Techniques for Determining Probabilities of Geologic Events and Processes

INTERNATIONAL ASSOCIATION FOR MATHEMATICAL GEOLOGY
STUDIES IN MATHEMATICAL GEOLOGY

1. William B. Size (ed.):
Use and Abuse of Statistical Methods in the Earth Sciences

2. Lawrence J. Drew:
Oil and Gas Forecasting: Reflections of a Petroleum Geologist

3. Ricardo A. Olea (ed.):
Geostatistical Glossary and Multilingual Dictionary

4. Regina L. Hunter and C. John Mann (eds.):
Techniques for Determining Probabilities of Geologic Events and Processes

Techniques for Determining Probabilities of Geologic Events and Processes

Edited by

Regina L. Hunter
Sandia National Laboratories

C. John Mann
University of Illinois

New York Oxford
OXFORD UNIVERSITY PRESS
1992

Oxford University Press

Oxford New York Toronto
Delhi Bombay Calcutta Madras Karachi
Kuala Lumpur Singapore Hong Kong Tokyo
Nairobi Dar es Salaam Cape Town
Melbourne Auckland

and associated companies in
Berlin Ibadan

Published by Oxford University Press, Inc.,
200 Madison Avenue, New York, New York 10016

Library of Congress Cataloging-in-Publication Data
Techniques for determining probabilities of
geologic events and processes /
edited by Regina L. Hunter and C. John Mann.
p. cm.—(Studies in mathematical geology ; 4)
Includes bibliographical references and index.
ISBN 0-19-507498-X
1. Geology—Mathematical models.
I. Hunter, Regina L. II. Mann, C. John.
III. Series. QE33.2.M3T43 1992 551'.01'5118—dc20 91-42353

2 4 6 8 9 7 5 3 1

Printed in the United States of America
on acid-free paper

Foreword to the Series

This series of studies in mathematical geology provides contributions from the geomathematical community on topics of special interest in the earth sciences. As far as possible, each volume in the series will be self-contained and will deal with a specific technique of analysis. For the most part, the results of research will be emphasized. An important part of these studies will be an evaluation of the adequacy and the appropriateness of present geomathematical and geostatistical applications. It is hoped the volumes in this series will become valuable working and research tools in all facets of geology. Each volume will be issued under the auspices of the International Association for Mathematical Geology.

Richard B. McCammon
U.S. Geological Survey
Reston, Virginia

Preface

The U.S. Environmental Protection Agency (EPA) has set a probabilistic standard for the performance of geologic repositories for the disposal of radioactive waste. This standard allows certain releases at large probabilities and greater releases at smaller probabilities. The EPA requires that repository performance be examined using assessments that consider uncertainty and that culminate to the extent practicable in the preparation of complementary cumulative distribution functions (ccdf's) showing the probability of exceeding any given release. The EPA has not suggested ways to determine probabilities of events or how uncertainty associated with the effects of property variability should be incorporated into the ccdf, but guidance appended to the standard states that numerical and analytical methods and expert judgment are probably all acceptable.

In the past, no consensus has existed about the best ways to determine the probabilities of the many events and processes that will be considered during the preparation of a ccdf. For this reason, the U.S. Nuclear Regulatory Commission (NRC) has undertaken the task of evaluating and selecting appropriate methods for determining these probabilities. The NRC funded the Waste Management Systems Division at Sandia National Laboratories to carry out the work. This volume describes some results of a part of the work. A slightly expanded version of Chapters 1, 5, 6, 7, 8, and 10 was published with limited distribution (Hunter and Mann, 1989). The manuscript for this study was originally completed in 1986, although it has been possible to revise some of the chapters in late 1989 and early 1990 to bring the references up to date.

This study is a literature review. It treats not only geologic events like fault movement, but also events that arise from the relation between human actions and geology, like drilling for resources, and some that arise from nongeologic processes that in turn affect a geologic process, like climatic change. In addition, it treats the relationship between property variability and probability of releases, e.g., in the chapters on ground-water flow and geochemistry. In preparing this volume, experts in various geologic and closely related fields were asked to compile techniques in use in their fields for probabilistic predictions and to evaluate the state of the art. The evaluations do not constitute endorsements; it is too early in the overall task to endorse one technique over another.

This work was supported by the U.S. Nuclear Regulatory Commission and by the U.S. Department of Energy and performed at Sandia National Laboratories, which is operated for the U.S. Department of Energy under contract DE-AC04-76DP00789. Research for Chapter 5, Climatology, was partially supported by grants from the National Science Foundation Climate Dynamics Program to Brown University (ATM 8713981) and the University of Oregon (ATM 8713980). Neither the U.S. government nor any agency thereof, or any of their employees, makes any warranty, expressed or implied, or assumes any legal liability of responsibility for any third party's use, or the results of such use, of any information, apparatus, product, or process disclosed in this report, or represents that its use by such third party would not infringe privately owned rights.

Albuquerque, New Mexico R.L.H.
Urbana, Illinois C.J.M.
February 1992

Acknowledgments

John Kutzbach provided critical comments on the manuscript of Chapter 5, Climatology. John C. Griffiths provided insight into URV analysis for Chapter 10, Resource Exploration. H. Einstein and G. Baecher contributed to the section on uncertainty in Chapter 1, Introduction. Chris Rautman provided helpful comments on Chapter 9, Geochemistry. Finally, George Barr and Rob Rechard of Sandia National Laboratories undertook the Herculean task of technical review, and Glenda Sweatt and Vilija Markunas of G^2 VanTel spent many hours checking references. We are grateful to these colleagues for their support.

Contents

Contributors

Gregory B. Baecher
ConSolve
Lexington, Massachusetts

Patrick J. Bartlein
Department of Geography
University of Oregon
Eugene, Oregon

Robert J. Budnitz
Future Resources Associates, Inc.
Berkeley, California

Jonathan F. Callender
Adrian Brown Consultants, Inc.
Denver, Colorado

W. Jay Conover
College of Business Administration
Texas Tech University
Lubbock, Texas

C. Allin Cornell
Stanford University
Stanford, California

Herbert H. Einstein
Department of Civil Engineering
Massachusetts Institute of Technology
Cambridge, Massachusetts

Carolyn Feakes
Somerville, Massachusetts

Allan L. Gutjahr
Department of Mathematics
New Mexico Institute of Mining and Technology
Socorro, New Mexico

John W. Harbaugh
Department of Applied Earth Sciences
Stanford University
Stanford, California

Heinrich D. Holland
Department of Earth and Planetary Sciences
Harvard University
Cambridge, Massachusetts

Steven Hostetler
Water Resources Division
U.S. Geological Survey
Denver, Colorado

Regina L. Hunter
Environmental Technology Programs Department
Sandia National Laboratories
Albuquerque, New Mexico

Alexander R. McBirney
Department of Geological Sciences
University of Oregon
Eugene, Oregon

C. John Mann
Department of Geology
University of Illinois
Urbana, Illinois

Malcolm D. Siegel
Geosciences Assessment and Validation Division
Sandia National Laboratories
Albuquerque, New Mexico

Gabriel Toro
Risk Engineering, Inc.
Golden, Colorado

Krishan K. Wahi
GRAM, Inc.
Albuquerque, New Mexico

Thompson Webb III
Department of Geological Sciences
Brown University
Providence, Rhode Island

Techniques for Determining Probabilities of Geologic Events and Processes

1

Introduction

Regina L. Hunter
C. John Mann
W. Jay Conover
and Robert J. Budnitz

The U.S. Environmental Protection Agency (EPA) (1985*) has published environmental standards for management and disposal of spent fuel (SF) and high-level and transuranic radioactive waste (HLW and TRU). The standard governs the allowable releases from potential geologic repositories in the United States for the disposal of nuclear waste. The containment requirement of the standard (U.S. EPA, 1985, Section 191.13) is probabilistic. It sets certain limits for contaminant releases that are estimated to have more than 1 chance in 10 of occurring within 10,000 years, and a less strict set of limits for releases estimated to have between 1 chance in 10 and 1 chance in 1000 of occurring within 10,000 years. No limits are set for releases of lesser probability. The standard requires performance assessments that estimate probabilities of events and processes that might cause or permit releases of radioactive waste from repositories, but it does not offer any guidelines for methods of determining probabilities.

*On July 17, 1987, the U.S. Court of Appeals for the First Circuit remanded Subpart B of the standard back to the EPA for further consideration. Specifically, the court held that Section 191.15, individual protection requirements, was arbitrary and capricious and that Section 191.16, ground-water protection requirements, had not met the notice and comment procedural requirements. This study of probabilistic techniques is useful primarily in determining compliance with Section 191.13, to which the court did not object.

No consensus has existed in the past in either the waste-management or geological communities as to how probabilities of geologic events and processes should be determined. Methods used by waste-management scientists have differed widely. In some cases, events and processes have been chosen for preliminary performance assessments without any published consideration of probabilities (Arnett et al., 1980). Some workers have assumed probabilities (Pepping et al., 1983). Some probabilities have been calculated with fairly sophisticated mathematical techniques but with limited or uncertain data (Claiborne and Gera, 1974; Crowe and Carr, 1980; Cranwell et al., 1987). Other studies have used a combination of these methods, depending on the availability of data (Hunter et al., 1983). The general geological literature also contains examples of probabilistic analysis of geologic events and processes that are based on a variety of techniques (Simpson, 1952; Rantz, 1964; Gretener, 1967; Clifton, 1978; Sadler, 1981; Carr, 1982). These methods have ranged from informed but unsupported guesses to elaborate calculations that may represent the best approximations that can be made today with available data. Methods commonly have contained a significant subjective component representing an individual's experience, judgment, or assumptions; this component may or may not be explicitly described. Few applications have involved adequate data bases, and therefore their results must remain suspect.

This literature review assembles and discusses existing techniques that can assign or can be helpful in assigning probabilities to events and processes of concern to long-term performance of nuclear-waste repositories. Much of the material may not seem directly applicable, at first glance, to repository performance assessment; the authors have drawn on a wide variety of materials, some more clearly applicable than others.

Some events and processes of interest in performance assessment are geologic, e.g., fault movement. Others arise from the relation between human activity and geology, e.g., exploration for resources, or between the repository per se and the geologic properties of the site, e.g., thermal expansion of rock. Finally, some events and processes are not geologic, but affect geologic events or processes, e.g., climatic change that alters ground-water flow. Some of the phenomena considered here are sudden, discrete events that may be rare, such as a volcanic eruption. Others are continuous processes that are certain to occur, such as ground-water flow. The probability associated with a specific ground-water flow rate is a function of the temporal and spatial variability of properties (such as permeability and hydraulic gradient) that control the fluid flow. Techniques appropriate for the probabilistic treatment of continuous processes or variables may differ from those that treat discrete events. Property variability introduces uncertainty into release calculations. It is in this context that the term *probability* is applied to parameter and process variability and uncertainty.

USING PROBABILITIES IN PERFORMANCE ASSESSMENT

A brief review of scenarios (potential sequences of events and processes leading to releases of waste) that might arise for geologic repositories in bedded salt (Bingham and Barr, 1979; Cranwell et al., 1990), domed salt (Harwell et al., 1982), tuff (Hunter et al., 1983), and basalt (Hunter, 1983) revealed moderate consistency in the events and processes considered capable of initiating releases from a closed repository. Bingham and Barr (1979) considered faulting, seal failure, igneous intrusion, drilling, mining, subsidence, thermal effects, and tectonics to be important, along with a few events specific to the Waste Isolation Pilot Plant (WIPP) site. Cranwell et al. (1990) considered drilling, seal failure, and dissolution to be important release phenomena and the effects of pumping, faulting, and igneous intrusion to be important transport phenomena. Harwell et al. (1982) examined the effects of a large number of geologic and human processes and events and concluded that only human intrusion was important enough to analyze. Hunter et al. (1983) considered effects of ground-water flow, tectonics, thermomechanical effects, igneous intrusion, subsidence, drilling, faulting, seal failure, and erosion. Hunter (1983) examined the effects of ground-water flow, tectonics, faulting, glaciation, erosion, thermomechanical effects, subsidence, seal failure, and drilling.

This study examines whether techniques are available to assign probabilities to these discrete events and continuous processes (hereafter called simply "events"); in contrast, probabilities of concern in performance assessments are those of release sequences, i.e., of scenarios. A scenario is composed of an initiating event and subsidiary events whose occurrence may or may not depend on the occurrence of the initiating events (e.g., Figure 1.1). To determine the probability of each scenario, not only must the probability of the initiating event be known, but also the conditional probability of each subsidiary event. A systematic examination of what these subsidiary events might be (i.e., scenario development) or of what the probabilities of any of the initiating or subsidiary events might be (i.e., probability assignment) is beyond the scope of this work. For the most part, only techniques for assigning probabilities to single events, independent of sequence, are discussed here.

GOALS OF THIS STUDY

The primary goal of this study has been to search out and evaluate existing quantitative methods for determining probabilities of events and processes in fields that seemed to be most closely related to the majority of the events considered important to long-term performance by earlier workers. These fields are thermomechanical behavior, mining engineering, hydrology, climatology,

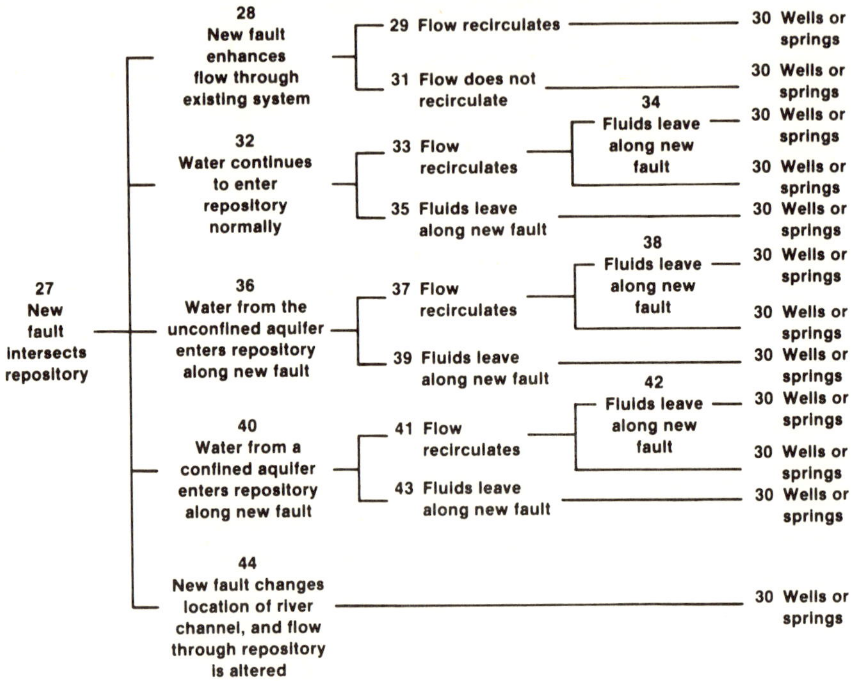

Figure 1.1. Example of an event tree containing several scenarios (Hunter, 1983).

seismicity and tectonics, seismic hazard, volcanology, geochemistry, and resource exploration. Most of these events and processes can initiate releases of radioactive waste from a geologic repository, although geochemical processes primarily affect transport of wastes after release. The quantitative methods for determining probabilities identified here are those that have been reported in the literature, and some that could be used but have not been reported. Merits and limitations of each method have been described, and the current availability of data bases adequate for determining accurate probabilities of events and processes has been appraised. A secondary goal has been to identify phenomena for which accurate probabilities cannot be determined now and areas of research that could materially improve our ability to make better probabilistic predictions in the immediate future.

Not all fields of geology or associated sciences known to bear upon long-term performance of a geologic repository have been considered here; nor have possible interactions between phenomena considered by the divers fields and their importance to long-term performance been evaluated. Comprehensive consideration of all possible contributors to performance of a geologic repository will probably be necessary in later stages of development of a probabilistic methodology for performance assessments of potential geologic repository sites.

Role of Probabilistic Calculations in the Regulatory Process

Calculations of probabilities of rare events and processes can and will play an important role in regulatory decision-making. This is true, in part, because certain probabilistic calculations must be performed to support regulatory decisions; but an important additional reason is that these types of calculations provide insights that cannot be obtained in any other way.

The EPA (1985, Appendix B) explicitly assumed that a number of techniques are appropriate for predicting likelihoods of events and processes that may disturb a repository. Complex computational models, analytical theories, and prevalent expert judgment were assumed to be appropriate techniques. In addition, the EPA assumed that, because of the uncertainties likely to be present, predictions may be supplemented with qualitative judgments.

APPROACH

Probabilistic risk assessment (PRA) offers a quantitative analytical technique for integrating diverse aspects of geologic events, processes, and other phenomena and their effects upon a potential geologic repository in order to evaluate what hazards they may hold for safe storage of nuclear waste. PRA calculates overall performance from the performance of individual, detailed components of a system. It also identifies what information is required for accurately analyzing generic and site-specific aspects of potential geologic repositories.

An ideal approach to PRA would model probability determinations in a completely general fashion for all possible geologic conditions. In practice, however, events and processes that are thought to be likely and of greatest concern at specific sites will be emphasized. Furthermore, a preliminary study such as this necessarily concentrates on the events and processes that have been treated in the existing literature.

One premise of this study was that most natural events and processes can be sorted into one of three groups when an attempt is made to assign probabilities of future occurrence. These groups are those for which probabilities can be determined (1) with confidence, (2) fairly accurately, or (3) with only limited confidence (i.e., containing inherently large uncertainty) (Hunter et al., 1986).

The timing and nature of some events or processes can be predicted with confidence, because the area of concern is well understood and because quantitative development of the subject is well established on either theoretical or empirical grounds. These events and processes are in the first group. Many of these natural events or processes are deterministic or virtually so. For example, heat production and changes in inventory in a radioactive source can be predicted with great accuracy, given data on the initial inventory. The physics of fluid flow through ideal porous media is well established theoretically (Davis and DeWiest, 1966; DeWiest, 1969; Marle, 1981); numerous quantitative models have been developed and applied to real geologic situations to predict ground-water movement; and hydrologists generally agree that several established methods and techniques provide acceptably similar results (Davis and DeWiest, 1966). Thus ground-water flow within some aquifers and basins also can be predicted, although the accuracy of the prediction depends on data that may not be readily obtained. Generally speaking, hydrologic data from sites currently being considered for waste repositories are sparse. Inadequacies in the data make uncertainty and sensitivity analysis of the predicted flow necessary. If a site is in a basin in which porous flow dominates, ground-water flow could be determined accurately after appropriate data are collected and ground-water flow is modeled.

Phenomena in the second group can be treated probabilistically with reasonable accuracy today. Adequate data bases exist, or may be easily obtained, to estimate their probabilities by establishing empirical frequency distributions and thereby estimating their probability density functions. This may be done even though we do not fully understand the event or process, particularly if the event occurs frequently. Phenomena might also fall into this group because data are inadequate to use existing deterministic models, because the process is too poorly understood to create deterministic models, or because the process can be considered random. An example might be volcanic eruptions in the western United States (e.g., Crowe and Carr, 1980). Thus phenomena that would normally be included in the first category when adequate data are available might be relegated to this second category in the absence of adequate data. Some phenomena can be predicted on the basis of general geologic knowledge to be nearly certain to occur in a given area during a long enough period of time. For example, Quaternary volcanism is entirely restricted to the western United States, where a lithospheric plate is overriding a subducting plate margin. Here, active volcanoes exist today (Williams and McBirney, 1979). Their frequencies of eruption during the Quaternary Period are generally well known (NOAA, no date) from geologic data. Because young volcanoes tend to be easily identified in the arid West, existing data are usually adequate. Even though it is not possible to predict deterministically when or where a volcano will erupt next, the process is well enough understood that it can be stated that some volcanism will certainly occur in the West. Moreover, enough data exist to determine the past rate of eruptions in the West, and therefore (assuming that the rate continues unchanged), the probability of new eruptions. Such a calculation has been made for a proposed nuclear-waste repository site in southern Nevada (Crowe and Carr, 1980); the probability of new eruptive vents was estimated from the known number of volcanoes and their frequency of eruption. Because the Gulf Coast salt domes previously considered for nuclear-waste disposal are well outside the area of Quaternary volcanism, volcanism probably would not have needed to be considered in salt-dome performance assessments.

Extreme values of some geologic parameters may be difficult to predict with great accuracy even when the parameters are well understood and available data are excellent. For example, whether future seismic events will exceed some stated level of energy release is uncertain. Seismologists have used Gumbel's theory of extreme values (Shakal and Willis, 1972), Markovian models (Patwardhan et al., 1980), and maximum entropy (Berrill and Davis, 1980), as well as assuming various probability distributions to estimate maxima that may be expected for future extreme values. Many such extreme values may have very small probabilities, however, and the EPA (1985, Appendix B) suggests that performance assessments need not consider events and processes that are estimated to have less than 1 chance in 10,000 of occurring over 10,000 years. For events of very small probability, a rough estimate of probability may be adequate to demonstrate compliance with the U.S. standard.

Phenomena that fall into the third and final category, however, present major difficulty because they can be predicted only poorly and with limited confidence. Mechanisms driving the event or process may be inadequately understood, or data may be inadequate to make accurate predictions. Geologic events and processes that are included in this third category of poorly predictable phenomena offer the greatest uncertainties to risk evaluation. Alternative techniques, however, such as consequence analysis, subjective probabilities (including expert opinions), sensitivity analysis, or establishment of acceptable bounds, may substitute adequately for probability density functions and normal probability estimates. Hunter et al. (1986, Figure 4) developed a flow chart offering a preliminary method for dealing with events and processes in the third group. The flow chart was arranged more or less in order of increasing uncertainty and greater undetermined risk.

METHODS OF DETERMINING PROBABILITIES

Commonly, the term *probability* is considered to be synonymous with likelihood and chance; the term is used here in this sense. Mathematically, the probability of a random event is the limiting value of relative frequency of occurrence reached in an infinitely long sequence of observations of that event. Even though individual outcomes of random events may not be predicted, they will show statistical convergence to a given value over several occurrences or predictions of occurrence. In contrast, individual predictions of other types of events may be made with certainty because they do not behave as random events; these types of phenomena are called "deterministic." When initial values of various variables are adequately known, only one result will be realized for deterministic events, and this result can be predicted with complete certainty. Conversely, for random events, even though initial parameters are known adequately, future events or results can be predicted only in probabilistic terms; that is, predicted only in an ensemble sense, without the ability to predict individual outcomes. Methods for determining probabilities can be divided into axiomatic approaches, frequentist approaches, modeling approaches, and expert judgment. Which of these approaches is most suitable in a particular application depends on the complexity of the system producing events whose probability is to be determined, how well the system is understood, and the number of data available.

Axiomatic approaches are theoretical in nature. Properties of a system are used to deduce possible outcomes of a trial and the probability of each outcome. For example, in a coin-tossing system, the analyst may determine that two outcomes are possible, heads and tails. No other outcome seems possible (neglecting the remote possibility that the coin may land on edge). A brief examination of the coin may show that it has no obvious bias. The axiomatic probabilist will therefore determine the probability of heads to be 0.5 and that of tails to be 0.5. The axiomatic approach assumes that the system is well understood; few geologic systems are simple enough or well enough understood to make an axiomatic approach practical.

Frequentist approaches depend on the analysis of existing data. Rather than determining what outcomes are possible in a system, frequentist approaches examine what outcomes have been recorded. The nature of the system need not be well understood if data are numerous enough to represent all possible outcomes. Unfortunately, "numerous enough" depends on the system; because the frequentist approach does not require any fundamental understanding of the system, an analyst can never know that one more trial will not produce a new outcome. In a coin-tossing system, the first few data might or might not produce a probability of 0.5 for heads. With three trials, the frequentist would assign a probability of 2/3 to heads, say, and 1/3 to tails. As the number of trials became larger, the frequentist normally would record a more equal number of heads and tails, so that the probability would approach 0.5 for heads and for tails. Frequentist approaches have been used in waste management to assign probabilities to occurrences of such events as volcanism (e.g., Crowe and Carr, 1980) and earthquakes of given magnitudes (e.g., Rogers et al., 1977).

Modeling approaches, often called Monte Carlo approaches, entail developing a model of the system, allowing the system itself to be subjected to numerous trials by proxy. A modeling approach to coin tossing might involve detailed measurements of the weights of various portions of the coin, measurements of forces applied to tossed coins, coin elasticity, and so on. After gathering many data and preparing a computer simulation that incorporates all parameters known or suspected to play a part in determining an outcome, the modeler samples the input data in some random way and executes numerous computer runs to determine the probability of certain events. The modeling approach has been used in waste management in complex computer models such as GSM (Geologic Simulation Model) (Foley et al., 1982).

The use of an axiomatic, frequentist, or modeling approach implicitly *assumes* that some part of an event or process under consideration is random. Geologic events and processes for the most part are not random phenomena. They differ from truly random occurrences in two major ways. First, geologic phenomena rarely occur repeatedly under identical conditions. Therefore, geologic phenomena rarely, if ever, can constitute an infinitely long sequence from which theoretically correct probabilities could be determined. (For this reason, Bayesian methods may be appropriate in many cases of interest to performance assessment [e.g., Jaynes, 1983, p. 163, 176].) In addition, slightly different conditions introduced for each subsequent occurrence of an event or process will generate greater variances than normally would be encountered if randomness and identical conditions truly existed. Second, most scientists would agree that many geologic events and processes actually are deterministic phenomena and could be treated as such if they were understood adequately and enough information existed to calculate these deterministic relationships. However, because of complexity in these phenomena and poorly understood relationships, many must be treated as probabilistic phenomena. This is completely satisfactory for scientific purposes, but it does mean that accuracy and precision are not as great as they theoretically could be if the events were handled deterministically. In final performance assessments, however, some expert judgment and irreducible uncertainty will probably be present, necessitating a final subjective decision on the acceptability of the residual uncertainty and risk presented by a repository.

Expert opinion in assigning probabilities is more subjective and uncertain than axiomatic, frequentist, or modeling approaches. In the absence of adequate data for objective probability determinations, probabilities estimated by experts may be used for performance assessment. Expert opinion may be completely adequate to determine that probabilities are so small that an event is not of regulatory concern, for example. Expert opinion may also be useful when site-specific data are sparse and would be prohibitively expensive or time-consuming to collect. Finally, great accuracy for a given probability may in many cases be unnecessary if consensus among experts about its possible range exists. Most commonly the basis for these estimates will be empirical Bayesian methods (Efron and Morris, 1973; Morris, 1983) or classical Bayesian methods (Box and Tiao, 1973; Louis, 1984). A disadvantage of all subjective probability determinations, which must be recognized, is that they are subject to manipulation or bias, either deliberately or unknowingly, because each holds varying amounts of individual judgment.

Bayesian probabilistic assessments may provide an evaluation of uncertainty associated with a probability estimate that is superior to that provided by objective probability determinations. This aspect commonly may be deficient in other probability determinations except for those derived from well established empirical or cumulative distribution functions.

ESTIMATING PROBABILITIES OF GEOLOGIC EVENTS AND PROCESSES

In determining probabilities of geologic events and processes for performance assessments, the question is not whether probabilities can be estimated or assigned to a specific event or process. Invariably probabilities can be and will be estimated. Rather, the correct questions are, "What is the most accurate means available by which realistic probabilities may be determined for an event at a specific site?" and "What is the probability that a variable parameter describing a property of the site will have a given value?"

Some means of assigning probabilities are more desirable than others. In general, objective (i.e., data-based), physically based probabilities are more satisfactory and more reliable than subjective, or judgmental, probabilities. A spectrum of quantitative measures exists in practice, however, ranging from virtually deterministic predictions with little associated uncertainty, through other objective predictions with increasing degrees of uncertainty or subjective probabilistic assessments with little uncertainty, to other subjective probabilities with associated great uncertainties. The following techniques for estimating parameter values or probabilities of occurrence of natural events are discussed in order of generally decreasing preference.

Objective Probability Estimates from Empirical Distribution Functions

Four functions (defined below) of interest to the discussion of probabilities of natural events and processes are the empirical distribution function (edf), the cumulative distribution function (cdf), the probability density function (pdf), and the complementary cumulative distribution function (ccdf). All are illustrated using the simple example of a sample of eight cars on a freeway. One car is traveling at a speed of 40 mph; one is traveling at 45 mph; three are traveling at 55 mph; two are traveling at 60 mph; and the last is traveling at 65 mph.

An edf represents relative frequencies of sample data less than or equal to x, for all $-\infty < x < \infty$. Its graph (Figure 1.2a) is a step function whose steps are of height k/n, where k = the number of observations at that sample value, $1 \leq k \leq n$, and occur at each observation in a sample of size n. The graph therefore increases from zero to one. It shows the proportion of sample observations $\leq x$ for $-\infty < x < \infty$.

For a random variable X, the cdf of X represents the true distribution of $X \leq x$. Because relative frequencies tend to approximate probabilities as sample size increases, the edf approximates the cdf as the sample size increases. In the example (Figure 1.2b), the graph of the cdf has been drawn on the assumption that the edf suggests that speeds of cars have a normal distribution. With such a small sample size, there is great uncertainty about the validity of this assumption; for example, a lognormal distribution may fit the data just as well or better.

For a continuous random variable X, a pdf is the first derivative of the cdf. If the random variable is discrete, the pdf of x represents the probability that the random variable X will equal x. Roughly speaking, it is a density function that indicates the likelihood of getting each value of x. The pdf at x of a continuous random variable X represents the relative rate at which the probability is accumulating in a small neighborhood of x. The graph of the pdf in this example (Figure 1.2c) is a bell curve, because the speeds were assumed in the construction of the cdf to be normally distributed.

A ccdf is equal to 1 - cdf and represents the probability that a random variable $X > x$. The EPA standard suggests that performance assessment contain a complementary cumulative distribution function (U.S. EPA, 1985, Appendix B). The ccdf for this example (Figure 1.2d) shows a probability of 0.5 that any given car will be in violation of the regulatory standard, i.e., the 55 mph speed limit. The sample showed only 3/8 of the cars to be in violation. Either the sample is inadequate or the cdf chosen does not adequately represent the edf.

Empirical Distribution Functions. Cdf's would provide exact probability determinations if they were available. Unfortunately, cdf's are rarely known in real applications: they must be estimated. The estimator for a cdf that is most free of assumptions is an edf derived from geologic data gathered locally. However, local data rarely will be adequate for this purpose and, hence, dependence will normally fall upon generic edf's derived from regional or, less satisfactorily, global data for the same geologic event or process. Generic data are less desirable than local data because geologic conditions rarely, if ever, will be identical for different geologic sites or times.

Both natural variations in an event or process and observational errors always will be present to some degree in edf data, and it may be impossible to distinguish between them. Whenever local, regional, or global edf's will be the basis for probability estimates, an investigator must decide whether a smoothing equation should be fitted to the edf (a cdf derived inductively) or whether an established family of cdf's (derived deductively) such as lognormal, gamma, negative beta, etc. (Elderton and Johnson, 1969; Johnson and Kotz, 1969, 1970a, 1970b, 1972; Patel et al., 1976), should be fitted to the edf. Many mathematical and geological reasons can be found why a fitted cdf should or should not be substituted for an

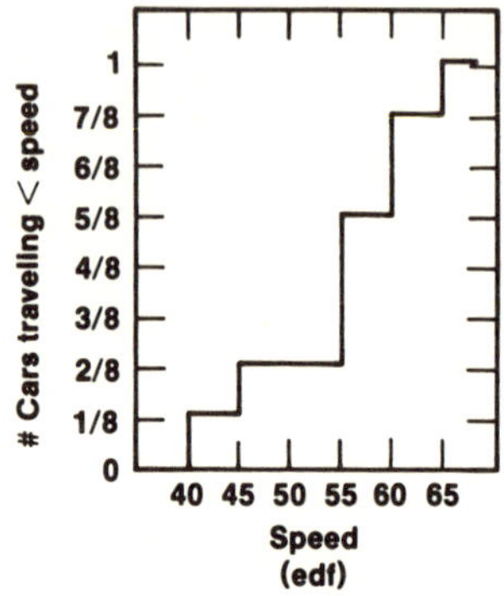

a. Edf for speeds of eight cars, traveling at 40, 45, 55, 55, 55, 60, 60, and 65 mph.

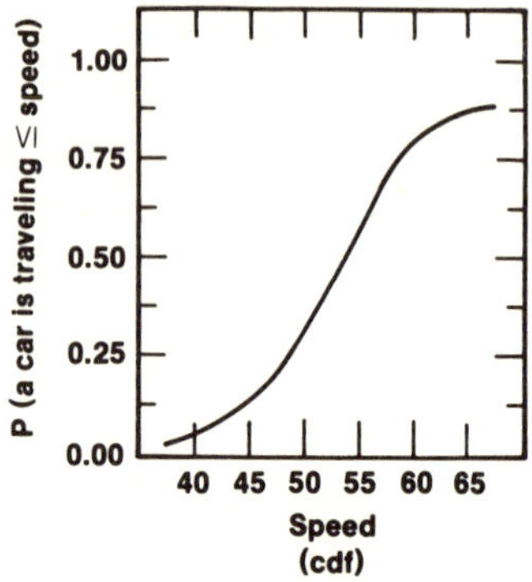

b. Cdf fitted to the edf shown in Figure 1.2a, assuming a normal distribution of speeds.

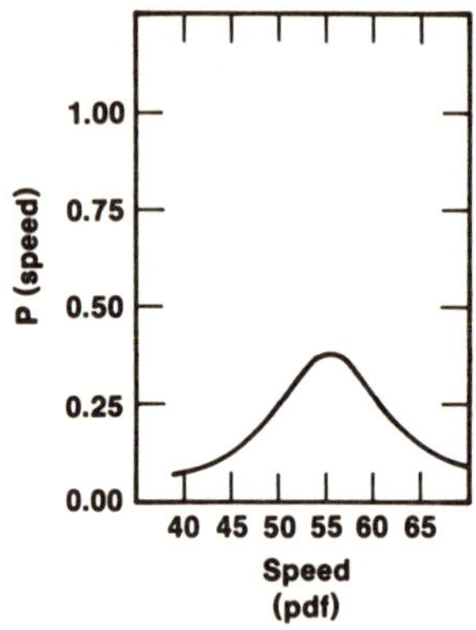

c. Pdf derived from the cdf shown in Figure 1.2b, showing the probability that a sampled car will be traveling at a given speed.

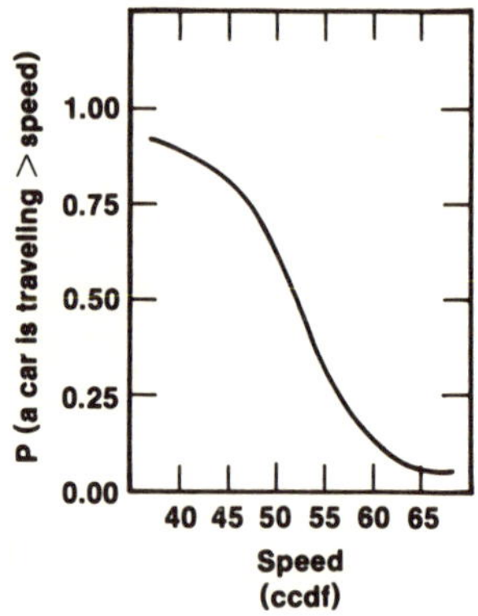

d. CCDF derived from the cdf shown in Figure 1.2b, comparing predicted speeds with the 55-mph speed limit.

Figure 1.2. Four mathematical functions used in determining probabilities.

edf. Substitution of a fitted cdf may be essential if one is concerned primarily with predicting rare events that are represented poorly, or not at all, by available empirical data. Commonly, large or infrequent events are so poorly represented in the geologic record that substitution may be essential for accurate probabilities. In such a case, the analyst must pay special attention to the tails of the distribution, rather than using a distribution that fits the main central part of the data and extrapolating it to the tails without thought. Observational data may fit two or more theoretical cdf's equally well, even though the cdf's differ in the portions of the curve for which data are unavailable.

Fitting Curves to EDF Data. Many mathematical techniques have been developed for fitting curves to edf's; some methods are more appropriate than others under various mathematical or geologic conditions, and some are totally incorrect under certain conditions. Therefore, care must be taken in deciding which method is to be used in any given case. Similarly, any technique used to determine a cdf or its associated pdf should be clearly described so that others may independently judge its validity for each specific situation.

Classical parametric methods of density estimation assume that certain mathematical conditions exist or are appropriate for data being fitted. These methods include maximum likelihood estimates (Fisher, 1925, 1934, 1974), weighted least squares (Sorenson, 1980), and moments (Pearson, 1902a, 1902b). Conversely, nonparametric techniques for density estimation (Wegman, 1972; Tapia and Thompson, 1978) have fewer mathematical assumptions regarding the form of the density function, but they require large sample sizes. These methods include kernel estimates (Rosenblatt, 1956; Parzen, 1962a; Singh, 1977), maximum entropy (Gokhale, 1975; Shore and Johnson, 1980; Gull and Skilling, 1984), projection pursuit (Friedman and Tukey, 1974; Friedman et al., 1984), splines (Boneva et al., 1971; Wahba, 1975), penalized maximum likelihood (Good and Gaskins, 1971; de Montricher et al., 1975), near-neighbor estimates (Loftsgaarden and Quesenberry, 1965), minimum distance (Wolfowitz, 1957; Woodward et al., 1984), incomplete means estimation (Houghton, 1978), orthogonal series (Schwartz, 1967; Wahba, 1971; Brunk, 1978), estimation using Poisson's distribution (Gawronski and Stadtmuller, 1980), and probability weighted moments (Greenwood et al., 1979).

Theoretical Distribution Functions. Lacking an edf or a cdf derived from empirical data, the next best estimate of a pdf or cdf normally will be derived from theoretical considerations based on knowledge of the geologic event or process. For example, seismologists long ago expected theoretically that frequency of seismic events would follow a Poisson distribution because of the nature of these processes (Epstein and Lomnitz, 1966; Cornell, 1971); observational data confirm that to a first-order approximation this is true for sites affected by several seismic sources. Historically, some distributions predicted theoretically have been reasonably close to subsequently observed data distributions; nonetheless, their accuracy directly depends upon our current understanding of a particular geologic event or process. In some instances, knowledge was poor at the time predictions were made, and theoretically derived cdf's have been significantly in error as a result.

Objective Probabilities without Distribution Functions

Often geologic data will be so incomplete that an accurate edf or cdf cannot be determined from either available data or our existing knowledge of the event or process. Even so, some estimates of objective probabilities of future occurrences still may be obtained on the basis of extremely limited data. These probability estimates will be objective determinations made without benefit of an edf or cdf. An example commonly encountered in repository performance assessments is frequency of movement along a recent fault; total data obtained from field studies may consist of only two or three values. Uncertainty in these objective probability estimates typically will be great.

Elimination by Consequence Analysis or Bounding

In absence of any probability determinations by the objective methods noted above, an event or process of concern for performance assessments of repositories possibly may be eliminated from further consideration either through consequence analysis or by establishing bounds on the probability. When an event or process can be shown through either of these techniques to have potential results that are sufficiently insignificant so as to pose no danger to a repository, it may be dismissed from further consideration for purposes of performance assessments.

Subjective Probability Methods

Even for geologic events and processes that can neither be assigned probabilities through objective means nor be eliminated from further consideration by consequence analysis or bounding, additional techniques embodying great uncertainty remain. Subjective probabilities may be established using expert judgment (Kyburg and Smokler, 1980; Uppuluri, 1983) employing several specialists from appropriate fields of geologic knowledge. Great uncertainties invariably are associated with subjective probabilities established in this manner because they represent an ill-defined basis that cannot be quantified easily and are not capable of experimental verification. Each expert's opinion depends on his or her understanding of an event or process, experience, and abilities, and that similarity which he or she perceives as existing between the particular geologic situation of concern and other situations that are known and are comparable to varying degrees. In some instances, expert-opinion predictions have been shown to provide accurate evaluations, and in other instances they have been demonstrated to be poor. Which will be the case for any given probability estimate obtained through expert opinion processes can rarely be assessed satisfactorily at the time those probabilities are assigned.

Closely associated with expert opinion are Delphi methods (Linstone and Turoff, 1975), which use opinions of numerous specialists from a variety of fields obtained by various methods to reach a group consensus on the probability of future events. Delphi techniques have been criticized, however, because the feedback inherent in them makes them suspect from a statistical point of view (Harris, 1984); collaboration can bias the results, and the expertise involved may have little bearing on the accuracy of such estimates.

Another technique is assigning relative probabilities to events and processes on the basis of general knowledge and individual or group understanding of them. These are subjective, order-of-magnitude estimates (Hunter, 1983). Although relative probabilities will have the greatest uncertainties associated with them and are the least desirable of all probability assessments, they may be sufficiently informative to be helpful in assessing performance risks.

In summary, the preferred order in which probabilities of geologic events and processes are to be estimated, or the event eliminated, for performance assessments for geologic repositories might be

- Objective probabilities obtained from
  - Edf's derived from local observations
  - Edf's derived from regional observations
  - Edf's derived from global observations
  - Pdf's or cdf's derived from theoretical considerations
  - Limited observational data (without benefit of edf or cdf)
- Objective probabilities (without distribution functions) established by
  - Empirical Bayesian methods
  - Bayesian methods
  - Expert judgment
- Elimination of an event or process from further consideration by
  - Consequence analysis
  - Bounding of probabilities
- Subjective probabilities established by
  - Delphi methods
  - Relative-probability estimation

The exact order of preference might be given differently by different authors. For most probability estimates to be used in performance assessments of geologic repositories for nuclear waste, the uncertainty associated with the estimate must be specified. In addition, the basis for both probability and uncertainty estimates and the methods by which they were obtained must be stated clearly and explicitly so that others are able to freely evaluate the validity of each probability and uncertainty estimate.

Even such ill-defined probabilities as those generated subjectively may be adequate for some parts of the performance assessment, for example, where the estimated probability is very small or large.

RELATIONSHIP BETWEEN UNCERTAINTY AND PROBABILITY

Geologic uncertainty may result from lack of detailed knowledge of the large-scale geometry of geologic units and major structural features such as faults and folds, or it may result from spatial and temporal variability in rock properties. These uncertainties typically include questions of geologic history, continuity and connectivity of geologic units, the potential existence of adverse features, and average property values. Geologic models are usually described subjectively and depend strongly on knowledge of regional conditions and relevant geological theory. Geologic uncertainties that arise from models are often dealt with by enumerating various possible models and assigning relative probabilities to each.

Model uncertainty results from abstracting and simplifying complex natural processes in order to develop workable analytical procedures. These procedures may be mathematical, numerical, or empirical, but in any case they only approximate reality. Presuming sufficient data from which to estimate parameter values, the difference between a model's description of a real process and the process itself is model uncertainty. Model uncertainty is unavoidable, and the more complex the real behavior, the greater the model uncertainty. Model uncertainty is estimated by comparing real performance with model predictions.

Parameter uncertainty refers to the precision and accuracy with which the values of specific model parameters can be estimated, typically from data. Estimates of parameter uncertainty are commonly based on statistical procedures and correlations between measured values and those calculated from scaled or prototype testing. The previous section describes the construction of a pdf that is related to parameter uncertainty.

Omissions pertain to elements that are inadvertently omitted from an analysis, even though they are important. No guarantee can be made that all important, relevant factors have been included. Omissions can be avoided only by diligence.

Most probabilistic modeling from which quantitative predictions are made emphasizes the effect of parameter uncertainty on predictions. Parameter uncertainty itself can be thought of as being composed of four parts, each of which has a distinct influence on the total uncertainty in performance predictions. The four principal components of parameter uncertainty are spatial variation, measurement noise, measurement bias, and statistical error. The first two give rise to data scatter. The second two give rise to systematic differences between best-estimate predictions and average performance.

A probability is a mathematical expression that embodies uncertainty about the future by definition: some workers think that to refer to the "uncertainty in a probability" is to misunderstand probability altogether. For most of the natural phenomena treated here, however, probabilities cannot be determined by performing repeated identical trials. Rather, probabilities must be calculated using data that are uncertain in the usual sense, that is, the exact value of a given parameter is unknown because of, say, measurement error or sparse data. Uncertainty in probabilities is used in this sense here. Uncertainty in probabilities may arise from two sources. First, the physical model may be inaccurate. For example, volcanism might be modeled as a spatially random process, even though it is known to be controlled by deep-seated crustal weaknesses. If the location of crustal weaknesses is unknown, spatial randomness may be the *best* model available, but it is not the *correct* model. Second, there may be uncertainty about values of parameters used to calculate the probabilities. For example, the probability of volcanism in a given area might be calculated partly on the basis of ages of existing volcanoes. Radiometric ages used to date volcanism always carry some uncertainty. Some estimate of the uncertainty in the probabilities of risks from proposed repositories is required.

The treatment of uncertainties in performance assessments involves a complex model of physical processes, which in turn consists of several submodels, each somewhat autonomous, and each characterized by inputs and outputs. Inputs may be known parameter values, estimates of unknown parameter values, or random quantities whose probability distribution may be known or estimated. Outputs of a submodel may serve as inputs to other submodels, and therefore the probability distribution of each output must be known or estimated. Alternatively, outputs of a submodel may go directly into determining the probability distribution of release of radionuclides into the environment.

For regulatory purposes, the latter probability distribution need not be known entirely, however. Perhaps only certain maximum limits need be determined; for example, the U.S. standard requires only that two quantities (0.1 and 0.001) be met. Nonetheless, in order to determine probabilities associated with those limits, the entire probability distributions of all preceding inputs and outputs must be known or estimated. Several methods have been proposed for combining uncertainties associated with inputs of the various submodels in order to achieve the desired estimates of released quantities, e.g., the Monte Carlo approach (Cranwell et al., 1987; Martz et al., 1983; McKay et al., 1979; Iman and Conover, 1980, 1982), the Maximus method (Maximus, Inc., 1980; Spencer and Easterling, 1984), the adjoint procedure (Cacuci et al., 1983; Dacol and Rabitz, 1983; Lewins and Becker, 1982), factorial designs (Campbell et al., 1980; Iman et al., 1978), and Bayesian methods (Martz and Waller, 1982). Iman and Helton (1985) thoroughly compared some properties of several of these methods and included an extensive bibliography.

SUMMARY

A number of techniques are available for calculating or estimating probabilities of various events that might occur during the lifetime of a radioactive-waste repository situated deep beneath Earth's surface. Most of these events have only a very small likelihood of occurring during the next 10,000 or even 100,000 or 1 million years, but cannot be excluded on the basis of what now is known. Related techniques are useful for considering the probability of release associated with spatially and temporally variable, and therefore uncertain, property values controlling processes such as ground-water flow. To understand whether the performance of the repository will be degraded by the possible occurrence of any of these events or processes, the best available probability estimates must be provided. However, each event or process is really a *category* of events of different sizes and different severities. Therefore, not only is an overall estimate of the probability necessary, but also a disaggregated estimate of the likelihood of various severities, recognizing that for each category under consideration the more severe events will probably be less likely to occur than events of lesser severity.

Inputs needed to make probabilistic assessments do not differ in principle from those needed for any other type of analytical method. The analyst requires a model of the phenomena under consideration, applicable data, a method for estimating uncertainties, and a means of explaining how the analysis was done so that it can be repeated and checked by others.

Several important limitations are inherent in the analyses discussed here. Perhaps the most important limitation stems from the character of the phenomena being studied: specifically, many are so rare that no way is known to validate the calculations by any actual experiments or experience. In this sense, the analyses will always be subject to varying opinions about the validity of the input information or the models used, and varying interpretations of the meaning of the calculated results will always be encountered.

Although the analytical methods under discussion are inherently quantitative in character, most of the broad conclusions are qualitative. This fact is not an artifact of probabilistic methodologies but rather an intrinsic characteristic of the regulatory issues and our incomplete knowledge. Because our experience is limited, the probability estimates cannot be based on experience data alone (such as an actual recorded number of volcanic events). Rather, they must be based on some combination of experience and analysis, specifically a combination of inferences based on the limited knowledge we do have, laboratory experiments, and theoretical considerations.

Inevitably, probability estimates will have uncertainties, and in many areas under study these uncertainties will be large. Whether the results of these calculations can be used as a basis for decision-making despite the large uncertainties is a question that cannot be addressed except case by case. To dismiss the results of these calculations without careful analysis of the decisions that are affected by them would be incorrect, however.

2

Thermomechanical Effects

Krishan K. Wahi

Consideration of deep geologic disposal of nuclear waste has created the need for a better understanding of thermomechanical processes in geologic media. Thermomechanical effects on underground structures or systems have not been considered probabilistically in the past. Unlike volcanic eruptions or earthquakes, thermomechanical processes are continuous, and the probability of the process occurring in a high-level waste repository is 1. Deleterious effects resulting from thermomechanical response are uncertain and must be addressed probabilistically in the context of repository performance. For example, what is the likelihood that thermoelastic uplift of the rock mass will fracture an overlaying aquitard and connect the aquifer to the host rock? How reliable is the predicted magnitude of thermoelastic uplift, given the uncertainties in the material properties and the simplifying assumptions of model calculations? Three important questions (Donath and Cranwell, 1981) related to the potential for deleterious effects resulting from thermomechanical effects are (1) what is the probability of an undetected fault; (2) what is the probability of new fault development or of movement along an existing fault for a given state of stress and the rock properties; and (3) what are the consequences for a repository system if a fault is present or develops? Perturbation of the in-situ stress field can cause movement along a fault that otherwise might have remained inactive. Stress perturbations can result from natural or man-made causes, such as the construction of a repository. Heat generated by emplaced waste introduces thermal stresses that are in addition to stress changes caused by excavation. Depending on the magnitude of these stress changes, development of new faults or fractures is also conceivable. Although substantial experience exists in mining and design of underground openings, unique complexities are introduced by heat generated by the emplaced waste. The thermomechanical response of the rock mass could significantly affect the overall performance of the repository system during the regulatory period.

The concerns resulting from rock movement in response to thermomechanical loads differ in nature during preclosure and postclosure phases of the repository. Prior to decommissioning, stability of roof, floor, and pillar walls is essential for operational safety. Following permanent closure, the concern shifts from stability of openings to long-term effects of thermomechanical loads that may increase or decrease the isolation capability of the host rock and the geologic setting. The deformational response of openings prior to decommissioning could also affect long-term performance. The rock-mass response to thermomechanical stresses and strains could activate processes that enhance permeability or create new connections to adjacent aquifers.

Intact rock does not contain joints, fractures, faults, shear zones, or other discrete discontinuities. A rock mass, in contrast, typically contains joints and fractures. A rock mass is essentially composed of pieces of intact rock with interspersed discontinuities. Soft rocks, such as salt, are more likely to contain larger volumes of intact rock than are hard rocks. Even in soft rocks, the rock mass is significantly weaker than intact rock. Laboratory-sized specimens tend to provide properties and characteristics of intact rock rather than those of the rock mass. Although the linear elastic behavior of intact rock is well understood and can be expressed analytically, rock masses in general exhibit more complex behavior. Numerical methods are available to describe thermal, mechanical, and thermomechanical responses of underground structures with material and geometric nonlinearities.

Wang et al. (1983) evaluated the thermal impact of waste emplacement in geologic media in analyses performed before 1983 and found that most modeling studies were based on thermoelastic calculations that neglected the effect of discontinuities. In identifying research needs, they stated that thermoelastic analyses could not predict the behavior of nonelastic, fractured rock masses and that rock failure conditions should be evaluated with fracture models. In spite of simplifications and inaccuracies of thermoelastic analyses, Wang et al. (1983) thought that such analyses could qualitatively identify zones of potential instability. Much developmental work on modeling of nonelastic or fractured rock masses has been done since 1983.

Each of the rock types that have been proposed in the United States as candidate host media for a repository (salt, basalt, and unsaturated tuff) has some unique characteristics that present difficulties to modelers. Although most rocks creep under stress, creep deformation is pronounced in salt subjected to differential stress. Creep is extremely difficult to quantify even at constant stress and temperature; incorporating creep into predictive models presents an enormous challenge. Both basalt and tuff are heavily fractured and jointed rocks, although fracture frequencies, joint orientations, and other joint characteristics differ for the two rock types. Numerical models must account for these discontinuities either directly or indirectly (i.e., by using continuum approximations). Because the unsaturated zone of tuff has been proposed as the emplacement horizon, additional modeling difficulties are presented in that medium. Thermal properties of tuff are sensitive to the degree of saturation. Single-phase heat transport calculations might be adequate for disposal at depths below the water table (e.g., in salt and basalt), but two-phase heat flow must be considered in unsaturated tuff. These medium-specific problems are in addition to modeling problems that are common to most rock types.

Many couplings among thermal, mechanical, hydrological, and chemical processes are conceivable. Not all couplings are equally significant, and different couplings may dominate at different times during the life of a repository. Thermomechanical, thermohydrological, thermohydromechanical, and thermochemical couplings are thought to be more important than others. A panel report (Tsang and Mangold, 1984) discussed 11 potential couplings and attached levels of significance to subprocesses within a given coupling. One dominant coupling during both preclosure and postclosure phases is between thermal and mechanical processes. Most analyses performed to date have not considered couplings other than thermomechanical or thermohydrological. Two primary reasons for the absence of such analyses are the relative lack of models and techniques to handle three- or four-way couplings and the general lack of appropriate input data (e.g., how does fracture permeability vary with normal stress?).

Chan et al. (1985) described the development of a three-dimensional finite element code, MOTIF, that models the coupled processes of ground-water flow, heat transport, brine transport, and single-species radionuclide transport in porous or fractured media. The model does not consider mechanical behavior, but developments are currently under way to enhance MOTIF into a coupled thermo-hydromechanical model. Elsworth and Goodman (1985) have presented coupled boundary element/finite element procedures to model hydromechanical behavior of fractured rock masses. Makurat and Barton (1985) reported on the application of their coupled model of shear-flow coupling in nonplanar rock joints. A coupled shear displacement-flow test conducted on a natural joint in granite (150-mm-diameter core sample) showed good agreement with their model calculations.

The hydraulic conductivity of a joint is sensitive to its aperture, and fluid flow and radionuclide transport depend on the deformability of the joints (Makurat and Barton, 1985). A model that can simulate coupled thermohydromechanical behavior in porous media has been developed (Ohnishi et al., 1987). The model is incorporated in a two-dimensional, finite-element code and has been verified or validated against three problems that pair two phenomena at a time. For a coupled hydromechanical problem, Terzaghi's one-dimensional consolidation problem was solved and compared with the analytical solution. Experimental results of a seepage experiment were compared with the numerical model results, providing a validation of that capability. The thermohydrological coupling was validated using test results of an experiment in a sand specimen. None of these is particularly relevant to the conditions in a repository, however.

Numerous analyses of very-near-field, near-field, and far-field thermomechanical effects have been made, and several conclusions can be drawn. In the very near field, peak temperatures in the waste, canister, and borehole wall are a function of waste type (e.g., reprocessed or unreprocessed spent fuel), thermal load per canister, ventilation, and thermal conductivities of the waste package components and host rock. Peak temperatures occur within 10 years of emplacement. The long-term near-field and far-field thermal response is insensitive to ventilation (or lack of it) during operation; however, ventilation does help reduce peak temperatures and thermomechanical stresses near the room. Convergences in unbackfilled rooms, in the absence of rock failure or collapse, are predicted to be small in basalt and tuff. Relatively large convergences are expected in openings in salt because of creep and a large coefficient of thermal expansion. Moreover, uncertainties in the parameters of empirical creep laws are large, making predictions unreliable. A benefit of creep deformation is that stresses and stress gradients are relieved. Experience indicates that at moderate and large confining stresses, creep is driven by the deviatoric stresses (i.e., the nonspherical component of the stress tensor).

ROCK PROPERTIES

Physical, thermal, and mechanical properties of the host rock and other strata affect the thermomechanical response of an underground structure or system. Similarly, the response of the geologic setting to a heat-generating repository is also a function of these properties. In general, perturbations are likely to be greatest in the host rock. With the exception of shafts, underground construction activity will be confined to the repository horizon. Because thermomechanical effects will be predicted primarily using Monte Carlo simulations with deterministic models, appropriate emphasis on characterizing the host rock and on obtaining ranges and distributions of values for critical parameters is important.

Thermal Properties

Thermal properties of the host rock and adjacent strata control the diffusion of heat into the rock mass. A low thermal conductivity means greater peak temperatures in the very near field and near field, that is, a slower dissipation of heat away from the heat source. Uncertainties in thermal property values imply uncertainties in predicting the temperature field. This would directly affect the ability to predict the magnitude of thermomechanical effects. Thermomechanical and thermohydrological responses would also become more uncertain; both of these coupled processes affect transport of radionuclides within the geologic setting. Therefore, the ranges and possibly the distributions of thermal-property data are important. The literature abounds in generic data on thermal properties of different rock types under consideration as repository host rocks (Table 2.1). Site-specific investigations have resulted in other data that are more recent and presumably have better quality assurance. Heat conduction is expected to be the primary mode of heat transfer from the waste package to the adjacent rock and surrounding rock mass. Conduction is controlled by thermal conductivity and volumetric heat capacity of the media. A larger thermal conductivity is desirable for host rock, because it permits heat to dissipate more rapidly and reduces temperature gradients and peak temperatures near the heat source.

Tien et al. (1985) compiled thermomechanical-property data for unsaturated tuff of the Topopah Spring Member at Yucca Mountain, Nevada (Table 2.2), and discussed the influence of temperature and porosity on these properties. Thermal properties of selected flows from the Columbia River Basalt Group at the Hanford Site have been studied by the U.S. Department of Energy (DOE) (1986b). Thermal conductivities range from 1.16 to 2.65 W/m-K for 47 samples from three flows; coefficients of thermal expansion range from 4.80 x 10^{-6} to 8.73 x 10^{-6}/°C for 49 samples from the same three flows. The coefficient of thermal expansion does not affect heat transfer but is typically reported as a thermal property; its effect on thermomechanical behavior is discussed below. Thermal property data for salt at the Richton Dome, Mississippi, have also been reported by the DOE (1986c). Thermal conductivities range from 2.66 to 4.17 W/m-K at 100°C for seven samples; coefficients of linear thermal expansion range from 37.5 x 10^{-6} to 46.5 x 10^{-6}/°C between 50 and 250°C for eight samples. Specific heats for these samples range from 913 to 933 J/kg-K. A more complete thermal-property data set for all former candidate salt sites was given by Lagedrost and Capps (1983). Tammemagi and Chieslar (1985) compiled thermal properties for generic crystalline rock (e.g., granite). They included ambient-temperature data as well as temperature dependence of these properties.

Generally, thermal properties of rocks depend on their mineral composition

Table 2.1 Thermal Properties of Rocks

Rock	Thermal conductivity[a] (W/m-K)	Specific heat[b] (J/kg-K)	Density (kg/m^3)
Salt	2.08-6.11	840-920	2128
Granite	1.99-2.85	880-960	2640
Basalt	1.16-1.56	710-920	2880
Shale	1.47-1.68	800-880	2560
Tuff			
(Welded)	1.2-1.9	800-900	2000-2400
(Unwelded)	0.4-0.8	800-1700	1500-2100

Source: Data from the DOE (1979) and Tyler (1979).
[a] Over the temperature range 0-400°C (except tuff).
[b] Over the temperature range 0-200°C (except tuff).

Table 2.2 Average Thermal and Mechanical Properties of the Topopah Spring Member, Yucca Mountain, Nevada

Property	Value
Thermal conductivity, saturated (W/m-K)	1.8 ± 0.4
Thermal conductivity, dry (W/m-K)	1.6 ± 0.4
Predehydration linear expansion coefficient (10^{-6}/°C)	10.7 ± 1.7 (to 200°C)
Transition-dehydration linear expansion coefficient (10^{-6}/°C)	31.8 (to 300°C)
Postdehydration linear expansion coefficient (10^{-6}/°C)	15.5 ± 3.8 (to 400°C)
Young's modulus (GPa)	26.7 ± 7.7
Poisson's ratio	0.14 ± 0.05
Unconfined compressive strength (MPa)	95.9 ± 35.0
Matrix cohesion (MPa)	28.5
Angle of internal friction (degrees)	26.0
Matrix tensile strength (MPa)	12.8 ± 3.5
Joint cohesion (MPa)	1
Coefficient of friction for initiation of sliding on joints	0.8

Source: Data from the U.S. DOE (1986a, Table 6-40).

and water content. Laboratory data from relatively pure and homogeneous samples should not necessarily be assumed to represent in-situ rock, which may contain impurities, heterogeneities, or anomalies. Theoretical techniques exist that permit calculations of thermal conductivities of mixtures, given the mineralogical content and conductivity of each component.

Thermal conductivity of salt has a strong, inverse temperature dependence. This inverse dependence is a disadvantage in the very near field, because the higher the absolute temperature, the more difficult it is for the heat to dissipate, resulting in greater buildup of heat in that region. However, even the "small" salt conductivity at expected peak temperatures is greater than that in other media. The temperature dependence of conductivity must be recognized and included to properly estimate the peak temperature in the very near field. The thermal expansion coefficient in salt is much larger than in other proposed host rocks. Heat capacity values tend to be similar in all rock types being considered, except that when porosity is significant, heat capacity becomes a strong function of saturation. Its dependence on temperature is relatively weak.

Mechanical Properties

A knowledge of mechanical properties (elastic moduli, nonlinear parameters, and strength parameters) permits predictions of the stress-strain response of a system that is subjected to mechanical or thermal loads. In particular, the onset of rock failure or structural failure cannot be predicted with confidence unless there is an acceptable level of confidence in the mechanical-property data used in calculating the state of stress and in estimating potential violations of strength criteria. Once mechanical properties are known within an acceptable confidence level, not only can relatively accurate stress estimates be obtained, but reliable strength criteria also may be developed. Mathematical failure surfaces can be constructed on the basis of strength criteria. Even when mechanical-property data are reliable, in practice rock failure may not occur as predicted; however, the probability of rock failure becomes a strong function of the proximity of the predicted stress field to the failure surface.

Uncertainties in mechanical properties stem from natural variation in the rock macrostructure, extrapolation of laboratory-scale sample measurements to field scale, and effect of discontinuities, among other factors. In the past, the predominant approach to dealing with such uncertainties has been to design with a large factor of safety. Regardless of the approach, the uncertainty in the mechanical property data must be known. The deformation of rock and rock structures is a function of the applied load(s), boundary conditions, and mechanical properties. When deformations are small, linear elastic theory (Hooke's law) can adequately describe the load-deformation response. In the elastic regime, elastic constants are the only mechanical properties needed to define the relationship between stresses and strains. For the most general kind of elastic material, 21 elastic constants are needed to relate stresses and strains (Sokolnikoff, 1956). But for isotropic materials only two independent elastic constants are needed to relate the stress and strain tensors completely. The most commonly used elastic constants are Young's modulus (E) and Poisson's ratio (ν). Other common forms of elastic constants are bulk (K_b) and shear (G) moduli. For transversely isotropic materials, five independent elastic constants are needed; many rock types (e.g., basalt and shale) exhibit transverse isotropy.

The term *failure* is used to indicate several types of deviation from elastic behavior, but not all deviations from elastic behavior involve failure. For example, creep deformation is nonlinear, inelastic, time-dependent deformation that may or may not result in failure. Brittle or ductile failure in a material can occur and reduce or eliminate its capacity to sustain load. A material is said to have "yielded" when the slope of the stress-strain curve shows an abrupt change and plastic deformation starts to occur. Yield strength of rocks is typically a function of confining pressure. Other common failure modes are rupture in tension or compression. The mechanical properties associated with these failure modes are yield strength, tensile strength, and compressive strength. In a rock mass that has preexisting fractures or joints, the effective moduli (stiffness) across and along the discontinuity can be strongly nonlinear. Design of openings in jointed rock must consider the orientation and spacing of joints in addition to the strength characteristics. Data have been accumulated and material models constructed to describe inelastic and failure responses of rock subjected to thermomechanical stresses.

Of the primary rock types being considered for a repository, only salt exhibits significant creep. Empirical laws have been proposed to estimate creep strain numerically. Wahi et al. (1977) formulated a generalized creep law, based on Starfield and McClain's (1973) one-dimensional model, that can account for transient and steady-state creep and was used in simulating the thermomechanical response of the Project Salt Vault experiment (Bradshaw and McClain, 1971). This experiment was conducted underground in a salt mine near Lyons, Kansas, to study the feasibility of geologic disposal and to measure the thermomechanical response of openings in salt. A few other creep models have since been developed, all of which are essentially empirical in nature. Of these, the "time-exponential creep model" is the most widely accepted. As a result, creep parameters reported in recent literature correspond to that model. Senseny (1983) reviewed several constitutive laws for creep. Pfeifle et al. (1983) reported experimental data and fitted parameter values for salts from several potential repository sites for the time-exponential creep law

$$\varepsilon_c = \dot{\varepsilon}_{ss} t + \varepsilon_a [1 - \exp(-\xi t)]$$

where

ε_c = axial creep strain
$\dot{\varepsilon}_{ss}$ = steady-state creep strain rate
ε_a = asymptotic transient strain
ξ = relaxation frequency
t = time

Further

$$\dot{\varepsilon}_{ss} = A\sigma^n \exp(-Q/RT)$$

where

σ = axial stress difference
T = absolute temperature
$A, n, Q/R$ = fitting parameters

Pfeifle et al. (1983) gave values for A, n, Q/R, ε_{ss}, and ε_a for Richton and Vacherie Domes and Cycles 4, 5, 6, and 7 of the Permian Basin. Updates and alternatives to creep modeling of salt were provided by Wawersik and Zeuch (1984) and Handin et al. (1984).

The DOE's final environmental assessments (EAs) contain data that are representative of the site-specific data available to date. For example, mechanical properties for the Richton Dome site (U.S. DOE, 1986c), the Cohassett flow vesicular zone (U.S. DOE, 1986b), and for the entire Cohassett flow (Basalt Waste Isolation Project [BWIP], 1983) are available. Mechanical properties for tuff have been presented in Table 2.2. Another recent compilation of mechanical properties of unsaturated tuffs was given by Tien et al. (1985). The property data have been obtained on laboratory-sized specimens and do not necessarily represent behavior in the field. The effect of specimen size on the strength of rock is a subject of controversy. Hoek and Brown (1980) thought that for most rocks a strength reduction occurs with increasing specimen size, a trend observed by many experimentalists. For unjointed, intact rock specimens between 10 and 200 mm in diameter, Hoek and Brown gave the following empirical relationship:

$$\sigma_{cd} = \sigma_{c50} (50/d)^{0.18}$$

where σ_{cd} is the uniaxial compressive strength of a specimen with diameter d(mm), and σ_{c50} is the compressive strength of a 50-mm diameter specimen.

TREATMENT OF DISCONTINUITIES

Before probabilities can be assigned to rock failure caused by thermomechanical effects, the potential mechanisms involved in rock movement along or across discontinuities must be understood. Slip along joints, closing and opening of fractures (aperture changes), rotation of a block with respect to neighboring blocks, and sliding along interfaces can induce rock failure or instabilities. Realistic predictions of site-specific behavior must include models that take into account geometrical and mechanical properties of discontinuities, unless a continuum approximation can be justified. In some cases where joints or fractures are randomly oriented, it may be sufficient to consider only the bulk mechanical properties.

The level of understanding of the behavior of intact rock is much higher than that for rock containing discontinuities. Mathematically, the elastic-plastic theory is often adequate to describe the behavior of intact rock, because the continuum approximation is generally valid. For a limited number of elastic-plastic problems, even analytical solutions are available where material property values could be substituted to obtain the desired answer. However, discontinuous behavior is difficult to characterize mathematically and difficult to measure in the field and laboratory. Unfortunately, most rock masses contain discontinuities, and the mechanical behavior is typically dominated by discontinuities. The state of knowledge and data bases for developing the constitutive relations for such rock masses are too limited at this time to be reliable. Application of models is possible and desirable, but uncertainties in the predicted behavior are large, especially when models have not been validated. The occasional claim by some analysts that thermoelastic calculations (numerical or analytical) are a conservative estimate of the thermomechanical response must be viewed with skepticism. Only some aspects of the thermomechanical response predicted by a thermoelastic analysis (e.g., surface uplift) are conservative. If processes that can lead to subsequent failure are not included in the model, whether and where failure would occur cannot be reliably determined.

Donath and Cranwell (1981) presented a method for assessing the probabilities of new fault development or of movement recurring on an existing fault. Determining these probabilities requires specifying the appropriate failure criteria and rock parameters as well as a knowledge of the state of stress. Donath and Cranwell (1981) stated that even though representative failure criteria and rock parameters can be determined with reasonable confidence, changes in stress levels cannot be predicted at the same level of confidence. Therefore, conditional probabilities will have to be calculated. Predicting changes in stress levels becomes even more difficult when temperature changes and thermomechanical effects must be considered.

Frequently, the effect of existing fractures is indirectly accounted for through the use of rock-mass properties that inherently represent a weaker rock compared with intact rock. This may be adequate for scoping, far-field calculations but may not provide the information needed to calculate stability of openings or onset of progressive failure. Hoek and Brown (1980) discussed strengths of rock with multiple discontinuities and heavily jointed rock masses and suggested that rock masses containing four or more sets of discontinuities should be treated as mechanically isotropic. According to Bray (1967), the overall strength of a rock mass with multiple discontinuities is estimated by the lowest strength envelope to the individual strength curves. Approximate equations for the strength of intact rock and heavily jointed rock masses have been developed by Hoek and Brown (1980). Parameters for these equations are estimated using subjective descriptions of the rock quality and composition. Hoek and Brown gave criteria and parameter values that are widely used by the rock-mechanics community.

If elastic solutions are used, they must be used cautiously in determining stress distributions around underground openings to prevent possible misapplications. In addition to stress levels, the stability of openings in fractured or jointed rock masses depends on the ratio of the excavation size to the size of the blocks in the rock mass (Hoek and Brown, 1980). Instabilities of underground openings can, therefore, occur because of either structural failure or strength failure. Elastic solutions also do not account for stability enhancement due to artificial support (e.g., rock bolts, shotcrete, steel sets, concrete liners, and steel lining), which is commonly used in underground mines. Increased stability can often be achieved passively by proper choice of shape and orientation of the opening for a given state of in-situ stress.

Attempts have been made to include the effect of joints in continuum models. Barton (1982) and Barton et al. (1983) proposed models to simulate joint behavior for repository applications. Based on Barton's (1973) earlier work, Wahi, Trent, et al. (1980) formulated a joint-slip model. Quasi-static and dynamic responses of underground openings in granite, salt, and shale were calculated using nominal property values reported in the literature. The joint-slip model was tested with simple test geometries to provide a comparison with direct shear experimental data. More recently, Zimmerman and Blanford (1985) reported results of model validation efforts in which in-situ measurements in welded tuff were compared with predictions from a numerical model. The effect of joints was numerically simulated with normal compliance and shear parameters. They concluded, in part, "that a jointed welded tuff can be represented under relatively simple stress states with a jointed rock continuum model using existing technology to provide model input parameters." Some modeling factors that could be significant but were not considered are joint orientation, joint dilation, hysteresis, and complex stress states with load reversal.

In analyzing the response of a hypothetical repository in granite, Ratigan (1977) first performed a thermoelastic analysis. By assuming that failure would occur along existing fractures and using a Mohr-Coulomb shear criterion, he identified regions of failure. The dependence of strength-based failure on joint orientation has been demonstrated numerically (Ratigan, 1977). Maxwell et al. (1982) demonstrated similar dependence on joint orientation in analyzing horizontal emplacement in granitic rock. Maxwell et al. (1984) reported the predicted joint response of Queen Grayburg siltstone due to excavation and construction of a repository shaft.

Large, single discontinuities such as faults or shear zones should not be treated the same way as ubiquitous fractures and joints. Some of the more sophisticated computer codes and models permit definitions of discrete discontinuities such that special properties along and normal to the discontinuity can be prescribed. For example, STEALTH (Hofmann, 1981), SPECTROM-32 (Fossum et al., 1989), and SANCHO (Stone et al., 1985) have been used in thermomechanical analyses where faults or sliding surfaces have been included.

NUMERICAL ANALYSES

Many thermal and thermomechanical analyses have been performed in the past decade in support of repository-related activities. The objectives included feasibility studies, preliminary design calculations, benchmarking, simulation of field tests, parametric and sensitivity analyses, model development, and model validation. To give the reader an appreciation for the variety of available thermomechanical modeling techniques (analytical or numerical), some recent analyses are described briefly here. The results have been included to put the magnitude of concerns in perspective. For instance, if a reliable thermal-response prediction shows that the peak temperature anywhere in the rock will be less than 200°C, collecting data and building models for failure mechanisms at much greater temperatures is unnecessary. On the other hand, if excessive closures are being predicted for an opening in salt, for example, many retrievability-related concerns come into play that may not be relevant in other media.

In thermal modeling, the two most common types of numerical methods used are finite-difference and finite-elcment methods. Many variations and options within each of these two methods are available. The numerical algorithms can be implicit or explicit, use successive over-relaxation, have time-step stability constraints, or be unconditionally stable. In mechanical or thermomechanical modeling, finite-element, finite-difference, displacement-discontinuity, and hybrid schemes are used most frequently. A majority of the thermomechanical codes require use of a separate thermal code to generate the temperature fields that are used as time-dependent input to the mechanical response code. The STEALTH code is an exception in that it simultaneously solves for the thermal response in each computational cycle and uses the deformed configuration (mesh geometry) at any given time. However, a major disadvantage of the STEALTH formulation is the time-step constraint for a stable thermal solution. The status of heat-transfer, thermomechanical, and coupled phenomena codes in 1984 has been summarized by Ubbes and Duguid (1985). Most thermomechanical phenomena can be modeled with existing codes, but fewer than half the codes had been verified and none had been validated in 1984.

Thermal Analyses

Generally, calculations of thermal response are easier and more economical than mechanical or thermomechanical calculations. The uncertainties in measured thermal property data are also small in comparison with mechanical properties. The simplest analyses assume heat conduction (only) in an isotropic, homogeneous medium with linear (constant) properties and compute the transient response to a heat source that can be described analytically (e.g., constant or exponentially decaying). In contrast, a complex numerical calculation might include stratigraphy, anisotropic conductivity, temperature-dependent thermal properties, arbitrary heat source(s), enclosure radiation, and convection along cavity walls. For repository applications, the scale of thermal calculations can be very near field, near field, or far field.

Rickertsen et al. (1982) used HEATING5 to calculate near-field thermal response for sequentially loaded spent fuel or HLW in salt. (Most other analyses assume instantaneous emplacement of all waste). HEATING5 is a three-dimensional, finite-difference, numerical code that solves the heat conduction equation to provide transient or steady-state temperature distributions in the most general type of media. It allows anisotropic thermal conductivities, layered media with nonhomogeneous properties, temperature-dependence of thermal properties, enclosure radiation, and an arbitrary thermal source description. Calculated maximum temperatures for a hypothetical spent-fuel repository in salt (Figure 2.1) were made for a maximum local loading of 21 W/m^2 and average thermal loading of 8 W/m^2.

McNulty (1987) used the TEMPV5 code to predict near-field thermal response at potential repository sites in domal and bedded salts. TEMPV5 is a semi-analytical, finite-line-source, heat-transfer code that can handle a temperature-dependent conductivity in an isotropic, infinite medium. The code models

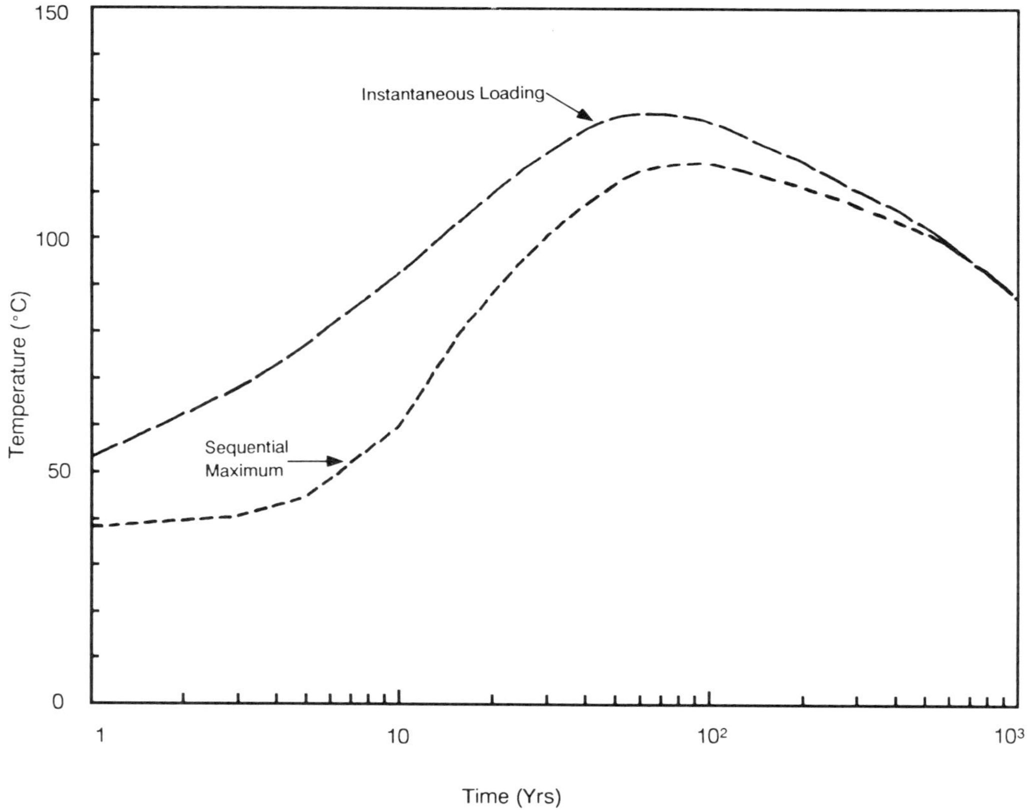

Figure 2.1. Temperature histories at selected locations in a hypothetical spent-fuel repository in salt (after Rickertsen et al., 1982).

individual waste packages as finite line sources and uses linear superposition of temperature contributions from individual finite line sources to calculate the temperature at a point. McNulty also benchmarked his results against numerical predictions with more sophisticated codes. Although site-specific thermal properties for salt were used, the laboratory thermal conductivity values were arbitrarily increased by 40% to account for sample disturbance and other speculated deviations. Calculated peak temperatures at salt-overpack interface for seven salt sites ranged from 130.3 to 416.7°C. The calculations show that the influence of using laboratory values of conductivity is significant. This important issue could impact the thermal load density in a salt repository.

Sundberg and Eaton (1982) performed a three-dimensional thermal analysis for a hypothetical high-level-waste repository in welded tuff. For a gross thermal loading of 100 kW/acre, peak emplacement-hole temperatures were predicted to be 225°C (when water boiling is allowed) and 188°C (no boiling). The difference is due to the lower thermal conductivity after boiling occurs. Even though some analyses have been conducted with two-phase flow, whether advection has a significant effect on total heat transfer is unclear. If total porosity and moisture content are small, advection can perhaps be ignored as far as heat transport is concerned. Attention must still be directed to possible hydrothermal or hydrochemical alterations that might affect mechanical properties of the rock. St. John (1985) performed another thermal analysis for a spent-fuel repository in welded tuff. Temperatures in the vicinity of a waste canister and an emplacement drift were calculated for ventilated and sealed drifts. Peak temperatures at the borehole wall were predicted to be 215°C (ventilated case) and 240°C (sealed case). The peak occurs at about 3 years for the ventilated case and 9 years for the sealed case. An important conclusion of that study was that two-dimensional analyses adequately predict the temperature field adjacent to the drift, provided that canisters are emplaced relatively close together.

Thermal analyses of various options of waste emplacement in basalt in the Canadian Shield (Tsui and Tsai, 1985) indicate far-field temperature profiles at a thermal-load density of 8.2 W/m^2 (about 33 kW/acre) (Figure 2.2). The effect of container pitch on maximum temperature of basalt and waste package components (BWIP, 1985) has also been studied.

Results of certain other thermal analyses are presented below under "Field Testing and Model Validation." The reliability of thermal-response predictions is relatively good. The largest uncertainty is in the thermal output from spent-fuel canisters. At one time, the thermal output of emplaced waste was expected to be known fairly precisely; however, the large-scale packing of spent fuel from different reactor sites and of different ages introduces uncertainties in the nature and thermal output of the inventory of a given package.

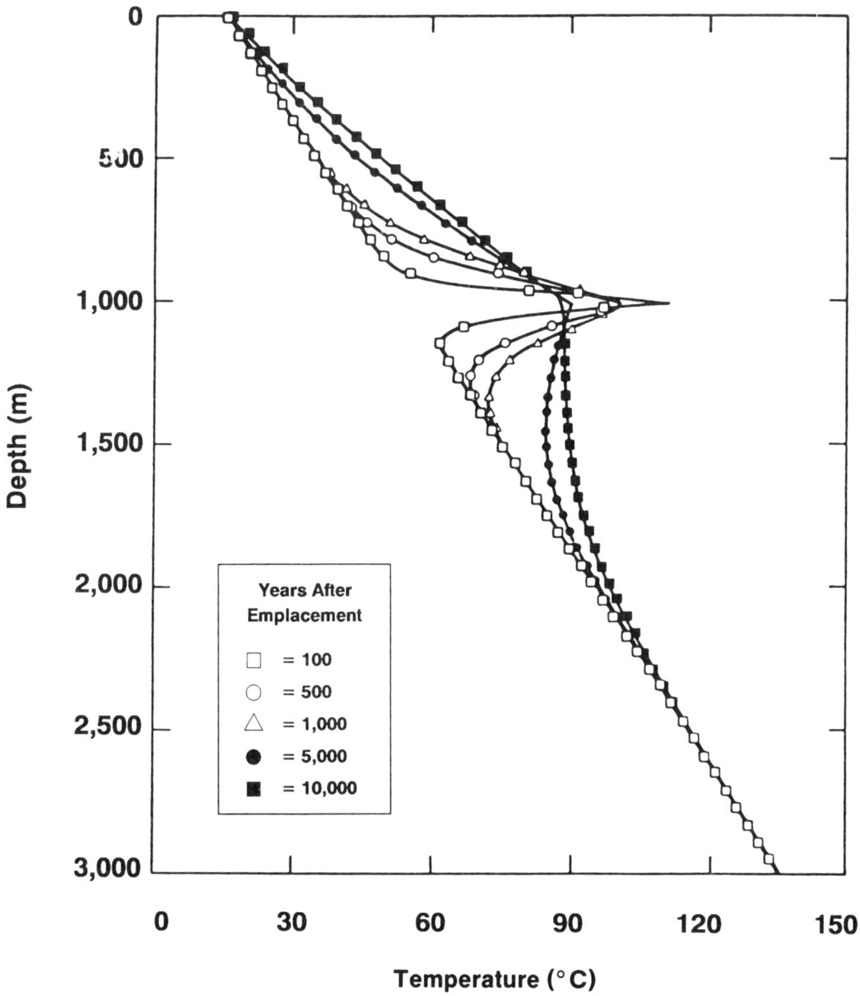

Figure 2.2. Temperature profile for a repository at 1000 m depth in basalt with areal thermal loading density of 8.2 W/m^3 (U.S. Department of Energy, 1984, Figure 6-5).

Thermomechanical Analyses

Before experiments at the WIPP were designed and conducted, a number of thermomechanical calculations were performed. Some calculations were benchmarking exercises with many participants, while others were pretest predictive calculations with only two participants. In one benchmarking exercise, nine structural codes were used to simulate the second WIPP Benchmark Problem, involving several nonlinear features that include a bedded salt

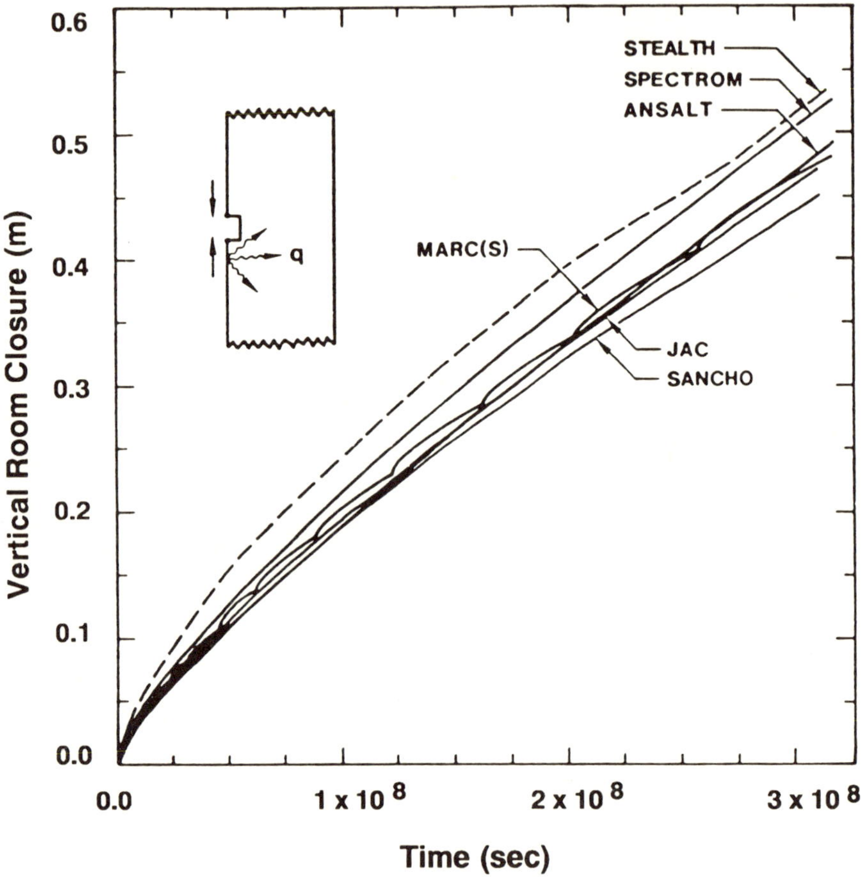

Figure 2.3. Revised vertical closure histories for the heated room, second WIPP Benchmark Problem (Morgan et al., 1981, Figure 52).

stratigraphy, slip along certain stratigraphic interfaces, salt creep, temperature-dependent conductivity, and enclosure thermal radiation (Morgan et al., 1981). Disparities among the results of different codes (Figure 2.3) were much greater than shown when initial comparisons of preliminary results were made. Subsequent checking revealed programming errors and incorrect problem formulation; corrections resulted in much better agreement.

Wahi et al. (1978) calculated the thermomechanical response of three hypothetical repositories in bedded salt. Near-field and far-field thermal and mechanical responses were obtained. Isothermal calculations for the 60-kW/acre, spent-fuel repository at 3323 years following emplacement (Figure 2.4) considered the appropriate heat source, but not the drifts (open or backfilled. Vertical displacement histories for a surface point on the axis of symmetry (i.e.,

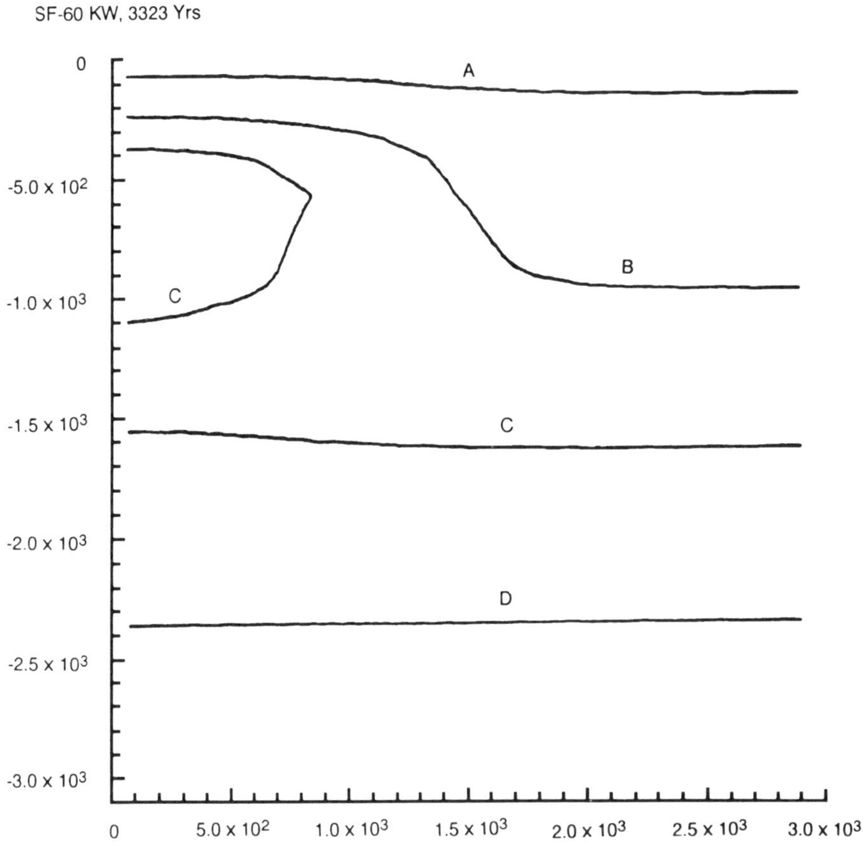

Figure 2.4. Isothermal map at 3323 years for hypothetical spent fuel (60 kW/acre) repository in bedded salt (Wahi et al., 1978). Contour levels: A, 20.0°C; B, 40.0°C; C, 60.0°C; D, 80.0°C.

point of maximum uplift) differed considerably for spent fuel and high-level-waste repositories.

Nelson and Fossum (1985) used numerical model calculations to investigate stresses exerted on waste packages emplaced in a high-level-waste repository in salt. Peak stresses predicted by these model calculations are substantially lower when an initial air gap exists between the waste package and the surrounding salt. Conditions of plane stress and plane strain, which bound the realistic case, were used in calculations with and without air gaps. The time at which the initial gap closes and the magnitude of stress following contact depend strongly on the creep parameters used. Nelson and Fossum (1985) stated that for early times the plane strain geometry better approximates the real situation. For plane strain conditions, stresses are substantially greater than for the corresponding plane stress cases. Moreover, an areal heat density of 21 W/m^2 was used; the proposed heat density in salt repositories in the final EA's is approximately 30 W/m^2. Greater stress levels should be expected for the 30 W/m^2 case.

A thermomechanical analysis of seismic response of heated repository tunnels used three different earthquake records as applied excitation to generic repository models in salt, granite, and shale (Wahi and Trent, 1982). The stability of openings in response to excavation, heating, and subsequent earthquake loading was predicted. Failure models included joint slip along existing discontinuities and creation of new fractures. Not surprisingly, granite and salt were found to provide a relatively stable environment and shale was not, assuming that the material properties and failure models used are representative of those rocks.

The final environmental assessment for the Richton Dome site reported results of room-scale calculations that predict a roof-to-floor centerline closure of 0.35 m in 5 years for 5.5-m x 5.5-m rooms (U.S. DOE, 1986c, p. C.8-7). DOE further asserted that "... thermomechanical calculations can reasonably predict the amount of room closure and far-field behavior due to creep in the salt host rock. However, these thermomechanical calculations are not very reliable when predicting the response of underground tunnels in the early stages of heating."

Thermomechanical analyses by Loken et al. (1987) predicted a maximum surface uplift of 1 m for conceptual repositories in bedded salt. These analyses were thermoelastic and ignored creep deformation. Although the final EA for Richton Dome (U.S. DOE, 1986c) implied that these estimates are conservative and overpredict the thermomechanical response, surface displacement alone is not a true indicator of potential problems caused by rock movement.

FIELD TESTING AND MODEL VALIDATION

In-situ test programs have been performed or are in progress in several countries in attempts to characterize rock behavior and measure in-situ thermal/mechanical properties (Table 2.3), but limited experience exists in the type of combined temperature and stress environment expected in a repository. Munson et al. (1985) described underground tests to determine the in-situ thermal/structural response of bedded salt at the WIPP facility. Progress and data from these tests are being reported continually. For example, Morgan et al. (1985, 1986) presented field data from WIPP and its use in evaluating drift response models. Wagner (1985) compared and evaluated field and numerical results from a heater test at Avery Island in Louisiana. BWIP conducted in-situ tests at the Near Surface Test Facility (NSTF) that have been analyzed and "predicted" with limited success (e.g., BWIP, 1981). Hocking et al. (1980) made numerical

Table 2.3 Repository-Related Field Tests

Test acronym	Location
PSV	Project Salt Vault, Lyons, Kansas
ASSE	Asse Salt Mine, West Germany
NSTF	Near Surface Test Facility, Hanford Reservation, Washington
STRIPA	Stripa Mine, Sweden
WIPP	Waste Isolation Pilot Plant, Carlsbad, New Mexico
Avery Island	Avery Island Mine, Louisiana
Block Tests:	
G-Tunnel (Tuff)	Nevada Test Site, Nevada
CSM (Gneiss)	Colorado School of Mines, Colorado
NSRF (Basalt)	Hanford, Washington
CURL	Canadian Underground Research Lab, Manitoba

predictions of two heater tests at NSTF. Hart et al. (1985) performed a numerical analysis of a triaxial loading test on a large basalt block. Their primary conclusion was that the complex deformation behavior may be attributed to the hexagonal shape of the basalt columns. The Yucca Mountain Project plans to conduct extensive in-situ testing as part of site characterization. A significant reduction in uncertainty of site-specific thermomechanical effects should occur if the tests are designed and conducted properly. An analytical tool that is starting to receive increasing attention in determining data needs as well as in defining the applicability of limited data to represent large rock volumes is based on geostatistics. This technique has the potential to guide geomechanics testing by optimizing the number of tests and sample locations. Buckley et al. (1986) discussed a geostatistical approach that could be used to evaluate the adequacy and representativeness of host-rock test data obtained during site characterization.

Zimmerman and Blanford (1985), in an initial effort in geomechanical model validation, compared numerical model results with data from the heated-block experiment in G-Tunnel at the Nevada Test Site. A joint annual report on the German-American brine migration tests at Asse Salt Mine in West Germany also reported thermomechanical data and measured and predicted horizontal room closure (Rothfuchs et al., 1984). Another experiment at Asse was numerically simulated by Wahi, Baltes, and Salz (1980). The temperature response to a heater inserted in a borehole and the borehole convergence were predicted using the power law of creep, which accounts for transient creep.

APPLICATION OF STATISTICAL AND PROBABILISTIC TECHNIQUES

Probabilistic assessments may be divided into two time periods: an operational (or preclosure) period and a long-term (or postclosure) period. Potential deleterious effects that can result from rock movement in response to thermomechanical loads differ in nature for the two periods. The amount of heat accumulated in the rock mass during the operational period is relatively small, but temperatures and temperature gradients in the vicinity of rooms and waste packages are large. Moreover, thermomechanical stresses local to openings are also large, causing convergence of the openings. If rock support is not adequate, the openings may collapse; ultimate damage could either be local or extend into the rock mass by progressive failure. Over longer times, the total heat in the rock mass will be large, but maximum temperatures and stresses will be less. After openings have been backfilled and the fill material consolidates, the net movement of rock is likely to result in an uplift followed by subsidence as cooling occurs. Several mechanisms could be triggered by the changing stress and temperature fields. Fracture apertures might be reduced due to larger compressive stresses, new fractures could form and propagate, and movement along faults or shear zones could be initiated by the accumulated strain energy.

The relatively small extraction ratio (<20%) planned for subsurface repositories makes surface subsidence unlikely during the first 50 years or so. Therefore, the primary concern before closure is room and pillar stability. Instability or failure can be structural (block movement) or stress related (stress in excess of strength). Rock strength decreases with increasing temperature, but thermal stresses increase; this synergistic effect is a new design concern for underground structural stability and is not well understood. Salamon (1974) and Salamon and Munro (1967) presented probabilistic concepts of pillar design for underground openings. The safety factor, η, in pillar design may be defined as

$$\eta = \frac{\text{pillar strength}}{\text{average pressure on pillar}}$$

Values of $\eta < 1$ indicate failure. If safety factors are predicted for a large number of cases where subsequent failure occurred, the η value will be near unity. The degree of scatter provides a measure of precision of predictions. Variation in the critical safety factor (i.e., safety factor at which failure occurs) can be described by a frequency distribution with a density function $f(\eta)$ (Salamon, 1974); $f(\eta)d\eta$ is the relative frequency of occurrence of pillars with critical safety factors in the interval $[\eta, \eta + d\eta]$. Empirical expressions are available in the literature to estimate the strength term, and the average pressure can be computed analytically or numerically. To estimate strength, Obert et al. (1946) recommended the expression $C[0.778 + 0.228\ w/h]$, where C is uniaxial compressive strength, w is pillar width, and h is pillar height. Holland (1962) and a number of other investigators have proposed the expression $Ch^{\alpha}w^{\beta}$, where α and β are determined from laboratory or in-situ tests. Although temperature is not a parameter in these empirical expressions, the concepts presented for the purely mechanical behavior are applicable to thermomechanical situations. The strength reduction at greater temperatures and the thermomechanical stress state in the pillar do warrant a development of more sophisticated expressions, empirical or otherwise.

Identifying mechanisms that can affect long-term repository performance is speculative. Yet a cause-and-effect relationship is thought to exist such that a random treatment could not be justified. Numerical models do exist that can be used to perform deterministic thermomechanical analyses of a given repository design for a given geologic setting. To account for uncertainties in material properties, model accuracy, model validity, and site stratigraphy, Monte Carlo simulations using deterministic models are recommended. For discrete discontinuities such as faults or shear zones, Donath and Cranwell (1981) used a probabilistic approach; however, deterministic analysis would still be needed to establish the state of stress that is conditional in this approach.

West (1983) compared real and predicted geology in five tunnels that ranged in length from 1.5 to 28 km with up to 300 m of overburden. Information from borehole logs, records, and data banks enabled a reasonably good prediction of the site geology. At one site during construction, however, the rock was found to have more joints than had been suggested from site investigation. This was attributed to the fact that vertically oriented exploratory boreholes are unlikely to intersect many of the vertical or steeply inclined joints. At another site, 64% of the tunnel length showed a match between predicted and as-found geology. Among West's (1983) suggestions to improve predictions are inclined boreholes, boreholes at the shaft sites, and logging with a borehole camera.

O'Rourke et al. (1986) have used statistical methods to analyze correlations of index test data with laboratory test data of rock properties for the same core in the Paradox Basin. Index tests are quick, economical, and easy to perform; those reported by O'Rourke et al. (1986) include point-load test, Schmidt hammer-rebound test, and abrasion-hardness test. Four types of variability affect the physical test results: (1) natural variability, (2) sampling variability due to sampling process, (3) sample-preparation variability, and (4) analytical variability introduced by the test methods. O'Rourke et al. further stated that test results tend to represent stronger samples, because weak samples can disaggregate during drilling, core retrieval, or specimen preparation, precluding testing or measurements. Fair-to-good correlations were obtained for predicting unconfined compressive strength and Young's modulus for sandstone and siltstone, whereas correlations were poorer for limestone.

The ultimate concern of performance assessment is the release of radionuclides from the waste containers to the accessible environment. To define the time-dependent source term (i.e., release at waste-package boundary), some knowledge of the times at which the containers will fail is necessary. The thermomechanical environment around the waste package can influence the waste-package failure rate. Not all containers are expected to fail simultaneously. Corrosion and structural deformation control degradation of canisters; however, fabrication and material defects are statistical in nature. Past experience with analogous components and accelerated testing can provide a reasonable data base for predicting canister failure in a repository environment. Probabilistic techniques have been used to estimate the distribution in time of canister failures.

Priest and Brown (1983) described a probabilistic approach for analyzing the stability of slopes excavated in fractured, variable rock masses. Though not specifically developed for repository applications, the method is potentially useful in performing similar stability analyses for the thermomechanical environment expected near repository openings. The method uses a Monte Carlo approach combined with empirical strength criteria to generate cumulative probability distributions of the factor of safety. Additional complexities that need to be considered are temperature dependence of strength parameters, in-situ stresses, and the length of time over which stability is desired. Two different probabilistic techniques were used by Nguyen and Chowdhury (1984) to study spoil-pile stability in coal strip mines. They found excellent agreement between the Monte Carlo simulations and the Rosenblueth method of estimating moments. Their study confirmed the suitability of normal and beta distributions in geomechanics in the context of slope stability.

Mrugala (1984) presented a statistical approach to determine mechanical model parameters for salt. Viscoelastic models consisting of springs and dashpots were assumed to represent the creep behavior of salt. Creep strain equations were written in terms of the model parameters and expressed in terms of equivalent regression models. A least-squares code, NLIN2, was used to fit the observed laboratory values of transient lateral- and longitudinal-creep strains. The deviations between the model and experimental results were expressed in terms of a "minimized function Φ" and the standard error of estimate. Mrugala (1984) showed that more complex mechanical models do not necessarily lead to more precise values of the associated model parameters.

SUMMARY

Rock as a material cannot be characterized with great certainty. Heterogeneities, local anomalies, joints, fractures, and interbeds all introduce uncertainties in rock behavior. These uncertainties extend into models and mathematical idealizations of thermomechanical response of a rock system. The data base on site-specific rock properties and their dependence on temperature is just starting to emerge. The state of the art in thermomechanical modeling permits nonlinear behavior (e.g., creep), discrete discontinuities (e.g., joints, faults), and coupled effects (e.g., thermohydromechanical) to be included. Such models have not been adequately validated and tend to be expensive to use, although limited validation has been achieved for some aspects of thermomechanical response. Not surprisingly, thermal-response validation has been considerably more successful than mechanical or thermomechanical validation.

Probabilistic treatment of thermomechanical effects in the context of subsurface structures has not been used or considered to date. In related applications, however, probabilistic techniques have been developed and applied to the analysis of rock-slope stability and coal spoil-pile stability (Priest and Brown, 1983; Nguyen and Chowdhury, 1984). Donath and Cranwell (1981) have developed an approach for probabilistic treatment of faults and fault movement. Their method is directly applicable in probabilistically analyzing some important failure mechanisms that can be induced by the thermomechanical response. A probabilistic approach for pillar stability, developed by Salamon (1974), seems appropriate for preclosure analyses but needs further development to account for elevated temperature and coupled effects. In general, however, the ability to assign probabilities to the potential disruption of the repository system or one of its components as a direct or indirect result of thermomechanical effects needs to be developed. Until appropriate methods are developed, these probabilities will likely be assigned on the basis of expert judgment.

A careful repository design along with a reasonable factor of safety can significantly reduce uncertainties in the expected thermomechanical response in the preclosure and postclosure phases of a repository. Proper site characterization is essential no matter what technique is chosen to assign probabilities. Identifying major structural discontinuities such as faults, brecciated zones, and brine pockets or gas pockets would help in making appropriate design adjustments and performing relevant scenario analyses.

3

Tunneling and Mining Engineering

Herbert H. Einstein
and Gregory B. Baecher

This chapter reviews probabilistic methods for predicting rock-mass and underground-opening performance pertinent to nuclear-waste isolation in geologic repositories. It identifies mechanisms of unsatisfactory rock-mass or opening performance, which in this context is any performance that increases radionuclide transport beyond that predicted as a design basis. It reviews present methods for assessing probabilities of unsatisfactory performance in general and with emphasis on applications in geotechnical engineering. Methods of performance prediction for underground openings are reviewed, and comments are made on combining probability assessment and performance prediction in probabilistic risk analysis.

An underground opening generally is called a tunnel if one dimension is more than about five times greater than the others, and if the principal axis of the opening is inclined 10° or less to the horizontal. An opening is called a shaft if similar dimensional relations exist but the axis is inclined more than 10° to the horizontal. Equidimensional openings are called caverns. In this discussion, tunnels, shafts, and caverns will all be called underground openings. Openings can be unsupported, can have distinct initial and final support systems, or can have only a final support (often called a liner). Supports can be made watertight in various ways. Szechy (1967) and Ontario Ministry of Transportation and Communication (1976) described basic tunneling technology in detail.

MECHANISMS OF UNSATISFACTORY PERFORMANCE

An underground opening can lead to unsatisfactory repository performance either through direct effects on waste package, backfill, or near-field rock mass or through interaction of the opening with the surrounding geologic environment. A variety of direct and indirect mechanisms of adverse performance exists. Direct mechanisms can be identified reasonably well in an abstract survey such as this chapter. Determining which mechanisms or combinations of mechanisms actually take place in the field is more difficult and usually is case-specific. Identifying indirect mechanisms is difficult.

Direct Mechanisms of Tunnel Performance

Two principal direct active mechanisms of opening performance may lead to loss of long-term waste containment at a repository. First, deformation of an opening may lead to failure of the tunnel support system and open existing discontinuities or create new fractures in the near-field rock mass. In the extreme, deformation could lead to collapse of the opening. Deformation will increase permeability of tunnel support systems and the near-field rock mass and may lead to interaction mechanisms discussed below. Long-term deformation of a rock mass that would normally not show such behavior may be induced by deformation during construction. In contrast, in swelling rock or in rock masses with swelling material in discontinuities, increased permeability may lead to increased swelling, which could have either favorable or unfavorable effects on repository performance depending on other circumstances. Second, deformation of an opening may affect backfill properties or the waste package without inducing fracturing or opening of pre-existing discontinuities. These deformations could reduce the volume and distort the shapes of openings, leading to shortened flow paths, openings in or erosion of the backfill, and possibly other consequences. All these mechanisms may be time dependent.

Deformation of openings can result from various causes. An opening changes the stress field. Construction may weaken wall rocks. Changes in ground-water flow may change cleft-water pressures and thus effective stresses. Temperature changes may cause anisotropic expansion or contraction. These causes are interdependent. Other causes not associated with the initially existing rock mass must also be considered, such as changes in the lithosphere or biosphere (e.g., earthquakes, glaciation) and man-made changes (e.g., large explosions, mineral exploration). Temperature effects, backfill compaction, swelling of backfill or sealing material, and chemical effects of the waste-package leachate may also lead to opening deformations.

Indirect (Interaction) Mechanisms

Indirect or interaction mechanisms resulting from stress changes may be associated with both intermediate and far-field effects of an underground opening. Stress changes can be the direct result of an opening, but in such cases usually are limited to the intermediate field. Stress changes also can be caused by permeability changes and subsequent modifications of cleft- or pore-water pressures. Temperature also influences stresses. Both permeability- and temperature-related stress changes may reach into the far field. During construction and operation, stress increases and decreases can occur at the same point in the rock mass (e.g., because of possible cooling during construction and subsequent heating during operation).

Stress changes can produce volume decreases or increases and shear deformations. The former express themselves in subsidence and heaving, the latter in movement on new or existing faults. Rock bursts are special mechanisms associated with stress changes. Both fault movements and rock bursts cause ground shocks with associated effects. For example, rock bursts can reach measured magnitudes of 3 or greater on the Richter scale (Obert and Duvall, 1967). Stress changes in rocks with swell potential can cause swelling, which is accentuated by direct temperature effects. Swelling also affects permeability.

UNCERTAINTY IN GEOTECHNICAL ENGINEERING

Both stochastic and subjective uncertainties affect performance predictions for underground openings.

Stochastic Uncertainties

Stochastic uncertainties, in this case, principally include spatial variation and its characterization. For example, the geometry of pervasive fracturing in a rock mass (e.g., jointing) is too complex to be measured or specified deterministically. Thus, stochastic representations of the frequency of fractures with particular geometric or mechanical properties are now widely used (Priest and Hudson, 1976; Dershowitz, 1984). These representations characterize fracture systems probabilistically. Values of parameters of stochastic representations are estimated from field data using statistical techniques. Probabilistic statements about the frequency of occurrence of specified conditions, e.g., adverse persistence or interconnectivity of fractures (Einstein, Baecher, and Veneziano, 1979), are deduced from the stochastic model. As a result, the assessment of uncertainty for the stochastic case follows explicit and well-established procedures. On the whole, uncertainties addressed in probabilistic modeling of the sort discussed above tend to be interpreted stochastically in practice.

Subjective Uncertainties

Subjective uncertainties in performance predictions for underground openings principally arise from geological or geomechanical hypotheses and the likelihood of undetected and possibly unexpected features in host formations. For example, the existence of an undetected shear zone is a one-time event that can only be assigned a stochastic probability through a leap of imagination. Estimating a probability for this event depends on geological theory, indirect evidence at the site, and reasoning about the chance that the executed site-characterization program would have failed to detect a disruption. As a result, the assessment of uncertainty for such cases rests upon subjective probability theory and the variety of methods that have evolved for quantifying expert opinion.

The practice of geotechnical engineering and engineering geology has about 15 years' experience with formal methods of subjective probability solicitation. Indeed, applied geology in the broader sense has been among the earlier disciplines to benefit from subjective probability theory (Grayson, 1960; Kaufman, 1963; Folayan et al., 1970; Wu, 1974; Baecher, 1972, 1983a; Einstein and Vick, 1974; Einstein et al., 1978; Kim et al., 1985; Call, 1985; Pentz, 1985; Bernreuter et al., 1984; Veneziano and Van Dyke, 1985; U.S. Geological Survey, 1975; Wight, 1980). Some difficulties remain with practical assessment techniques, however (e.g., Capan, 1976; Hogarth, 1975; Dalkey, 1975; Dalkey et al., 1970; Winkler, 1968; Morris, 1974).

Updating Uncertainties

Performance predictions for underground openings are sensitive to results of site characterization and to information collected during preliminary phases of construction. For this reason, updating of uncertainties as new information becomes available is an important operational part of any probabilistic analysis of underground-opening performance. This updating is accomplished by means of Bayes' theorem or similar mathematical techniques, in which the probability distribution $f_X(x)$ over some uncertain parameter X is modified according to the relation

$$f_X(x|\text{data}) \propto f_X(x)\, L(\text{data}|x) \tag{3.1}$$

where $f_X(x|\text{data})$ is the modified or updated distribution of probability over X, given a set of data; $f_X(x)$ is the original distribution before the data were collected or incorporated; and L(data|x) is the likelihood (i.e., conditional probability) of the observed data given x. This equation states that probabilities inferred after observing certain evidence should be proportional to the product of those probabilities held before the observations were made and the likelihood of the observations conditioned on possible hypotheses or parameter values.

Updating in this form plays a central role in the geologic prediction and updating model of Einstein and his colleagues and in formal approaches to what geotechnical engineers call the "observational approach" (Casagrande, 1965; Peck, 1969a; Asaoka, 1978). It also plays a central role in search theory (discussed below) and in formalized site-characterization strategy analysis (Baecher, 1972). In recent work, probabilistic updating by Kahlman filtering has been applied to the performance of underground openings by use of in-situ monitoring data (Jurina et al., 1977).

SITE CHARACTERIZATION

Geologic features and engineering parameters required for tunnel-performance predictions are described during site characterization. These descriptions can be quantitative or qualitative and should reflect associated uncertainties. The objective of site characterization for tunneling or underground construction is to develop information on regional and local geology, structural features, depth and character of overburden, ground-water conditions, intact rock characteristics, and in-situ stresses. These features or parameters are described from information obtained in outcrops, boreholes, pilot shafts, and tunnels and through geophysical methods. Descriptions can be qualitative or quantitative; both are useful in tunnel-performance and interaction prediction.

Site characterization is evolutionary. It begins typically with air-photo analysis and geologic mapping during prefeasibility studies and progresses through exploratory borings, laboratory testing, and large-scale in-situ tests during the design phase. The purpose is to develop an operational concept of the geology of a site, to estimate geometries and material-parameter values for engineering analysis, and to ensure that details of the site geology that might adversely affect facility performance have been identified. Because not all potential geotechnical problems can be identified no matter how extensive the exploration program, site characterization must continue throughout construction. Foundation grouting, rock support, and amount of reinforcement in tunnel linings are some examples of construction items requiring definition during construction.

A convenient organization for site characterization based on functional tasks is reconnaissance, mapping, search, and parameter estimation. During reconnaissance, qualitative information is collected and a geologic concept of the site is developed. Mapping recognizes geologic forms and reconstructs zonations of the site. Search looks for geologic details that might affect facility performance. During parameter estimation, field or laboratory testing determines parameter values and boundary and initial conditions for engineering analysis.

The reconnaissance function remains, as before, an attempt to develop initial hypotheses about the geologic structure of a site, to establish an initial outline of the geometry of site formations, and to create a list of potential details of site geology that might later cause problems. The remaining three functional areas address specific data needs, testing strategies, and data analysis.

Considerable work has been done on probabilistic analysis of mapping, search, and parameter estimation for geotechnical site characterization. Although only part of this work specifically considers underground openings, most is pertinent to waste repositories. Some of the work is theoretical, some practical. Each of the three problem areas is considered separately below, but the general approach to probabilistic analysis is similar across all three.

Probabilistic Models in Site Characterization

As a first approximation, the natural variability of rock masses is of three types: stratigraphic variability among homogeneous zones, spatial variability within zones, and geologic anomalies or details. Site characterization attempts to describe this variability (e.g., Dowding, 1978). Variability of rock masses is not "random," it is spatial. In principle, sufficiently many observations can characterize such variability in detail. This is not possible in practice, but only for operational reasons. Data are limited, because not every point can be observed; therefore, properties of unobserved locations must be inferred. Probabilistic models are used to analyze the uncertainties associated with those inferences and to evaluate how effort might best be allocated to reduce uncertainty. The uncertainty about subsurface conditions results from lack of information and not from inherent randomness. Probabilistic models characterize the rules of inference adopted for interpolating between observations.

Most probabilistic models used in site characterization are based on the artifice of a generating process. The actual spatial variability of the subsurface is assumed to be one realization of this process, like one flip of a coin. By hypothesizing some function by which spatial properties of the geology are generated, inferences can be placed within a logical context, and conclusions can be drawn from statistics. The suitability of a particular generating function is empirically verified from data.

Because site characterization is a fundamentally subjective undertaking, much of the literature is based on a Bayesian view (Tang, 1971; Wu, 1974; Meehan, 1984). Within this view, inferences are made according to Equation 3.1. The inference can be divided into two parts: a prior distribution that summarizes existing geologic information (usually subjective), and a likelihood that summarizes relevant information in the exploration program and observations (Figure 3.1a). With little observational evidence, the posterior inference (Equation 3.1) mostly depends on prior information. As observational evidence grows, the weight of evidence shifts and the likelihood becomes more and more important until it predominates.

In Equation 3.1, the statistical model of exploration is entirely contained in the likelihood function. Typically, this likelihood has two parts, a geologic model and an observational model (Figure 3.1b). The geologic model is an abstract generation model of the spatial variation of the rock mass. The observational model characterizes the way measurements are made on the geologic model. Dividing the likelihood this way leads to an expression of the form

$$L(z|\theta) = \int f(G|\theta)\, f(z|G)\, dG \tag{3.2}$$

where the first term on the right is the geologic model, expressed as a probability distribution over geologic states G given a set of parameters θ. The second term is the observational model, expressed as a probability distribution over the set of observations z given the geologic state G. The integral is taken over the set of possible geologic states G (e.g., Veneziano, 1979). Geologic and observational models address different facets of the exploration problem. The geologic model describes the generating process for the subsurface formations, and the observational model describes the operating statistics of the sampling devices and the spatial array of observations.

Uncertainty about a set of parameters being estimated, θ, is determined from Equation 3.2 by combining prior information on θ expressed as a probability distribution with the likelihood function of the observed data $L(z|G)$. An exploration program is optimized by modifying the type, number, or spatial array of tests, which in turn define $f(z|G)$. Changing $f(z|G)$ correspondingly changes $L(z|G)$ and thus increases or decreases the uncertainty in θ expressed by the posterior probability distribution $f(\theta|z)$ (Eq. 3-1) (Veneziano and Faccioli, 1975).

The problem in planning site-characterization programs is that the actual observations z are not all known until exploration is complete. During the planning stage, z's are uncertain quantities. Therefore, planning an efficient site-characterization program involves projecting the range of observations that might be made, ascertaining the uncertainty in the estimate of θ that would result from each possible set of observations, and then weighting possible outcomes by their probabilities. In statistical decision theory, this procedure is called evaluating the expected value of sample information (e.g., Baecher, 1972).

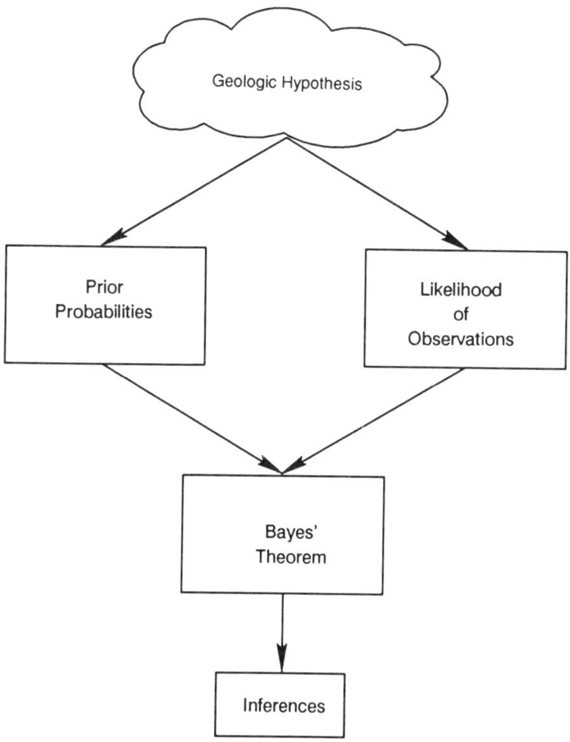

Figure 3.1a. Inference of geologic conditions from observation using Bayes' Theorem (Baecher, 1985). Copyright 1985 by the Engineering Foundation.

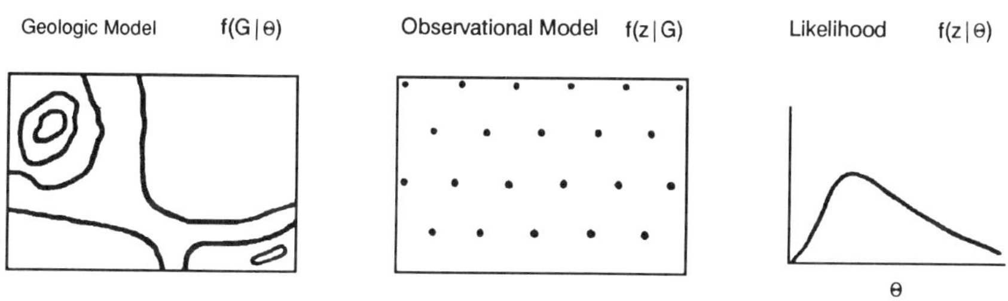

Figure 3.1b. Abstraction of the information obtained in exploration into a geologic model and an observational model (Baecher, 1985). Copyright 1985 by the Engineering Foundation.

This entire procedure reduces to contingency planning: identifying what possible observations might be obtained with a particular strategy, what the consequences of those observations might be on inferences and predictions, and how the worth or utility of the observations might be altered by changing the exploration strategy (e.g., Einstein et al., 1978; Einstein et al., 1983; Einstein and Baecher, 1983).

Mapping

The mapping task in site characterization recognizes patterns in site geology and reconstructs the large-scale zonation of formations at the site. Zonation refers to separation of the site into homogeneous zones for the purposes of engineering modeling. Mapping is usually directed toward one or two types of variability: differentiation of a profile into strata (e.g., Dowding, 1978), and differentiation of a site into zones (e.g., Wu et al., 1975).

The inference of stratigraphic variability is among the most important tasks of subsurface exploration, but little probabilistic work has been done on the problem. The only basis upon which strata can be interpolated is a clear understanding of the geologic processes leading to their formation. Some geologic work has attempted to fit simple curves or surfaces to strata boundaries. The most common approach is trend surface analysis (e.g., Krumbein and Graybill, 1965), which fits a polynomial surface by regression analysis. Fourier series as approximations to boundaries or topographic surfaces (J. C. Davis, 1973) and kriging of strata contacts (Journel and Huijbregts, 1978) also have been used.

If certain types of rocks are observed to follow mathematically describable geometries in the field, a spatial model can be justified. This has been suggested for fold geometry in structural geology (Busk, 1929). However, work by Wahlstrom (1964), Robinson and Lee (1964), and Dowding and Miller (1975) casts doubt on the practicality of this approach.

Interpolating between borings is difficult because the geometry of natural units is controlled by geologic processes that are difficult to model. If geologic control is ignored, however, discrete random fields can be used to model both vertical stratigraphy and horizontal zonation. If geologic processes leading to the units are simple enough to be modeled, random process models of the genesis of a profile might be used. This approach has been applied in sedimentology (e.g., Merriam, 1972) and led to the development of models of sedimentary-sequence development. Finding a place for such techniques in geotechnical exploration, either for a particular site or at the research level of establishing satisfactory trend models, seems far away.

Related work in statistical geology has been done on sedimentary sequences and transformations between strata perpendicular to bedding planes.

Transformations between sedimentary facies seem to have Markovian properties represented by a transformation matrix that seems to remain constant over large geographic areas (Vistelius, 1961; Krumbein and Sloss, 1963). Whether such findings can be exploited in geotechnical exploration is unclear. In petroleum exploration, the transition matrix has been used in conjunction with pattern-recognition techniques to correlate boring logs over separation distances on the order of kilometers. Much work in the earth sciences and in pattern recognition and image processing has specifically addressed the two-dimensional problem of mapping horizontal zonation and reconstructing discrete patterns. The following remarks pertain only to the two-dimensional case, although certain of the comments made can be extended to three-dimensional variability.

As in all exploration analysis, the fundamental question, whether the data are analyzed statistically or intuitively, is how to specify the geologic model. Abstracting the problem somewhat, statistical mapping assigns a probability distribution to each point at the site, defined over the set of possible geologic units or rock types.

Mapping of nonstructurally controlled units can be modeled by proximity rules, much like the concept of Thiesen polygons. Probabilistically, the units are idealized as discrete correlated random fields. The process is specified by a probability mass function over potential rock types (states) and an autocovariance function. Models of this form have been applied to rocks (Matheron, 1971; Switzer, 1967, 1975) and to soils (Wu et al., 1975; Nucci, 1978). Switzer (1969) also considered the case of random boundary lines partitioning the two-dimensional plane into random polygons. Miles (1969, 1971) introduced related mathematical developments on so-called Poisson planes. From models of this type, error rates in mapping can be determined as a function of the number and spatial array of observations and of correlation distances of the process. Little work has been done on statistical uncertainties in the probabilistic map itself. Nevertheless, for the simplest, two-color case, optimal classification rules that reduce misclassification errors substantially over nearest-neighbor rules can be generated. With either nearest-neighbor rules or optimal classification rules, the percentage of an area misclassified can be surprisingly large. Depending on the ratio of areas, expected misclassifications of 20% to 30% of a map are not uncommon, even for sample sizes of 50 to 100 (Nucci, 1978). In principle, optimal (or at least improved) observation patterns can be developed, but such patterns have not appeared in the literature.

The geometry of structurally controlled units is difficult to model for the same reasons as discussed above. In principle, spatial models can be proposed, but they are difficult to validate. Dahlberg (1975) suggested that such maps may be better inferred by direct intuition, which rests on a more immediate interpretation of geological experience. This is consistent with the strong influence of geological hypothesis, and therefore subjectivism, on mapping structurally controlled formations (Harrison, 1963).

Search for Geologic Details

A geologic detail is a local anomaly in a geologic unit, such as a solution feature, rock-mass alteration, or shear feature, that may adversely affect engineering performance. A convincing argument can be made that details at a site are the most important determinants of geotechnical performance (Terzaghi, 1929). Thus, finding such details, should they exist, is an important part of exploration. If design performance is sensitive to details above a certain size or of a certain character, effort must be expended to find that size or type detail or to reduce the postexploration probability of their existence below some acceptable level.

Exploration for details is usually analyzed as a search problem. A geologic model characterizes the shape, size, and number of details as random variables, and an observational model relates their size, shape, and number to probabilities of finding the details during the search program. Thus, modeling results are expressed initially as probabilities of a find, and then as posterior probabilities of size, shape, and number of details through Bayes' theorem. The target model usually abstracts details as random point processes with associated size, shape, and orientation distributions (Baecher, 1979). For example, lenses might be assumed to be elliptical in horizontal projection, with lognormal size distribution, uniform orientation, and random location (i.e., Poisson distributed). The adequacy of this sort of modeling, which de facto is the type used, is a central question in the application of search theory to geotechnical exploration.

The most common assumptions about location are that the target centers have Poisson (i.e., independent) or negative binomial (i.e., clumped) distributions. Several other point process models have also been used (Rogers, 1974; Ripley, 1981). Poisson models are chosen if the details are thought to be spatially independent. Negative binomial models are chosen on the basis of empirical evidence in certain mineral exploration applications (e.g., DeGeoffroy and Wu, 1970; Uhler and Bradley, 1970). Empirical verification in geotechnical applications is difficult to find. From a stochastic process point of view, negative binomial locations might be generated by assuming spatial independence at the local level and nonuniform distribution of spatial density across the site (Greenwood and Yule, 1920).

The size, shape, and orientation distributions are typically assumed to follow common univariate pdf's. Lognormal assumptions on size are widely verified by geologic statistics (e.g., Agterberg, 1973; Kaufman, 1963) and have been observed in certain rock-mechanics applications (e.g., Baecher et al., 1977). As a first approximation, the assumption of lognormality in most cases seems accurate but may be influenced by the common sampling bias that the probability of detecting objects is proportional to their size (Baecher and Lanney, 1978; Barouch and Kaufman, 1975).

Orientation has not received much attention in geotechnical applications, but distributional forms have been investigated, and statistical results are available (Mardia, 1972). Empirical analysis of rock-joint orientations (e.g., Mahtab et al., 1972; Dershowitz, 1979) has failed to identify consistent trends in distributional forms. Whether similar difficulties will be encountered with other geologic inhomogeneities is an open question.

The assumption of simple shapes for target geometry can be inaccurate (Baecher, 1979). Phenomena such as sink holes and previous channels are poorly modeled by ellipses or other regular forms. In principle, any geometric assumption that is representable mathematically can be made; however, which of these are appropriate in particular cases has yet to be investigated.

The observational model for search is divided into two parts, one based on geometry and one based on conditional detection. For a detail to be observed, it must be first intersected and then recognized (i.e., detected given that it has been intersected), that is,

$$\Pr(\text{finding detail}) = \Pr(\text{intersection}) \times \Pr(\text{detected}|\text{intersection}) \tag{3.3}$$

The first probability follows from geometric probability theory (e.g., Kendall and Moran, 1963; Harding and Kendall, 1974). Mathematical complications arise, but for regular geometries these can be handled. For problems like grid search, closed-form solutions exist or tables and charts have been developed (e.g., Savinskii, 1965; Singer and Wickman, 1969). As with any problem in geometric probability, paradoxes can arise unless the definition of randomness is invariant with respect to transformations of the reference (Kendall and Moran, 1963). This is embodied in results of geometric probability (stochastic geometry) theory (cf. Harding and Kendall, 1974).

An important question in analyzing the results of geometric search is whether biases arise in the way observations are made. For example, in observing details of varying size, the probability of intersection and therefore the likelihood is a function of detail size. Thus, inferences about size must account for this bias. Barouch and Kaufman (1975) developed a search model for petroleum based on intersection probability linearly proportional to size. Baecher and Lanney (1978) showed that intersection probabilities may be proportional to the square of size in sampling for rock joints. Terzaghi (1964) described biases as a function of orientation. Specific censoring and truncation biases have been discussed by Epstein (1954), Cruden (1977), and Priest (1985), among others. If biases are not considered adequately, inferences of subsurface properties will be in error.

Less work has been done on detection models than on geometric models, but they are no less important. Detection for borings, the easiest case, might be expressed as a simple probability. For example, the probability of detecting a shear zone of certain description in an NX core might be described as, say, 20% (Geotechnical Engineers, Inc., 1976). Such a number could in principle be inferred from past data, but in general would be assigned subjectively. Detection would be a function of, at least, detail depth, size, and contrast with host formations, but might also depend on lateral distance from the geophysical traverse and other aspects of the particular situation.

Optimization of search allocations is possible for most problems once geologic and observational models are obtained. For uniform prior location probabilities, the optimal allocation of effort is also uniform. As the prior distribution becomes nonuniform, optimal allocations resolve to putting more effort where details are more likely to be found. Depending on the extent of nonuniformity and the amount of effort available, the optimal allocation may be to concentrate effort on the more probable locations and not to search at all in locations with small or even moderate prior probability. Baecher (1979) discussed uniform allocations. Optimal allocation over nonuniform priors was discussed by Koopman (1956a, 1956b, 1957), Stone (1972), and Morse (1974). Applications of search strategies and inference to mineral exploration have been considered by Allais (1957), Slichter (1955), Brown (1960), Griffiths (1967), and Engle (1957), among others. In geotechnical engineering, analyses of search allocations have been made by Baecher (1972), Drake (1976), and Grant (1973).

Parameter Estimation

More work has been done on the spatial variation of engineering parameters within homogeneous zones than on any other type of subsurface variability. Much of the early work was based on classical sampling theory, although more recent work addresses problems of sampling from correlated random fields. The former approach assumes a generating model from which each observation is an independent identically distributed (IID) random variable; estimates of parameter values of the parent distribution are inferred. The latter approach assumes a generating model in which the actual profile is one realization of a stochastic field; estimates of the parameter values of the stochastic process including autocorrelation are inferred.

Because a rock mass is physically continuous, elements near one another might be expected to have properties that are somewhat similar. On the average, the smaller the separation between elements of a rock mass, the more similar the properties are. Thus, from a single observation one would extrapolate confidently to nearby locations. As the distance increased, confidence would decrease. To the extent that this is reasonable for rock masses, stochastic processes are better models than simple IID sampling.

Sampling with IID Observations. Common sampling theory applied to geotechnical exploration is based on an assumption of IID observations from some parent population or generating process. Whether relative frequentist or Bayesian, sampling theory results in inferences about the spatial mean and variance of rock-mass properties, but no inferences about spatial correlation. The primary use of simple sampling theory has been in conjunction with random lumped parameter models, in which properties are taken to be spatially constant but uncertain (e.g., Harr, 1977; Yucemen, 1973; Krizek et al., 1977; Hoeg and Murarka, 1974; McMahon, 1971; Wu and Kraft, 1967). To the extent that observations are widely spaced, and thus independent, simple sampling becomes a special case of sampling from a correlated random field.

The theory of relative-frequentist sampling is well developed (e.g., Kendall and Stuart, 1973, 1976, 1977), allowing confidence to be established for a broad range of specific sampling problems. Although many results are distribution dependent, the robustness of estimators and confidence limits has been extensively studied for common cases. Nonparametric methods are also available (e.g., Puri and Sen, 1971).

Optimization of sampling plans within relative-frequentist approaches struggles with the difficulty of relating confidence limits to risks in performance predictions. Because relative-frequentist approaches do not admit probabilities directly on states of nature, that is, directly on in-situ properties, probabilities cannot be directly associated with performance predictions. For a given exploration cost, programs of observations can be determined that minimize estimator variance. However, the criteria for balancing local exploration cost against confidence limits are not obvious. One common practice is to specify a hypothetical "risk cost" associated with different confidence limits, and then find an optimal balancing. How such a risk cost is quantified is also not obvious.

Bayesian sampling techniques (based on Equation 3.1) are mathematically similar to maximum likelihood estimation. In general, Bayesian inference is based on distribution assumptions in order to specify the joint likelihood function of the observations, $f(z|\theta) \equiv L(\theta|z)$. However, limited nonparametric results have been developed (e.g., Jewell, 1973; Ferguson, 1967). Because Bayesian techniques follow from a subjective view of probability, probabilities can be directly placed on states of nature (i.e., parameter values) and can be directly propagated through geotechnical analysis to yield probabilities on performance predictions. Thus, sampling plans and amounts of effort can be optimized directly, without considering confidence limits and so on (Grayson, 1960; Folayan et al., 1970; Baecher, 1972; Wu and Kraft, 1967).

Full expression of the uncertainty in in-situ properties is obtained by integrating over uncertainty in θ to obtain the predictive distributions. Aitchison and Dunsmore (1975) presented predictive distributions associated with common parent populations. Veneziano (1974) developed predictive distributions for stochastic processes.

Correlated Random Fields (Stochastic Models of Spatial Variation). Elements of the rock mass that are spatially contiguous would be thought to have similar engineering properties, and elements that are widely separated would be thought to be approximately independent. Therefore, the introduction of spatial correlation (i.e., autocorrelation) is an obvious extension of simple IID sampling.

Geotechnical work on sampling from correlated random fields (Lumb, 1974, 1975; Alonzo, 1976; Veneziano and Faccioli, 1975) is related to analyses of sampling problems in geology (Agterberg, 1970; Matheron, 1971), hydrology (Rodriguez-Iturbe and Mejia, 1974; Bras, 1975), ground-water flow (Bakr et al., 1978; Wilson and Dettinger, 1974), time-series analysis (e.g., Box and Jenkins, 1976), and other fields (e.g., Matern, 1960; Whittle, 1963). Veneziano (1979) has presented a comprehensive overview of statistical sampling from correlated random fields.

The statistical theory of sampling from correlated random fields is mathematically tractable for special classes of problems (Veneziano, 1979). In particular, those cases most extensively studied are second-order stationary (i.e., nonparametric) and Gaussian (strictly stationary) fields. Many problems remain, however. Among the more important questions still requiring work is inference of the autocorrelation function (or variograms) for correlated random fields and uncertainties in the inferences. Similarly, sampling networks can be optimized in special cases (e.g., Veneziano and Faccioli, 1975; Rodriguez-Iturbe and Mejia, 1974). Several workers have recently addressed sampling theory for variograms (e.g., Journel and Huijbregts, 1978).

Joint Surveys

The stochastic nature of jointing has received considerable attention in both statistical assessment of jointing parameters and in subsequent use on probabilistic mechanical and hydrologic models. The principal geometric jointing parameters are orientation, spacing (location), and size (trace length, persistence). They can be described in a disaggregate manner by collecting data on each individual parameter and representing each parameter and its distribution separately. Extensive research and practice have shown that spacing and size are statistically well behaved. Spacing is usually exponentially and sometimes lognormally distributed (Priest and Hudson, 1976; Call and Nicholas, 1978;

Einstein, Baecher, and Veneziano, 1979). Trace lengths are exponentially (e.g., Call and Kim, 1978; Robertson, 1970), lognormally (Einstein, Baecher, and Veneziano, 1979; Bridges, 1975; McMahon, 1971), or hyperbolically (Segall and Pollard, 1983) distributed. Joint orientation tends not to be so well behaved statistically. Fisher (1953), bivariate Fisher (Dershowitz, 1979), and Bingham (1964) models have been applied to joint-orientation data, but none appears to provide adequate descriptions of empirical data except in peculiar circumstances. Some evidence suggests autocorrelation in joint properties (e.g., Miller, 1975, 1979; LaPointe, 1980; Einstein, Baecher, and Veneziano, 1979), but the evidence is contradictory.

Joint geometry can also be described by stochastic geometry. Dershowitz (1984) thoroughly surveyed stochastic-geometry models for joint systems. The orthogonal (e.g., Snow, 1968a; Irmay, 1955), disk (Baecher and Lanney, 1978; Barton, 1977), polygonal (Veneziano, 1978; Dershowitz, 1984), and mosaic block tessellation models (Dershowitz, 1984) are most common.

Mechanical properties of joints, notably friction angles, can also be represented statistically. Work remains to be done on the variability of mechanical properties before definite conclusions can be drawn on generally valid distributional forms (Einstein, Baecher, and Veneziano, 1979). Data scatter caused by different testing methods and different rock properties makes regression analyses of joint-friction data inadvisable except for focused sampling (shear tests with the same equipment on the same joint set in a locally well-defined formation). Some geometric properties on this scale, e.g., roughness, have been treated statistically to a certain extent (Swan, 1983; Roberds, 1979).

Much work has been done on statistical sampling plans for joint surveys (e.g., Kiraly, 1969; Baecher, 1972; Call, 1984). Because jointing is a geometric property, sampling plans must be designed with care to avoid severe sampling biases. Most joint surveys designed according to statistical principles involve two stages; first, specific outcrops or locations are chosen for sampling, and second, a predesigned sampling geometry is applied to the chosen outcrop. Typical sampling geometries include multiple scanlines, box-shaped windows, and circles. The geometry of the sampling plan, presuming a stochastically homogeneous (stationary) population of joints, has received more attention than the selection of outcrops across a site and the associated question of mapping trends in jointing parameters (Call, 1985).

Exploration Planning

Statistical and probabilistic methods have been applied to site characterization for two purposes: interpreting test data (considered above) and planning exploration programs. Exploration planning selects types, locations, and numbers of tests and phases site characterization so as to obtain the most information from a fixed budget or to balance the cost and effort of site characterization against the benefits obtained from resulting information. As a rule, statistical and probabilistic methods have been less used for exploration planning than for interpreting field data or drawing inferences. Exploration planning has had to face more directly the issue of subjective uncertainty, the problem of combining different types of uncertainty, and the problem of complex objective functions for optimality.

The planning of site-characterization programs for underground construction is a decision-analysis problem. A set of objectives specifies the use to be made of site-characterization data. For repositories, for example, site characterization forms a basis for judging a repository's compliance with specifications. Once the objectives are defined, each alternative exploration plan is evaluated according to how well it is expected to achieve the stated objectives. In the case of waste repositories, achieving the objective of forming the basis for performance predictions means providing sufficient information that a prediction with specific confidence can be made.

Approaches on Tools for Exploration Planning. The formal methods used for exploration planning are, in general, variations on the basic statistical decision-analysis procedure. Analyses are based on what might be called "virtual exploration." The range of possible outcomes and subsequent implications of an exploration program are forecast, probabilities are assessed for each outcome, and some form of expectation is taken over the range. In decision theory, such procedures are called a calculation of the expected value of sample information (EVSI) (e.g., Raiffa and Schlaifer, 1961).

Einstein and a series of collaborators (e.g., Einstein and Vick, 1974; Kim et al., 1985; Einstein et al., 1978; Ledezma, 1983; Ioannou, 1984; Chan, 1980) have developed a tunnel cost and scheduling model based on a combination of decision theory and simulation. This model relates site-characterization decisions to construction cost and scheduling for tunnel construction. The model is based on a one-dimensional discretization of a proposed tunnel alignment. A matrix of engineering, geologic, and construction-related parameters is defined for each element along the alignment. Probabilities are addressed over the elements of the matrix. Monte Carlo simulation of the tunnel construction is used in conjunction with construction management techniques to evaluate the uncertainty in construction cost, advance rate, and schedule as a function of uncertainty in site characterization.

TUNNEL-PERFORMANCE PREDICTIONS

General Deterministic and Probabilistic Aspects

Performance predictions relate geologic, construction, and operating conditions to tunnel performance. Tunnel performance can be either the construction process, particularly time and cost to completion, or physical behavior of the tunnel (opening) during construction, at the end of construction, or in operation. Physical behavior more specifically means stress-deformation-time performance of tunnel supports and surrounding rock.

The information necessary for performance predictions should represent all possible influences. For this reason, not only geologic features and parameters, but also construction and operating effects must be considered. Site characterization provides geologic input and certain information on waste-package performance and operating conditions. Operating conditions also include external, long-term, man-made effects (exploration, explosions). Input representing construction effects is more difficult to obtain. Much of this is construction related and may have to be obtained from modeling the construction process or from empirical estimates. The relations between input and construction process or performance predictions are incorporated into models representing physical mechanisms. Such models can be divided into four classes: empirical methods, analytical methods, scaled physical (laboratory) models, and numerical methods.

Most of these methods are deterministic. Uncertainty is incorporated by propagating variances of probability distribution as input parameters through the models, typically by using first-order propagation of moments; approximation by point estimates, response surfaces, or similar techniques; or Monte Carlo simulation. Uncertainty or error-propagation techniques are not widely used in engineering practice for tunneling, but they are methodologically well developed and verified in practice.

A difficulty in making any prediction of tunnel performance is specifying what constitutes unsatisfactory behavior. Stresses, deformation, and so forth can be calculated, but their effect is often unclear. Usually a level of unsatisfactory behavior is simply defined. For example, unsatisfactory behavior can be deformation exceeding a certain level or the occurrence of cracking. The reliability of a tunnel design is expressed as a probability. When only moments of input uncertainties are used in a calculation, and thus only moments of a prediction are calculated, such a probabilistic statement is not possible. In this case, standard practice is to use a first-order reliability index (β). This reliability index is a measure of the separation of mean performance from a predicted threshold of unsatisfactory performance, measured in numbers of standard deviations.

Empirical Methods

Empirical methods have a long and well-established standing in tunneling. Steiner and Einstein (1980) reviewed empirical methods and their history. Hoek and Brown (1980) reviewed better-known methods and presented their own combined empirical-analytical approach. Empirical methods range from qualitative (e.g., Terzaghi, 1946), to quantitative single-parameter (Deere et al., 1969a, 1969b; Merritt, 1972), to quantitative multiparameter (Barton et al., 1974; Bieniawski, 1973, 1974, 1976, 1979) methods. Barton and Bieniawski also reviewed the application of their methods and some extensions (International Symposium on Engineering Geology and Underground Construction, 1983). Empirical methods characterize a rock mass in a way that is related to tunnel support requirements. Information on these parameters is usually collected through observations in the tunnel. Because of the importance of jointing, at least one if not most parameters reflect geometric and mechanical properties of joints. Some methods do not rely on a fixed set of parameters and fixed relations between parameters and tunnel supports, but adapt them to particular geologic conditions. The New Austrian Tunneling Method (NATM) (Rabcewicz, 1969; Einstein, Steiner, and Azzouz, 1980; Einstein, 1977) uses this approach, as does the MIT Tunnel Cost Model (Einstein and Vick, 1974; Moavenzadeh et al., 1974); the latter is actually a construction-process prediction that includes parameter-support relations. Empirical methods have found increasing application over the last decade. Consequently many case histories have been reported and compared (Rutledge and Preston, 1978; Barton, 1977; Einstein et al., 1983; Cameron-Clarke and Budavari, 1981; International Symposium on Engineering Geology and Underground Construction, 1983).

Empirical methods relate rock-mass parameters to support dimensions (and possibly also to excavation methods) that in the past have led to satisfactory tunnel performance. Rarely is the deformation of supports included in assessing satisfactory behavior (except for Terzaghi's method and the NATM). Otherwise, the existence of an undisturbed support is simply taken as evidence of satisfactory performance. The support might, on the one hand, be several-fold overdesigned, or on the other hand, fail over longer periods of time. The latter case is particularly difficult to detect, because base cases for many of the empirical methods have only been observed for short times.

Probabilistic empirical methods exist mostly as a part of the MIT cost models.

Parameter trees with associated probabilities or the probabilistic parameter profiles are related to support requirements and excavation performance. Because these are models of construction processes, however, the resulting probabilities pertain to the need for specific support types rather than to unsatisfactory performance. A further step, but still not a complete probabilistic performance prediction, is found in the Decision Support System for Tunneling (DSST) (Kim, 1984; Kim et al., 1985), which calculates the probability of selecting an inadequate support prior to excavation of the next round. Panet (1978), in the French ground classification for tunnels, and Nicholas and Marek (1981), for feasibility studies, included limited statistical aspects. Related applications of probabilistic empirical methods are the study by Kaiser and Harr (1984), in which a number of empirical methods were statistically correlated with each other, and the attempt to use fuzzy sets for rock-mass classifications by Zhen and Zu-Zeng (1983).

Analytical Methods

Analytical methods exist for a variety of material conditions and tunnel shapes. Two groups are distinguished in this discussion: analyses based on continuum or quasi-continuum assumptions and analyses for discontinua.

Continuum analyses for unsupported openings range from linear elastic plane-strain or plane-stress solutions for circular openings (Timoshenko and Goodier, 1951; Obert and Duvall, 1967) to elastic analyses for more complex shapes (Davis and Poulos, 1974) and elastoplastic analyses (usually for simple shapes). Time-dependent behavior also can be analytically expressed through creep or relaxation expressions. Using conformal mapping, essentially any shape and material behavior can be modeled (Savin, 1970); the mathematical effort required, however, is usually not justified given the capabilities of numerical methods and computers. Elastoplastic analyses range from those assuming no volume change, through those including empirical volume-change expressions, to those explicitly expressing volume change (Schwartz and Einstein, 1980; Schwartz et al., 1980).

Because these methods are based on continuum assumptions, they use parameters that often are not directly measurable. Jointing, for instance, is introduced by changes to elastic constants (Deere et al., 1967) and direction-dependent shearing resistance (Mueller, 1963; Goodman and Duncan, 1971). Specific discontinuities cannot be considered in this manner.

The effect of water pressure can be considered in all these analyses through the usual geotechnical consideration of water-pressure-dependent strength. Essentially any material behavior (e.g., temperature or chemistry dependence) can be formulated, but appropriate testing is required to establish validity.

Analytical solutions for tunnels with support also exist, although they are usually limited to simple material behavior and shapes. An elastic solution, including delayed support installation and a variety of internal and external loads, has been developed by Ushijima (1982) based on preceding work by Schwartz and Einstein (1980), Schwartz et al. (1980), Ranken et al. (1978), Hoeg (1968), Burns and Richard (1964), Muir Wood (1975), Curtis (1976), Morgan (1961), Daemen and Fairhurst (1972, 1973), Egger (1973), Panet and Guellec (1974), and Panet (1976). Schwartz et al. (1980) and Detournay (1983) developed methods for considering plastic behavior. Time-dependent effects have been considered, e.g., by Ladanyi (1974), Ladanyi and Gill (1984), Carter and Booker (1983a, 1983b), and Semple et al. (1973, 1974). Duddeck and Erdmann (1982) critically reviewed these and other methods. Simpler solutions include Peck's (1969b) approach and multilayer analyses, which through iterative procedures also treat elasto-plastic behavior (Newmark, 1969; Hendron and Aiyer, 1972). Three-dimensional effects as they occur around an advancing tunnel have been considered by Lombardi (1973), Panet and Guellec (1974), Panet and Guenot (1982), Laabmayr and Swoboda (1978), and Muehlhaus (1985). Muehlhaus (1985) and Detournay (1983) considered nonuniform stress fields. A wide variety of methods exists to consider different support types (e.g., Deere et al., 1969a, 1969b, for shotcrete; Panek, 1962, and Larsson et al., 1985, for bolts).

The convergence-confinement method is philosophically related to analytical approaches for ground and ground-structure behavior. This method represents ground-structure interaction by characteristic curves (Peck, 1969b; Kerisel, 1979, 1980).

Discontinuous rock masses are usually treated analytically through limit equilibrium analyses of blocks or wedges isolated by joints. Such methods in slope stability analysis are also applicable to underground works (e.g., Hendron et al., 1971; Einstein, Baecher, and Veneziano, 1979; Hoek and Bray, 1981), although special tunnel applications also exist (Cording et al., 1971; Hoek and Brown, 1980). New developments by Goodman and Shi (1985) and Warburton (1981) verge on numerical methods. The constrained conditions in tunnels and the associated effect on stresses require special considerations in limit equilibrium analyses (Glynn, 1979; Crawford, 1982). The action of the support must be included through simplified force assumptions, e.g., of bolt forces (Cording et al., 1971) or shearing through the support (Rabcewicz, 1969).

Probabilistic analytical methods both for continuum and discontinuum assumptions can be developed by propagating moments or point estimates through deterministic analysis. Few such approaches exist. Matsuo and Kawamura (1983) presented the background for a reliability analysis assuming continuum conditions and the reasonably complex support structure (shotcrete, bolts) of the NATM. Probabilistic analyses of block stability, particularly for slope-stability problems, however, have been extensively developed. Many of these approaches use lumped-parameter methods (Glynn, 1979), but some consider spatial variability (Glynn, 1979; Einstein, Baecher, and Veneziano, 1979; Call and Kim, 1978; Call and Nicholas, 1978; Crawford and Lau, 1983; Chan and Goodman, 1983). Simulation often is used if stochastic characteristics are considered. These methods are applicable to tunneling.

Scaled Experiments

Physical model experiments have been extensively used for tunnel-performance predictions, particularly before the advent of numerical methods for the analysis of complex shapes (Keneshige et al., 1960; Nigam et al., 1976; Obert and Duvall, 1967). Model tests on tunnels in jointed rock have been conducted under plane-strain and triaxial conditions and with base friction models (e.g., Worotniki et al., 1976; Barton and Bandis, 1982). Support and excavation procedures have been investigated through models (Korbin and Brekke, 1975). Centrifuge testing has also been applied to overcome scaling problems of the other procedures (Atkinson and Potts, 1977; Kimura and Mair, 1981).

Model tests still serve an important purpose, because they are often the only way to validate complex numerical models. Conditions that cannot be tested in reality, but for which tunnels have to be designed (e.g., blast protection structures with large overpressures), can be physically tested only in this manner (Kennedy and Lindberg, 1978a, 1978b; Lindberg, 1982).

Probabilistic methods in the sense discussed above (propagation of uncertainties) can practically not be applied in model tests. However, statistical approaches are often essential in test evaluation (Einstein, Baecher, and Veneziano, 1979; Baecher and Einstein, 1981).

Numerical Methods

Numerical methods, including finite-element, distinct-element, boundary-element, and displacement-discontinuity-element methods and their hybrids, can handle any combination of material behavior and opening geometry. Computer size or computation time has limited the use of these methods, but even this has been overcome by hardware development. The literature abounds with descriptions of numerical methods and parameter studies; summary reviews by Einstein, Schwartz, et al. (1980) and the U.S. National Committee on Rock Mechanics (1981) are dated because of recent developments. The most up-to-date collection is probably in the hands of the International Society for Rock Mechanics' Committee on Computer Programmes, in draft form. Given the amount of information available, it serves little purpose to list all methods. Only two aspects of special interest to repository analysis are mentioned here: consideration of discontinuities and lack of field validation.

Numerical methods, even though based on continuum mechanics, can be augmented to treat discontinuities. Finite-element methods (FEMs) have long included discontinuity elements (Goodman et al., 1968; Goodman and St. John, 1977), and recently three-dimensional, arbitrarily shaped contact elements have been introduced (Bathe and Chaudhary, 1984, 1985; Bathe, 1985; Mehlhorn et al., 1985; Zienkiewicz, 1977). Quasi-continuum considerations are used also (Goodman and Duncan, 1971). The most general (three-dimensional, elasto-plastic, time-dependent, slip or contact elements) FEMs are ADINA (Bathe, 1976, 1977, 1985; Langer and Stockmann, 1985), BMINES (Balachandra, 1978; Ewing and Raney, 1976; Van Dillen, 1978; Agbabian Assoc., 1981); Nike 3 (DYNAD3) and ABAQUS (Hibbitt, Karlsson and Sorenson, Inc., 1981). They are limited, however, with regard to representation of geostatic conditions. Applications of three-dimensional FEMs have been reported since 1973 (Descoeudres, 1974; Gartung et al., 1979; Agbabian Assoc., 1981; Nigam et al., 1976; Tanimoto, 1978; Wittke and Pierau, 1976 [Face Effect]; Ranken and Ghaboussi, 1975). Chandrakant and Christian (1977) reviewed FEM applications to tunneling. Special FEM developments to consider rock bolts (Heuze and Goodman, 1973; St. John and Van Dillen, 1983) also are interesting.

Discontinuity-element, direct boundary-element (Brebbia and Walker, 1978; Crouch, 1976; Crouch and Starfield, 1983; Hocking, 1978; Brady and Bray, 1978; Brady, 1979; Beer and Meek, 1980; Moore and Brooker, 1982), and distinct-element (Cundall, 1971; Voegele, 1979; Dowding et al., 1983; Lemos et al., 1985) methods open an unlimited number of possibilities to represent discontinuities and interfaces. Elastoplastic character has been incorporated in boundary-element (BE) methods (Telles and Brebbia, 1981) and three-dimensional aspects (e.g., Hocking, 1976; Hocking et al., 1976; Niwa et al., 1977; Brown and Hocking, 1976; Ushijima and Einstein, 1985; Ushijima, 1983). Various element types have also been hybridized for tunnel analysis (e.g., Lorig and Brady, 1984).

Although numerical methods make it possible to analyze underground openings, there is a surprising lack of supporting field validation of them. The number of methods and theoretical parametric studies far exceeds the number of field or even laboratory validations. Case studies have compared FEM results with field or laboratory measurements (Fischer, 1985; Nord and Stille, 1983; Einstein et al., 1983; Ushijima and Einstein, 1985; Borsetto et al., 1983; Sweeney and Hovland, 1981; Rescher et al., 1973). Ushijima and Einstein (1985) also included a comparison with a hybrid BE-FE method, and Rescher et al. (1973), with photoelastic studies.

Probabilistic numerical analyses have been conducted in a number of areas. Baecher and Ingra (1981) developed a first-order stochastic FEM for soils problems, and many stochastic FEM techniques have been developed in ground-water modeling (e.g., Wilson and Dettinger, 1974). Through simulation, essentially any uncertainty can be propagated through a numerical analysis, although validation can become ever more difficult. Somewhat simpler is the use of probabilistic quasi-continuum parameters like the reduced modulus of elasticity as a function of joint stiffness and orientation (Dershowitz et al., 1979) or stochastic analysis of wedge/block instability (Call and Kim, 1978; Call and Nicholas, 1978; Einstein, Baecher, and Veneziano, 1979). So far these approaches have not been used in tunneling.

Time-Dependent Performance

Temporal effects can be predicted to some extent with empirical methods (e.g., Myer et al., 1981) and can be included in analytical or numerical methods (e.g., Semple et al., 1973, 1974; Carter and Booker, 1983a, 1983b; Ladanyi and Gill, 1984; Liedtke and Bleich, 1985). Validation is scarce, however. Laboratory model experiments can accelerate time-dependent effects (e.g., Korbin and Brekke, 1975) and thus indicate what might occur.

A special time-dependent effect, swelling, has received considerable attention, and a number of empirical, analytical, and numerical methods exist for the prediction of swelling effects on tunnel supports (e.g., Terzaghi, 1946; Gysel, 1977; Grob, 1972; Einstein and Bischoff, 1976; Einstein, Bischoff, and Hofmann, 1979; Wittke and Rissler, 1976; Fritz, 1981). These methods are notoriously weak in relating time to the predicted swell phenomena.

Although time-dependent performance seems to be difficult to predict, disturbing evidence suggests that time-dependent effects may be severe. Many old railroad tunnels show considerable decay. Tests on intact rock reveal that stress at the start of crack propagation and long-term strength are similar; both may be considerably smaller than "intact strength." In contrast, natural caverns in a variety of rocks have existed for hundreds or thousands of years. Temporal effects clearly introduce major uncertainty, but little basis for estimating the magnitude of this uncertainty is recognized.

Other Effects (Water Pressure and Flow, Dynamics, Temperature, Collapse)

The effect of water pressure on the ground and tunnel supports is usually included in the prediction methods discussed above. This includes the effect of cleft-water pressure on ground and support and the effect of possible internal pressures on support and ground. If a perfect seal exists, cleft-water pressure can be handled through effective stress considerations, and internal pressure can be considered as internal pressure load. Problems arise if the seal is not perfect. Static ground instabilities are known to occur (Terzaghi, 1962), and transient conditions may have to be considered (Deere et al., 1969a, 1969b). Flow from or into the opening is part of the larger ground-water hydraulics/hydrology problem.

Recent developments go beyond continuum approaches. Numerical models incorporating stochastic descriptions of discontinuities have recently been developed by Long et al. (1982) and Rouleau (1984) for two dimensions and by Dershowitz (1984) for three dimensions. Samaniego and Priest (1984) have published a specific application for tunneling. More general discussions on stochastic discontinuity modeling of flow have also taken place (Comite Milieux Poreux, 1985). A somewhat different direction is taken by coupled methods that consider flow and pressure, that is, effect of pressure on joint aperture and thus rock-mass deformation (Kafritsas et al., 1984; Lemos et al., 1985).

Numerous studies have been conducted on dynamic effects. These range from surveys of damage to tunnels from earthquake shaking (Rozen, 1976; Dowding, 1977; Owen et al., 1979; Owen and Scholl, 1985; Pratt et al., 1979) and blast effects (Labreche, 1983), to development of numerical methods (Dowding et al., 1983; Kobayashi and Nishimura, 1982a, 1982b), to overviews on numerical methods (St. John and Zahrah, 1985). More general discussions on the engineering of tunnels under dynamic effects were given by Owen and Scholl (1985), Hendron and Fernandez (1983), and Dowding (1985).

A case history reported by Rehbinder (1984) predicted effects of temperature variation from hot water storage (40 to 95°C) and compared them with measurements. The coincidence of predicted and measured results was considered to be largely fortuitous, however, thus illustrating the practical difficulties of prediction and validation.

Collapse of an opening could lead to chimney caving to the surface. Such events are known to occur in mining. Goel and Page (1982) studied collapse and developed a method for computing the probability of a chimney reaching the surface. O'Riordan et al. (1984) addressed the collapse of old underground limestone mines in England; their study was not statistical, however.

OBSERVATION AND EVALUATION OF TUNNEL PERFORMANCE

Observations of tunnel and mine performance can be traced into the last century (e.g., Steiner and Einstein, 1980). Empirical methods are to some extent based on actual performance measurements. Extensive performance-observation programs were made for the Washington, D.C., subway system (e.g., Van Sint and Cording, 1984). Such programs are most useful if they are combined with back analyses. Additional studies, particularly related to mining applications, were discussed by Brown and Hudson (1984). In mining, statistical approaches may be applied in evaluating performance observations. In some cases, failure may be considered statistically, because the extraction process either purposely includes failure or is intended to be as close to failure as possible, thus inducing some failures. In such cases, performance of various pillar geometries (Wagner and Madden, 1984) or chimney caving (Goel and Page, 1982), for example, can be statistically compared.

Prediction methods can be validated through performance observation and back-analysis; however, given the variability of ground conditions and construction procedures, each validation holds only for a particular case. Extrapolation to other ground conditions, opening geometries, and construction procedures is questionable. This can be remedied to some extent by using observational approaches, in which encountered ground conditions are used to update the original information on geologic features and parameters, and performance observations during construction are used to calibrate the prediction models. Deformations of the unsupported opening or of an opening with minimum initial support are often of such a magnitude that they can be observed, particularly right after excavation. With more sensitive instruments, observation can be extended to later construction stages and to the final tunnel (Kovari and Amstad, 1979; Kovari, 1983).

Observational methods in geotechnical engineering have been advocated by Terzaghi (1961), Peck (1969a), and Casagrande (1965). Observational methods assist in choosing a design that is adequate and not overly conservative. Essentially a number of design alternatives are developed, and their performance is predicted. During construction, performance is observed (e.g., deformation of the initial support). If this performance does not correspond to that predicted for the particular design construction alternative, another alternative is chosen (e.g., heavier support is chosen if the deformation is greater than a certain value). Observational methods in tunneling have found increasing acceptance. The NATM and other methods that include adaptable excavation procedures and support techniques are examples. (Some misunderstanding of the term *observational methods* exists; sometimes it is used to express a procedure in which one "designs as one goes"; i.e., without predesigned alternatives. "Observational method" is used here as defined above and by Einstein [1977].)

Using observational approaches has significant advantages in tunneling. Consequently, a number of techniques have been developed to put these approaches on a more formal basis. The geologic prediction and updating model (Chan, 1980; Ashley et al., 1981) incorporates a Bayesian updating procedure to include observations made in the tunnel or in boreholes ahead of the excavation. Rollin (1979) developed the concept and some techniques by which the prediction model can be updated on the basis of observed performance. Cividini et al.'s (1983) parameter-updating procedure for E-modulus is a more detail-oriented technique. DSST (Kim, 1984; Kim et al., 1985) uses geologic updating and information on construction performance to choose the optimum construction procedure in the next construction step (e.g., a tunnel round). Procedures developed by Sakurai (e.g., Sakurai and Shinji, 1984) are more design-adaptation oriented.

Among other references on quantified approaches to the observational approach for geotechnical practice are those by Asaoka (1978) and Whitman (1984). Most of this work is based on a Bayesian view of inference, and Bayes' theorem is used as the principal vehicle for combining prior information with facility performance to draw inferences about subsurface conditions. Kahlman filtering approaches have been used by Jurina et al. (1977).

PROBABILISTIC RISK ASSESSMENT FOR UNDERGROUND WORKS

Probabilistic risk assessment (PRA), as used here, is a systematic, quantitative evaluation of the uncertainties affecting a complex engineered facility. The objective of PRA is an assessment of overall risks associated with the operation of a facility. The greatest use of PRA and its principal methodological devices, namely fault and event trees, has been in the aerospace, electronics, and nuclear-power industries. These industries use PRA to analyze the aggregate performance of a system, on the basis of failure rates of individual subcomponents and the interaction among subcomponents. Although a variety of probabilistic analyses have been made of the behavior of underground works, only a limited number of comprehensive PRA-type efforts have been attempted. The more successful of these focus on specific classes of performance.

In application to underground works, the purpose of PRA is to systematically account for uncertainties entering a prediction of facility performance, and by so doing to obtain probabilistic statements on the safety of the facility throughout its service life. PRA differs from other probabilistic or statistical analysis in that it intends to be comprehensive. Whereas a probabilistic analysis of, say, thermomechanical effects in the rock mass surrounding a repository might lead to predictions of induced fracturing and associated probabilities, a PRA uses such probabilistic results as a starting point, combines the results across all physical mechanisms that in consort could lead to releases of radionuclides to the accessible environment, and attempts to calculate aggregate probabilities of release. A PRA combines the results of other probabilistic analyses to estimate overall risk, and is therefore an ambitious undertaking. The result of a PRA is twofold: first, an identification of important chains of events, and second, a quantification of the probabilities associated with those chains of events.

The primary applications of PRA to underground works has, in fact, been in association with the nuclear-waste program. Early applications (Logan and Berbano, 1978; Schneider and Platt, 1974) addressed generic repository sites using variations of event-tree probabilistic analysis. A more detailed study (Arthur D. Little, Inc., 1980) considered four geologic media and primarily used a fault-tree analysis, similar to that used in the U.S. Nuclear Regulatory Commission (NRC) Reactor Safety Study (U.S. NRC, 1975). More recently, a proposed PRA-type safety study of the BWIP facility (Golder Associates, 1985) was based on a combination of fault- and event-tree techniques.

A difficulty with the early studies has been the ambition to comprehensively model a physical system as large, complex, and poorly defined as a repository, and to include both mid-term physical processes (e.g., thermomechanics, rock-mass behavior) and long-term geologic events (e.g., historic changes) in the analysis. This has led to large bushy fault or event trees and a lack of assurance that important chains of events have been adequately represented. Unlike a power plant, aircraft, or electronic circuit, a repository is an open system. Boundaries of the system are ill-defined for purposes of analysis. Even within the near field, not all components of the system may be recognized or identified. Thus, representational and completeness issues cloud the results of comprehensive PRAs for geologic repositories. Current approaches to PRA for repositories, as evidenced by Golder Associates (1985) and others, are focused on narrower classes of repository performance, and thus may provide more concrete results.

An added problem of PRA studies for repositories is that they are forced to deal explicitly with subjective uncertainty and expert opinion. In other probabilistic analyses, subjective uncertainties often may be conditioned or referred to qualitatively. PRA, in contrast, has the goal of being comprehensive. Subjective uncertainties, to the extent they influence risk, must be incorporated. This means that subjective uncertainties must be quantitatively evaluated, usually as subjective probabilities, with the attendant difficulties.

PRA analyses of underground works deal with natural systems, which are poorly defined, not engineered systems. Boundary and initial conditions are imprecise, because they must be estimated from site-characterization data. Natural mechanisms of performance often are incompletely identified or represented. As a result, basic tenets of PRA as applied to power plants or industrially produced goods, for example, that the system be bounded and all subcomponents be specifiable, are violated.

The present state of the art of PRA for geotechnical aspects of constructed facilities is such that practical benefit can be derived from such analyses, but that the profession has only limited experience with large-scale PRA's in practice. To a large extent, the nuclear-waste program itself is forcing the development of PRA in geotechnical applications, while other applications, for example, dam safety or reliability analysis of offshore platforms, will benefit from the experience gained in the repository projects.

SUMMARY

Basic understanding of mechanisms of tunnel performance and of uncertainty in geotechnical engineering exists, and a number of engineering methods are available for predicting tunnel performance. Similarly, practical methods exist to determine geotechnical uncertainties, to propagate them through engineering analyses, and to probabilistically assess risks associated with particular designs. Although no satisfactory probabilistic performance-assessment method for tunneling has been developed, probabilistic tunnel cost and time estimation methods are used in practice.

Detailed consideration of PRA techniques for underground construction suggests that certain questions must be answered and additional developments made before probabilistic methods can be applied to the tunneling aspects of radioactive waste repositories. First, tunneling is a complex geotechnical problem area, characterized by the interaction of many physical mechanisms. Identifying relevant mechanisms in a particular case and modeling combinations of mechanisms are not satisfactorily resolved, particularly if long-term effects are important. Such model uncertainty has yet to be incorporated into tunnel-performance predictions. On the other hand, adaptive design approaches, in which tunnel design is modified to accommodate encountered conditions, is in widespread use, although adaptable methods have not yet been developed for waste-repository applications. Second, subjective uncertainties are of great importance in tunneling, and assessment methods do exist. The principal limitation on using these methods is the lack of practical tools for applications to tunneling. Third, in site characterization a large body of methods has been developed and applied to more traditional geotechnical projects. Experience is lacking, however, with the application of such methods to site-characterization programs as complex as those associated with repositories, and further development of practical tools is necessary. Finally, probabilistic risk assessment for tunnel performance has not been attempted in a systematic way, at least as reflected in the literature. Probabilistic time and cost estimate methods aimed more at project management and control, however, are fairly mature, and could provide a basis for developing performance PRA methods. Any attempt to use large-scale PRA for a repository system would have to incorporate subjective uncertainties and would have to deal with site-characterization questions, as noted above. None of these problems is insurmountable, and as a result, practically applicable approaches appear within reach.

4

Hydrology

Allan L. Gutjahr

Dispersion of nuclear waste from a repository most likely will occur within ground-water systems. Methods and procedures available in ground-water hydrology for probabilistic predictions fall into three major categories:

1. Deterministic porous-media models, where the probability component enters primarily through parameter variations.

2. Stochastic porous-media models, where the probability component enters through the treatment of the medium itself as well as through parameter variations.

3. Fractured-media models, where both stochastic and deterministic features may appear.

These groups are not mutually exclusive, and combinations of the classes may be encountered, but a single dominant classification will usually be made. Any possible interaction will be noted in the model description. The probabilistic framework and kinds of data needed can vary considerably between the classes; in many cases the differences appear in the analysis and interpretation of the data.

Even though the models share a common physical basis, the probabilistic frameworks may differ conceptually. Therefore, a clear consensus as to which model is correct or best is not always possible. The debate on how to incorporate probability into predictions about assessments focuses on two aspects:

1. On the subjective level, how are probabilities assigned, and how can data be incorporated? To some degree this is a debate between classical and Bayesian views of statistics.

2. On the objective level, is the process itself best modeled as a particular realization of a stochastic or random process or as a deterministic process that only needs a certain number of parameters for complete determination?

If complete data were available for a model, no probabilistic component would be necessary, although *all* parameters at *all* locations would need to be known. Such a complete data set cannot realistically be expected. In consequence, the probabilistic approach can be envisioned as a way to model both the uncertainty inherent in nature and the lack of information about ground-water flow and dispersion. As more data become available, the probabilistic model can be "conditioned" by the data and the uncertainty in predictions reduced.

Several excellent and useful reviews are related to the topics discussed here. Neuman (1982) reviewed stochastic and statistical methods in hydrology; a more up-to-date version of stochastic hydrology was given by Dagan (1986) and Gelhar (1985). M. P. Anderson (1979) extensively reviewed dispersion. Koplick et al. (1982) included a good discussion of hydrologic models and subjective probabilities. DeMarsily's (1983) review is an excellent recap of many of the strategies and issues involved in nuclear-waste disposal, emphasizing probabilistic and hydrologic details.

DETERMINISTIC POROUS-MEDIA MODELS, WITH PARAMETER VARIATION

The probabilistic assessments done with deterministic parameter models are either sensitivity analyses or Bayesian in nature. In each case, a criterion function $G(\underline{\alpha})$ is studied by generating realizations for $\underline{\alpha}$ and using an estimate for associated probabilities. If the basic models discussed below are used for flow and dispersion, the output will correspond to $G(\underline{\alpha})$; it will be a function of the input vector. Within this framework, several questions arise: What do the joint probability distributions for $\underline{\alpha}$ represent? How can these joint probability distributions be obtained or estimated? How can data be incorporated into estimates of the joint distributions and the models?

Joint probability distributions are often taken to reflect factors like measurement uncertainty and estimation uncertainty. Ultimately, however, they represent subjective prior probabilities of the kind used in a Bayesian statistical framework (DeGroot, 1970; Luce and Raiffa, 1957; Luce and Suppes, 1965). They are not classical sampling distributions, which refer to estimates and generally still include the parameter estimated. For example, the distribution for the sample mean from a population, $\bar{Y}$, will still include the true unknown mean, μ, in a sampling distribution framework. Hence the joint distribution of $\underline{\alpha}$ is a subjective prior distribution reflecting information about the parameters. This distribution reflects a degree-of-belief statement rather than a relative-frequency statement, though the latter may be used to estimate the prior distribution. Methods have been described for obtaining such prior distributions for one variable (Benjamin and Cornell, 1970; DeGroot, 1970; Einstein and Baecher, 1983; Luce and Suppes, 1965; Winkler, 1985). However, the choice of joint priors for several variables is not as well defined. One procedure followed for single parameters, e.g., mean transmissivity (Cranwell et al., 1987), has been to use generic distributions obtained from the literature. For example, many studies (Freeze, 1975; Freeze and Cherry, 1979; Hoeksema and Kitanidis, 1985) have indicated that transmissivity has a lognormal distribution for any given material.

How to obtain joint probability distributions is not completely solved in the Bayesian framework. Generally, generic joint information for all parameters (transmissivities, porosities, etc.) has not been obtained; to do so is almost impossible. Instead, individual distributions could form the starting point, and correlations between variables (again, subjective prior estimates) could be used to obtain joint information. Iman et al. (1979) used this approach, assuming variables are jointly normal. Iman and Conover (1982) also showed how one can induce correlations between ranks, that is, between the ordered values, for use in Latin hypercube sampling. Although this does not get at the issue of correlation between parameters, it may yield a suitable working model especially useful in sensitivity studies.

The inverse methods discussed below give one procedure for incorporating data into estimates of joint distributions and models. Data can be incorporated either into the prior distributions or directly into estimates for $G(\underline{\alpha})$. In either case, similar results should be obtained. In effect, the data tie down or "condition" the model. If complete data on all variables were known, no subjectivity and no probability would be necessary. The probabilistic approach models the uncertainty and lack of information.

Models

All deterministic porous-media models are based on standard flow and dispersion equations, which may include radioactive decay and adsorption terms. Some models include heat transport and other characteristics. Parameters such as hydraulic conductivity, transmissivity, flow velocity, dispersivity, and so on are treated as constant, deterministic values within various zones. Constant parameter values, however, are generally not known exactly. Several procedures, including sensitivity and probabilistic analysis, are used to examine this uncertainty. In addition, efforts have been made to use data in inverse procedures to increase the accuracies of flow predictions and to aid in predicting concentrations or discharges. One major probabilistic aspect thus comes from the uncertainty associated with constant but unknown parameters. Another probabilistic aspect comes from dispersion itself, which is generally treated as a random mechanism.

For saturated media, the basic equations (for a single species) are of the form (M. P. Anderson, 1979; Bear, 1972)

$$q_i = -K \frac{\partial \phi}{x_i} \qquad \text{[Darcy equation] (4.1)}$$

$$\sum_{i=1}^{m} \frac{\partial q_i}{\partial x_i} = 0 \qquad \text{[continuity equation] (4.2)}$$

$$n \frac{\partial C}{\partial t} = \sum_{i=1}^{m} \sum_{j=1}^{m} \frac{\partial}{\partial x_i} D_{ij} \frac{\partial C}{\partial x_j} - \frac{\partial}{\partial x_i} (Cq_i) - \lambda K_d C \qquad \text{[transport equation] (4.3)}$$

where K = hydraulic conductivity

ϕ = head

q_i = specific discharge, direction x_i

m = dimension (1, 2, or 3)

n = porosity

C = concentration at location (x_1, x_m) and time t

D_{ij} = dispersion coefficient tensor

λ = decay constant

K_d = adsorption coefficient

t = time

These equations can be generalized to include heat flow and brine transport (Bear, 1972; Reeves and Cranwell, 1981) and decay and transport of several species (M. P. Anderson, 1979; Campbell et al., 1981; DeMarsily, 1983; Gureghian and

Jansen, 1985; Koplick et al., 1982). The equations are well-accepted models for flow in homogeneous media and, with appropriate modifications, can be applied to heterogeneous media if the heterogeneity is modeled deterministically. The role of the dispersion terms is still being actively studied (Dagan, 1986; Gelhar et al., 1979; Gelhar and Axness, 1983; Gueven et al., 1984; Matheron and DeMarsily, 1980; Pickens and Grisak, 1981; Sudicky, 1985) and will be explored in the section on stochastic methods.

The complexity of the models dictates that solutions to Equations 4.1 through 4.3, or their counterparts, must be obtained numerically by finite-difference (Reeves and Cranwell, 1981) or finite-element techniques (Daus and Frind, 1985; Huyakorn et al., 1985). Procedures are available for analyzing possible instabilities and inaccuracies that occur when numerical techniques are used.

One difficulty encountered in uncertainty modeling is that if parameters are assumed to be random variables because of uncertainties and if probabilistic predictions are based on the models, many computer runs are required for an uncertainty analysis. Using the full numerical model will make such calculations expensive. For sensitivity and probabilistic analyses, simpler flow and dispersion models have been developed (Campbell et al., 1981; Koplick et al., 1982). NWFT/DVM (Campbell et al., 1981), a typical representative of such models, is a one-dimensional network flow and dispersion model used in conjunction with a full finite-difference model, SWIFT (Reeves and Cranwell, 1981). The full model is used to find the dominant flow paths, and then NWFT/DVM is used to perform probabilistic parameter variation and sensitivity studies economically.

Alternatives to numerical models also have been studied. Gureghian and Jansen (1985), for example, presented an analytical procedure using Laplace transforms to solve dispersion equations in layered media for a three-member chain. Their procedure is exact for nondispersive cases and might yield an alternative to parts of NWFT/DVM. NWFT/DVM incorporates the effects they studied as well as aspects like mixing and multimember chains.

With the exception of finding appropriate dispersion coefficients (Gelhar et al., 1979; Matheron and DeMarsily, 1980), flow and dispersion models are well-defined models with many parameters whose values are unknown. Research continues on dispersivities and dispersion coefficients. Although some authors suggest that dispersion models and coefficients are, at best, empirical (Gillham and Cherry, 1982), most researchers think these models can be adapted to real cases with appropriate parameter definition.

Remson et al. (1982) pointed out that porous-media models apply to cases where hydraulic conductivity is not small. At waste-management sites, conductivity values may be less than values where Darcy flow occurs. Freeze and Cherry (1979) suggested that for such small conductivities a cutoff (no-flow) condition may exist. In any case, assuming that the porous-media models hold, flow and dispersion should yield conservative results.

Significant controlling factors for radioactive-waste problems are distribution (K_d) or retardation (R_d) coefficients (M. P. Anderson, 1979; Cranwell et al., 1987; Waste Isolation Systems Panel, 1983). These can vary widely and have great associated uncertainty. Not only are reported K_d values extremely variable, but models using these values assume linearity. The K_d values used play a critical role in retardation. Field experiments are needed both to measure K_d values and to examine their role (M. P. Anderson, 1979; Remson et al., 1982; Waste Isolation Systems Panel, 1983).

Probabilistic and Sensitivity Aspects

In probabilistic analyses for these models as they apply to repositories, a variable like integrated discharge for species i is designated as G_i ($\underline{x}$, t; $\underline{a}$) at location $\underline{x}$, time t, and a vector of parameters $\underline{a}$ (hydraulic conductivities, porosities, retardation coefficients, etc.). (Underlined quantities denote vectors.) A probabilistic analysis entails choosing specific numerical values from some joint probability distribution of the parameters $\underline{a}$ and calculating integrated discharge for that set. This procedure is repeated many times (N sets of parameters $\underline{a}_1$, $\underline{a}_2$, ... , $\underline{a}_N$ are chosen and the calculations performed for each set) and then the probability that G_i at $\underline{x}$ and t exceeds a given level of g^*_i is estimated by

$$\hat{p} = [\text{number of times } G_i\,(\underline{x}, t; \underline{a}_j\,) > g^*_i \tag{4.4}$$

This approach was followed by DeMarsily (1983), Campbell et al. (1981), and Cranwell et al. (1987). Sensitivity analyses also require multiple runs (Iman et al., 1978; Iman et al., 1979; Iman and Helton, 1985).

The models may be combined further with efficient sampling techniques for the parameter values like Latin hypercube sampling (LHS) (Iman and Shortencarier, 1984; McKay et al., 1979). Latin hypercube sampling is a method to combine parameter values in an efficient manner, so that unbiased results with small variances are obtained for the output (G_i or $\hat{p}$ in the example above). LHS can be extended to incorporate rank correlations (Iman and Conover, 1982; Iman and Shortencarier, 1984). LHS yields unbiased estimates of mean values and the distribution function; furthermore, for monotonic functions variances are smaller than those obtained for random sampling (Iman et al., 1978; McKay et al., 1979). It does yield slightly biased variance estimates (Iman et al., 1978; Iman and Helton, 1985). Similar procedures are followed for sensitivity analysis.

Sensitivity analysis is useful for probability assessment, because it can indicate the most influential parameters for a detailed probabilistic study. Procedures for sensitivity analysis other than LHS have been proposed (Metcalfe et al., 1985; Oblow, 1983; Sykes et al., 1985). Oblow outlined a general approach based on a linearization scheme that used an automated procedure to carry out adjoint sensitivity analysis. The adjoint sensitivity method was also discussed by Sykes et al. (1985) and DeMarsily (1983).

Probability and Data Incorporation

If certain a priori distributions are used for various parameters, incorporation of data is still necessary for a specific site analysis. How many more data are needed, where data should be taken, etc., are also vital questions. Sensitivity analysis can help with these questions; inverse methods have also been recommended (Waste Isolation Systems Panel, 1983).

Most inverse procedures focus on flow problems and the use of hydraulic-head values to estimate conductivities or transmissivities. They also are useful for transport and dispersion phenomena, because heads and transmissivities or conductivities drive the velocity fields, and the latter, in turn, determine transport (Neuman, 1984). Early methods (Cooley, 1977, 1979) for deterministic two-dimensional problems used nonlinear least squares. Flow equations were incorporated numerically, and zoned transmissivity values were estimated. Cooley (1982) showed how "prior" information could be taken into account in the inverse model. The statistical aspects of this approach primarily reflect measurement error. The parameters presumably are deterministic, and the main objective is to get estimates for parameters that yield a good predictive model for the deterministic output.

Later works by Neuman (1984) and Clifton and Neuman (1982) have combined the inverse method with geostatistical procedures. Again, a zoned approach is taken, but now correlations are allowed between zoned values. Best unbiased estimates of transmissivity are obtained in various zones by kriging. Similar prior estimates for heads are obtained. These estimates are used along with flow equations and a weighted least-squares approach to find new parameter (transmissivity and head) estimates. This procedure seems to yield smoothed predictions with zonal intercorrelations and prediction variances for the parameters. It may be a fruitful approach for the deterministic case, although it is quasi-stochastic. Clifton and Neuman (1982) referred to this as a Bayesian method, even though initial distributions are not updated. This method will yield precise "conditioned" (on the data) fields useful for transport predictions if many data points are available.

Data Bases

Although data and data bases have improved since the study by M. P. Anderson (1979), many of her comments are still valid. As she pointed out, the models proposed require more detailed and extensive field data than previous groundwater models; this is especially the case for dispersion models. The data base for transmissivities given by Freeze and Cherry (1979) and the discussion by Freeze (1975) are still useful. Cranwell et al. (1987) summarized other data bases for parameters needed in modeling a hypothetical bedded-salt site. Thus, although some significant generic data bases exist, they do need expansion, especially for quantities like dispersivities and retardation coefficients.

STOCHASTIC MODELS OF POROUS MEDIA

In recent years, widespread interest in stochastic models of heterogeneous porous media has grown (Freeze, 1975; Dagan, 1986). The demarcation between stochastic models and the deterministic parameter-variation models discussed above is not always clearly defined. In general, the randomness inherent in the parameter-variation (nonstochastic) approach is associated with errors in measurements or uncertainty about hydrological or physical parameter values. In contrast, stochastic models consider randomness to be an inherent feature of flow and dispersion. Media properties like hydraulic conductivity are viewed as if they were stochastic or random processes in space, and consequently derived quantities like head or concentration are also stochastic properties.

Work in this area involves both flow and transport and consists of several types: spectral and perturbation methods; Monte Carlo methods; and Green's functions and perturbation methods. In addition, various geostatistical approaches have been associated with this class of models, especially insofar as data analysis and incorporation are involved. These approaches will be examined, because they relate to basic questions about probability estimates in waste-site studies.

In the stochastic case, parameters are statistical parameters like the mean of the hydraulic conductivity in a particular field. The process itself is still random with given parameters even after the mean value is chosen.

In sensitivity analyses, sensitivity to various covariance function forms also should be examined. Some sensitivity to covariance forms exists (Mizell et al., 1982; Dagan, 1982), but the full three-dimensional model is usually less sensitive to the forms. Gutjahr (1985) showed that for a one-dimensional dispersion model analogous to NWFT/DVM, results for means and variances are insensitive to covariance form for velocity.

One limitation of stochastic models is that they have often been based on perturbation analyses that assume input variances are small, when in fact such variances can be large (Freeze, 1975; Hoeksema and Kitanidis, 1985). Flow-perturbation models have been tested against more exact, but time-consuming, Monte Carlo models (Gutjahr and Gelhar, 1981; Mizell et al., 1982) and found to be accurate. Similarly, for flow models theoretical analyses (Dagan, 1985a;

Gutjahr, 1984) also showed that perturbation results are accurate even for large variances. For transport models, such analyses remain to be made. Smith and Schwartz (1980) did a Monte Carlo study of robustness in concentration models, but their results must be tempered by the fact that little spatial correlation was used in the cases they analyzed.

Many stochastic analyses refer to ensemble results. For waste-site analysis, these results must be conditioned to apply to specific cases. In effect, such conditioning ties down the possible realizations or possible spatial reality for a particular domain by using the associated data. Thus one could use generic data of the type given by Hoeksema and Kitanidis (1985) and then condition the domain with site-specific data. This approach most likely would require many conditioned Monte Carlo runs. As in the deterministic case, equivalent one-dimensional models could be developed to study the probabilistic aspects.

Because concentration and related quantities can fluctuate widely, many runs would be needed for probabilistic analyses. Conditioning can greatly reduce the concentration variability (Dagan, 1984, 1985b). Such large concentration fluctuations may be a true feature of reality.

If stochastic models are taken to model variations that exist in place of zoned, deterministic models and if further generic results like those of Hoeksema and Kitanidis (1985) are used, stochastic models may require fewer data than deterministic models. Furthermore, the added information available from different data sources can be incorporated into stochastic models via co-kriging and geostatistical approaches and the worth of such information evaluated (Gutjahr, 1981; Gutjahr and Wilson, 1985). The indications are that stochastic models do simulate features observed in the field (Dagan, 1986; Gelhar et al., 1979; Sudicky, 1985). Another issue of practicality, mentioned above, is that a stochastic representation may be more realistic and useful than a zoned deterministic model. Computer runs needed for probability calculations can be costly and time consuming, however.

How do stochastic models compare with deterministic models? Parameter-variation features are similar; however, if the stochastic approach is used to model spatial variability, fewer parameters like mean transmissivities may be required. If one looks at conditioned values and if the measurements are extensive (e.g., within one or two correlation scales of each other inside the domain of interest), stochastic and deterministic results are similar.

In stochastic models, variability is inherent in the model description; probabilistic calculations arise within a natural context. That aspect of the probabilistic calculation is objective, but again, the choice of statistical parameters is subjective. The dispersion phenomenon, in both stochastic and deterministic models, is also a random inherent feature that is not completely understood.

Models

Essentially all the models use the basic flow and dispersion Equations 4.1, 4.2, and 4.3 as a starting point, except that now the hydrologic parameters and variables are viewed as spatially varying stochastic processes. Thus, in Equation 4.1, $K = K(\underline{x})$ is a stochastic process, as are all the other variables. In most of the models, $K(\underline{x})$ is modeled as a second-order stationary process (also called a random field), that is, as having a constant mean and covariance behavior depending only on spatial separation. Because various studies (Freeze, 1975; Hoeksema and Kitanidis, 1985) have indicated that $\ln K(\underline{x})$ is approximately normal, $\ln K(\underline{x}) = F + f(\underline{x})$ is often used instead of $K(\underline{x})$, where F (the mean term) is constant and $f(\underline{x})$ is the stochastic term. The covariance function is

$$R_{ff}(\underline{\xi}) = \mathrm{cov}[f(\underline{x}+\underline{\xi}), f(\underline{x})] \tag{4.10}$$

and the associated spectral density for the spectral methods is

$$S_{ff}(\underline{k}) = \int_{-\infty}^{\infty} \frac{R_{ff}(\underline{\xi}) \exp(i\underline{k} \cdot \underline{\xi})\, d\underline{\xi}}{(2\pi)^m}$$

where $\underline{k}$ is called a wave-number (or vector) and the integral is over m dimensions.

$R_{ff}(\underline{\xi})$ measures the strength of the statistical relationship between two points that are $\underline{\xi}$ units (vector) apart. It measures, in a sense, the statistical continuity of the field. Other quantities used are integral or correlation scales, which measure average distances over which significant correlations occur for the process in question. Stochastic analyses have studied several aspects of these models. Initially, relationships between input (e.g., log of conductivity) and output (head) variances were derived (Bakr et al., 1978; Chirlin and Dagan, 1980; Dagan, 1979, 1982; Mizell et al., 1982; Smith and Freeze, 1979a, 1979b). These studies indicated that the correlation scales are important features of the models, that heads are correlated over larger distances than conductivity correlations, and that statistically isotropic inputs can yield nonisotropic outputs.

Effects of boundaries on the calculations were further explored by Gutjahr and Gelhar (1981) and Smith and Freeze (1979a, 1979b). Conditions for which stationarity of heads holds and the effects of using perturbation models for flow were also explored (Chirlin and Dagan, 1980; Dagan, 1985a; Gutjahr, 1984; Gutjahr and Gelhar, 1981). Relationships yielding effective flow parameters are important because they show how averaging should be treated (Dagan, 1979, 1982, 1986; Gutjahr et al., 1978).

These results are all relevant to predicting probabilities, because they indicate the kinds of parameters that must be estimated. Furthermore, the results are needed for probabilistic predictions for solute transport.

Stochastic solute-transport models have helped to delineate several important aspects of dispersion phenomena. One aspect is macrodispersion, which is additional dispersion due to correlation between the stochastic velocity and concentration (Dagan, 1982, 1984, 1985b, 1986; Gelhar and Axness, 1983; Gelhar and Gutjahr, 1982; Gelhar et al., 1979; Gutjahr, 1985; Matheron and DeMarsily, 1980; Simmons, 1982). These papers show how to obtain macrodispersivities in terms of input covariance functions and their parameters and also show that the approach to a Fickian diffusion model is time dependent and can be slow.

The anisotropic nature of concentration (Dagan, 1984; Gelhar and Axness, 1983) is an important feature of the multidimensional stochastic solute-transport problem. Relationships between dispersivities and effective conductivities (Gelhar and Axness, 1983) can be useful (Hufschmied, 1985).

Variances of concentrations (Dagan, 1982, 1984, 1986; Vomvoris and Gelhar, 1985) can be large. Dagan's calculations show that, especially on a regional scale, concentration has a very large variance, indicating that any predictions made on the basis of mean concentration equations have a large uncertainty. Also, concentration itself is extremely unpredictable insofar as quantities like arrival times are concerned. Gelhar and Gutjahr (1982) and Gutjahr (1985) noted a similar result for variances in a one-dimensional dispersion problem.

Dagan (1984, 1986) and Matheron and DeMarsily (1980) used a generalized Brownian-motion approach similar to the particle-tracking methods used in classical dispersion problems. Gelhar and Axness (1983) and Gelhar et al. (1979) instead used a Eulerian description. Uffink (1985) used a particle-tracking and random-walk technique for dispersion with a Runge-Kutta numerical procedure to get results analogous to those of Gelhar et al. (1979). Brownian-motion and random-walk methods possibly may be incorporated in models like NWFT/DVM that include other features like retardation and decay.

Some stochastic developments more directly relevant to radioactive waste disposal were given by Gelhar and Gutjahr (1982) and Gutjahr (1985), who studied a one-dimensional concentration process for a single decaying species. Both mean and variance equations were developed for concentration. Gutjahr et al. (1985) found the effect of different covariance functions to be small. In addition, comparisons with a parameter-variation model that had the same mean and variance for the input showed that the stochastic model predicts arrival of mass at a location more conservatively than the parameter variation model.

The Monte Carlo approach to the solute transport model was illustrated by Schwartz and Smith (1981) and Smith and Schwartz (1980). Their conclusions were similar to those given above. The analytical approaches described above yield insights into controlling mechanisms and the importance of various parameters. Implementation for actual sites will most likely require numerical approaches. If a Monte Carlo modeling approach is coupled with a similar Monte Carlo approach for parameter variation, the calculations required could become prohibitively expensive and time consuming.

The role of scale-dependent dispersion (Pickens and Grisak, 1981) is not resolved for either the stochastic or deterministic models. Gillham and Cherry (1982) viewed dispersivities as purely empirical with little relation to the underlying equations. Because using macrodispersion coefficients yields more realistic results, authors like Gueven et al. (1984) claimed larger values than those indicated by Gelhar et al. (1979). However, what the actual scale values should be in the work of Gueven et al. (1984) is not clear, and that may significantly influence their results. Sudicky (1985) reported measured dispersivity values that agree well with the model developed by Gelhar and Axness (1983).

Other questions concerning stochastic models involve appropriateness of scales and their roles, averaging, and ergodicity (Dagan, 1985b, 1986). The stochastic modeling approach is becoming more widely accepted, however.

Probabilistic Aspects

The probabilistic content of stochastic models includes variation of the statistical parameters (means, variances, scales, etc.) similar to that in the parameter-variation models. Stochastic models also include probability in a much more direct way: the model itself is stochastic so that even if the parameters were known, randomness would remain as part of the description of the medium. This spatial stochastic approach can replace the deterministic parameter-variation method. For this reason, stochastic models may actually need fewer data for their implementation than deterministic models. Sensitivity and uncertainty analyses similar to those previously discussed could be used to study probabilistic aspects of the parameters and to study effects of covariances and scales.

Probability and Data Incorporation

Many techniques used to incorporate data into stochastic models involve the geostatistical ideas of kriging, co-kriging, and conditional simulation (Delhomme, 1978, 1979). In addition, inverse modeling has been combined with geostatistical methods to study uncertainty in stochastic models (Carrera et al., 1985; Townley and Wilson, 1985). Alternatively, Dagan (1982, 1984, 1985b) used conditional probabilities, assuming that quantities like conductivity and head are jointly normal. Some of the approaches are not fully stochastic. Although most of the work has focused on flow, it can be extended to solute transport.

Gutjahr (1981) combined co-kriging with stochastic analysis to reconstruct smoothed head and conductivity or transmissivity fields. Kitanidis and co-workers used this concept in conjunction with numerical methods to find covariances and cross-covariances (Hoeksema and Kitanidis, 1984; Kitanidis, 1983; Kitanidis and Vomvoris, 1983). In their approach, head and transmissivity (smoothed) are reconstructed by

1. Estimating covariance relations by maximum likelihood or weighted least squares for transmissivity or conductivity.

2. Connecting input (transmissivity/conductivity) and output (head) covariances and cross-covariances using the stochastic equations and transfer functions.

3. Reconstructing the transmissivity and head fields using co-kriging and the relationships derived in Steps 1 and 2.

Gutjahr (1981) and Gutjahr and Wilson (1985) did Step 2 analytically, assuming a particular covariance function from Step 1. Hoeksema and Kitanidis (1984) and Kitanidis and Vomvoris (1983) performed Step 1 as indicated and Step 2 numerically. In both cases, the equations used were linearized.

This approach is especially useful for deciding where additional field measurements should be taken and for comparing different measurement networks. It is appealing because it combines the stochastic and statistical approaches. In each case, a kind of conditioning is carried out in the use of the data.

Dagan (1982, 1984, 1985b, 1986) used conditional normal distributions to achieve results similar to those discussed above. Dagan's results are especially important because he studied concentration conditioned on head and transmissivity. He concluded that for a two-dimensional problem (depth averaging used), conditioning on transmissivity can significantly reduce concentration variances if measurements are within two correlation scales of each other. He further concluded that head measurements did not reduce variances greatly or aid in predicting concentrations. These conclusions about spacing also apply to flow problems. Gutjahr and Wilson (1985) concluded that head or transmissivity measurements should be within one correlation scale of each other for accurate predictions.

The predictions referred to above are for smoothed or mean values. In waste disposal, variation is also of great importance. In an early attempt to get at the issue of variability and conditioning, Delhomme (1978, 1979) performed conditional Monte Carlo simulations. The results used the stochastic behavior of transmissivity along with a numerical model. This was methodologically sound: it not only attempted to incorporate all data, but it also allowed probability statements to be made for the output. Unfortunately, the joint conditioning was only carried out ad hoc. Such an ad hoc procedure would be costly to complete for more complex models. The work of Dagan (1984, 1985b, 1986) could be used to perform conditional simulations.

Devary and Doctor (1982) used geostatistical methods to estimate errors in velocities. Their model, however, was not fully stochastic and was inconsistent because correlations between conductivities and heads were not considered. They treated the estimation variance of the velocity, which differs from the actual variance. The latter is the controlling factor in concentration-fluctuation calculations.

Other authors have used geostatistics in conjunction with inverse techniques to reconstruct head and conductivity fields that influence transport. Clifton and Neuman (1982) used conditioning and an inverse estimation procedure and concluded (in contrast to Dagan) that conditioning on heads played a significant role. Their model was not for concentrations and, furthermore, was a smoothed model combining stochastic and deterministic features. Carrera et al. (1985), Sagar and Clifton (1985), and Townley and Wilson (1983, 1985) have used first- and second-order methods in conjunction with stochastic models and inverse procedures to address issues concerning reconstruction of heads and conductivities. No one seems to have extended these methods to probabilistic predictions about concentrations or integrated discharge.

In addition to the kind of data required on parameter uncertainty in the deterministic case, stochastic models also require information about covariances and associated parameters like scales. The remarks made in the previous subsection on probability and data incorporation for the deterministic case also apply to the stochastic case.

Data Bases

In addition to the data-base information given for the deterministic case, Hoeksema and Kitanidis (1985) gave useful results for stochastic models. They used innovative statistical procedures to obtain both distributional information and information about scales. Data on scales yield surprisingly large values, indicating a large degree of continuity. Delhomme (1979) also reported large scale values for the two-dimensional problem. Such large scale values would auger well for predictive modeling, because for large scale values a few samples can significantly reduce prediction variances.

An alternative promising approach to estimating quantities like the scale has been to try to relate such quantities to readily observed geologic features (e.g.,

MacMillan and Gutjahr, 1986). MacMillan and Gutjahr found that the average thickness of layers for different classificatory systems corresponded well to calculated vertical integral or correlation scales.

Sudicky (1985) demonstrated how macrodispersivities in a sandy aquifer can be derived using models by Gelhar and Axness (1983) and Dagan (1982). Dagan (1984, 1986) carried out a similar comparison and also concluded that field observations on dispersion do seem to be in accord with the stochastic theory results.

Hufschmied (1985) has performed an interesting experiment using flow-meter and pumping tests to estimate effective conductivities and other parameters like scales needed in stochastic analyses. This procedure shows promise for gathering data needed for stochastic analyses.

FRACTURED-MEDIA MODELS

Research efforts have been made recently in defining flow and dispersion in fractured media, and modeling and experimental work are still developing (Brown, 1984; Brown and Gelhar, 1985). Models of single fractures or parallel sets of fractures are common. Random aspects included have been assumptions of random apertures (widths), fracture lengths, spacing, and orientation. Many of the models have looked at flow, generally assuming that Poiseuille flow occurs within fractures. Several tried to see if flows and dispersion could be modeled using equivalent porous media. A few models have included stochastic features, incorporating correlation.

Models

Early work by Snow (1968a, 1968b, 1969, 1970) dealt with infinitely long fractures and included work on effective permeabilities and some data analysis to obtain necessary distributions. That work and virtually all work in this domain assumes Poiseuille flow. If Q is the flux, this assumption means

$$Q = \frac{gb^3 J}{12\nu}$$

where b is aperture width, J is hydraulic gradient, g is the gravitational constant, and ν is kinematic viscosity.

Single-Fracture Models. Models with single discrete fractures (Grisak and Pickens, 1980; Neretnieks, 1980; Neretnieks et al., 1982; Sudicky and Frind, 1984) are probably the simplest, but even these yield complex results. These studies examined flow and dispersion mechanisms that are controlling in single fractures. In addition, they indicate that matrix diffusion and sorption are important. Neretnieks (1982) and Neretnieks et al. (1982) noted that matrix diffusion is especially important in retarding contaminants.

Neuzil and Tracy (1981) looked at a single fracture where aperture varied in the direction perpendicular to flow. In the flow direction, flow was still assumed to take place in parallel channels. Neuzil and Tracy used their model to derive a modified Poiseuille equation involving the moments of the aperture, $E(b^3)$ and $E(b)$, and applied the model for lognormally distributed apertures. They studied effects of compression and showed that average fluid velocity could be obtained if the aperture-width moments were known. Their modified Poiseuille equation agreed well with experimental results when an approximate aperture-width distribution was used.

Sudicky and Frind (1984) examined transport of a radionuclide within a single fracture, including molecular diffusion into the porous matrix and adsorption both on the rock face and in the porous medium. They derived an analytical solution for two-member chains and showed that the parent nuclide must be included even if it is short-lived. In addition, the daughter may advance ahead of the parent even if the parent's half-life is longer.

Schwartz and Donath (1982) looked at effects of a single fracture in a nuclear-waste facility design. The effects of such a fracture are complex; small changes in geologic conditions can result in large differences in the pattern of contaminant transport.

Brown (1984) used a perturbation and spectral analytic approach to analyze flow and transport in a single fracture, where aperture is a stochastic process, $b(\underline{x})$. The analysis for flow is similar to that of Mizell et al. (1982) for stochastic porous media. He derived the effective aperture for flux and found results similar to those of Gelhar and Axness (1983). For solute transport, the mean concentration distribution, analogous again to the porous-media case, can be described by effective advective and dispersive transport components; a retardation effect on the solute velocity occurs due to the random media properties. Dispersion includes a macrodispersion term similar to that obtained for stochastic porous media.

Neretnieks (1983) examined an extension of a single-fracture model to stratified flow in parallel channels with random apertures. He showed that dispersion in such a situation is not Fickian. Use of a Fickian model will give misleading results about dispersion and diffusion. His results are typical of any medium that is perfectly stratified with no transverse dispersion (Brown and Gelhar, 1985).

Multiple-Fracture Models. Sudicky and Frind (1982) modeled multiple, parallel fractures and showed that greater contaminant-penetration distances occur along closely spaced fractures than along a single fracture. Grisak and Pickens (1980) considered a single set of periodic fractures and included matrix diffusion, distribution coefficients, and so on in a numerical model. They concluded that matrix diffusion was especially important to retardation for flow in fractured media. Grisak et al. (1980) also compared the above model with experimental results.

Equivalent-Porous-Medium Models. Several studies have modeled a fractured, porous medium by assuming that an equivalent porous medium could be defined. Duguid and Lee (1977) used a dual-porosity model for flow in fractured media. The fractures were "lumped" and modeled using an anisotropic secondary porosity. They concluded that even at small permeabilities, the permeability of the porous medium could not be ignored. Narasimhan (1982) also used an equivalent-porous-medium approach and suggested using a dual-porosity model for studying flow in fractured media. He indicated that macroscopic parameters needed for such models are hard to estimate and have a large degree of uncertainty, and that interactions between large- and small-scale phenomena are not well understood.

Both of the previous models treat flow rather than transport. Barbreau et al. (1982) used a two-dimensional equivalent-porous-medium approach to study transport in fractured granite. They used an apparent porosity for fractures and incorporated adsorption and variation of permeability with depth into the model. They concluded that regional distribution of permeabilities played an important role. In addition, they stated that retention within the rock was due to both a static factor related to K_d and a kinematic factor related to transport velocity in fractures and diffusive porosity.

Recent work has focused on models with multiple fractures and fracture sets and on discovering when a fractured medium could be modeled as an equivalent porous medium with respect to flow and dispersion.

Sagar and Runchal (1982) examined the symmetry of the permeability matrix. They claimed that this matrix is asymmetric and nontensorial and depends on the size of rock volume and fractures. Brown and Gelhar (1985) pointed out that Sagar and Runchal's model is inconsistent because, for finite fractures, continuity is violated. Sagar and Runchal (1982) also performed an uncertainty analysis using a second-order Taylor series analysis to examine the mean and variance of the permeability matrix entries. They indicated that aperture-width uncertainties lead to large permeability uncertainties.

Long et al. (1982) studied fracture domains with finite fracture lengths, random orientation, and random apertures. The primary focus was on two-dimensional flow, and their intent was to see whether an equivalent-porous-medium approach could be applied. Using Monte Carlo methods, their results show that the samples do not have a symmetric conductivity tensor and that large variations exist in conductivity from sample to sample. Thus, for moderate fracture density, the medium cannot be modeled as an equivalent porous medium. As fracture density increases, the hydraulic behavior does approach that of a homogeneous, anisotropic, porous medium. They observed further that

1. Fracture systems with random (distributed) apertures behave less like porous media than those with fixed apertures.

2. The size of the representative elementary volume (REV) (Bear, 1972) may need to be large to get an equivalent porous medium.

3. Fracture systems with random orientations behave more like porous media than those with fixed orientations.

Brown and Gelhar (1985) pointed out that the results of Long et al. (1982) imply that two important scales in fractured systems may occur: the length of fractures and the spacing of fractures. Endo et al. (1984) extended the study of Long et al. (1982) to transport properties. They examined deterministic fracture systems, one with two sets of parallel, equispaced fractures 30° apart, and another with orthogonal fracture sets. Individual sets had constant apertures, but each set had different apertures. For this model, flow can be modeled by an equivalent porous medium as determined by examining the permeability tensor and the flux, but transport cannot be modeled.

Other Models. Smith and Schwartz (1984), Schwartz et al. (1983), and Schwartz and Smith (1981) have studied random fracture networks, including flow, dispersion, and travel times. Schwartz et al. (1983) studied two perpendicular discrete fracture sets with the flow gradient arbitrarily oriented in the domain, using a two-dimensional Monte Carlo study. A particle-tracking procedure was used to study transport and dispersion. Controlling factors were orientation of fractures, differences in mean velocities in the two sets, and standard deviations of the aperture widths for the two sets. They observed significant longitudinal dispersion and transverse dispersion that was relatively independent of longitudinal dispersion. They concluded that classical dispersion models did not apply to the medium. The concentration distributions were strongly skewed, with long tails, and considerable variation existed between realizations.

Smith and Schwartz (1984) examined factors determining permeability of individual fractures and connectivity of sets. They found that mass transport was related to the amount of direct connection in the system and that dispersion increased with the mean and variance of the velocity in the primary fracture set and with the average path length in the secondary fracture set. Apertures of both sets also affected macroscopic dispersion and travel time distributions. Reducing the number of fractures in the primary set delayed breakthrough but increased macroscopic dispersion. Smith et al. (1985) extended this work to three dimensions and included roughness of apertures.

Rouleau and Gale (1985) studied fracture interconnectivity as it relates to the permeability tensor in continuum models. They carried out numerical simulations using a stochastic discrete-fracture model. They related an interconnectivity index to porosity of the system and suggested its use for percolation studies in networks of fractures.

Andersson et al. (1984) used infinite fractures in a two-dimensional flow model. They looked at effects of conditioning on flow characteristics. Additional fractures not observed were generated and effects of added measurements on flow predictions shown. Andersson (1985) generalized the work to cases where apertures can change in size between intersections, leading to a kind of correlation scale for fracture length.

Dershowitz et al. (1985) explored three-dimensional models. Their model used Poisson plane processes to simulate fracture domains with polygonal fractures that terminate at intersections with other fractures. The resulting fracture configurations were similar to actual observed fractures. The resulting fracture networks are used to study flow properties like effective hydraulic conductivity using a finite-element program. These three-dimensional models show great promise for the study of realistic fracture networks and represent an important first step.

Limited work has been done on unsaturated-flow models. Wang and Narasimhan (1984) hypothesized that fractures will be essentially dry and that flow will occur across fractures rather than within fractures. Montazer and Wilson (1984) used a double-porosity model with hysteresis included. Both studies were modeling approaches and represented important first steps.

Results from fracture flow models tend to be confusing; there is no unanimity as to which factors are important. Unsaturated or partially saturated flow models require added study. More experimental work would be useful in both cases.

Probabilistic Aspects

The probabilistic content of fractured-media models includes the parameters and distributions of aperture widths, fracture length, orientation, density, variability and uncertainty in boundary conditions, and properties of the solid matrix. Furthermore, if fractures are modeled as stochastic processes, the correlation structure must be included. Parameters of the distributions could be uncertain, and uncertainty analysis, similar to the previous two cases, should be performed. Fractured-media models are not nearly as well validated experimentally nor as well accepted as the porous-media models. Fractured-media models can be segregated into deterministic parameter-variation models and stochastic-process models.

Probability and Data Incorporation

All methods suggested in the previous two classes can be used for fractured media. The conditioning approach suggested by Andersson et al. (1984), for example, is analogous to the approach of Dagan (1984, 1986) for incorporation of data into porous-media models. Generic distributions can be used for the various parameters required.

Data Bases

Bianchi and Snow (1968) found apertures to be lognormally distributed, with means decreasing with depths; they also examined aperture orientation. Baecher (1983b) pointed out that biasing may occur in orientation and length sampling. Spacing and length have been studied by Baecher et al. (1977) and Baecher (1983b). Lognormal distributions for lengths (Baecher, 1983b) are generally accepted, although some authors have used exponential distributions.

For a single fissure, Neretnieks et al. (1982) studied tracer movement experimentally. They observed short residence times and little diffusion into the matrix and suggested independent channels. Gale (1982) examined flow in parallel plate fractures and discussed various measurement systems for fractured systems. Brown and Gelhar (1985) also summarized available data for fractured systems. In addition, they discussed various measurement systems and work that remains to be done in that area. Intera Environmental Consultants, Inc. (1983) provided information and data relevant to fractured-media parameters.

Other measurement techniques useful for fractured-domain studies include geophysical methods (Nelson, 1982) like acoustic and electrical cross-borehole methods. These procedures use borehole probes with tomography to study fractures and other subsurface features. Work on measurements of fractures and fissures has been performed in France using geostatistical methods (DeMarsily, 1984). For example, Razak (1985) found the variogram of fracture density in a region and used that to calculate a random field of fractures. Such procedures can be useful for flow and dispersion studies of a stochastic nature.

CRITICAL COMPARISONS

Models

In all models discussed, the role of adsorption coefficients, K_d, is not completely resolved. The validity of the linear isotherm approach, the possible field values of K_d, the degree of variability of K_d, and the probability law for K_d all need further study. Once generic values for K_d are determined, site-specific values must be incorporated.

Deterministic parameter-variation models require a large number of parameters for reasonably realistic zoned values. Even then, how correlations between different zones or fluctuations from mean values (which are important for transport and dispersion) could be incorporated is not always clear. For any particular site, the application of deterministic models would require many data points and perhaps a large number of zones; such a complex numerical model still may not represent the site adequately.

Stochastic models include variability inherently within the model calculations. Data are becoming available on scales and variabilities (Gelhar, 1985; Hoeksema and Kitanidis, 1985) and on dispersivities (Gelhar, 1985; Sudicky, 1985). Gelhar (1985) pointed out how stochastic models can be calibrated and how dispersion tensors can be estimated. These models have been used to explain field data, but more experimental work is required, especially on transport and dispersion processes.

Generic results from stochastic models must be combined with site-specific information (Dagan, 1985b) and problems of appropriate scales resolved (Dagan, 1986; Gelhar, 1985). Many generic results rely on perturbation analyses, and although some progress has been made in applying these results to flow problems (Dagan, 1985a; Gutjahr, 1984), the robustness of transport results needs further study. Stochastic models may require fewer data than parameter-variation models for a particular site-specific case if used with generic values, but better procedures for estimating scales and variances, perhaps based on geologic features (MacMillan and Gutjahr, 1986), must be developed.

Stochastic-model development using conditioning and Monte Carlo approaches is applicable for making probability predictions at a particular site. The theoretical underpinning for such modeling exists, but the actual incorporation into computer models may be expensive and awkward (Dagan, 1984, 1986; Delhomme, 1978, 1979; Gutjahr and Wilson, 1985).

For both deterministic parameter-variation models and stochastic models, the role and value of dispersivities are being debated. Evidence is mounting that macrodispersivities account for departures from classical models observed in the field (Dagan, 1986; Gelhar et al., 1979), but additional studies are needed. For both types of models, dimensionality must be further examined. Many of the models used are two-dimensional, but in some cases three-dimensional models are more appropriate. Such three-dimensional models can be developed, but they are costly both to develop and to run. Again, deterministic models require many parameter values for specific application. The validity of equivalent one-dimensional models should also be examined in any specific application.

Fractured media require more study before model results can be applied confidently. From a practical point of view, the stochastic approach would be more amenable for realistic applications. Location of all specific fractures and their exact parameters would be formidable, if not impossible. Results obtained regarding equivalent porous media and scales as well as results of the type given by Brown (1984) again are useful from a generic viewpoint. Almost all fracture models are two-dimensional. Full three-dimensional models now under development could yield significantly different results.

Probabilities

The models discussed require inputs like conductivities and dispersivities and yield outputs like concentration. One major aspect of probabilistic studies is the uncertainty or probability assigned to various inputs. In addition, behavior of parameters through time are also implicitly modeled in the input data.

All models discussed assume unchanging geologic conditions (DeMarsily, 1983). Disruptive events and processes could be incorporated. Although this adds some complexity, it does not seem to add any fundamental problem. DeMarsily (1983) describes other approaches that model changes in parameter values; such changes, if made an inherent feature, again require a fair amount of subjective judgment. For time intervals of thousands of years, the fixed geologic model seems reasonable unless disruptive events occur. Near-field changes imposed by site placement may need to be included.

The probabilities discussed above are for input parameters. For the output obtained from the models, interest is often in the tail probabilities, that is, probabilities for relatively rare events. Model outputs include mean values, variances, and in some cases estimates of rare-event probabilities. In the case of deterministic inputs, output reflects the probabilistic content of the input. In the case of stochastic models, output further includes the random media effects. Means and variances can be used to bound extreme event probabilities with Chebychev's or Markov's inequality (Ross, 1985), but such bounds may be too coarse. Probability estimates based on Monte Carlo methods, on the other hand, may require too many runs for reasonably accurate estimates. As DeMarsily (1983) pointed out, estimating events that have probability 0.001 would require 10^4 runs, unless one assumes a form for the probability law (e.g., lognormal) for the output.

Estimating rare event or tail (extreme) probabilities is, in general, difficult without making a variety of assumptions. The actual tail probabilities can depend strongly on the specific probability functions and the correlation structure of the stochastic processes involved. Both of these features may be difficult to obtain for ground-water phenomena. Although possible procedures for finding the extreme probabilities have been described, no complete answer to questions related to these probabilities and their estimation is known.

Koplick et al. (1982) suggested that subjective probabilities assigned by experts may simply obscure the role of judgment; however, judgment seemingly cannot be avoided. Any subjective procedure used should be explicitly stated and explained with justification. The use of methods like the Delphi method is also suspect, and a more straightforward, informed approach may be preferable. If extensive data are available for associated generic cases, using data-based distributions is the recommended approach. Generic data should be combined with site-specific data via conditioning to calculate output values and probabilities. When fewer data are available, sensitivity analysis of different assignments could be used to give information about relative weights. Sensitivity analyses should be separated from uncertainty analyses. Sensitivity studies (Iman et al., 1979; Iman and Helton, 1985; Sykes et al., 1985) are useful for determining important parameters for more detailed probability studies. They further indicate which parameters require accurate determination in order to make accurate output predictions.

UNPREDICTABLE AND LESS DEVELOPED AREAS

Some studies have appeared for unsaturated flow (Yeh et al., 1985a, 1985b, 1985c) and transport (Mantoglou and Gelhar, 1985) in porous media. A more crucial need for waste-disposal problems would seem to be in the area of unsaturated fractured media, because fractured media in general pose many unresolved problems. Although development is proceeding rapidly, predictive efforts are still in their infancy. The modeling approach may help to give some general guidelines. As important as modeling, however, is the effort to gather experimental and field data to check various models. Even though current models may be useful, their predictive worth requires evaluation. Other areas for development include isotope methods, studies of adsorption, and new data-gathering methods described in the previous sections.

SUMMARY

Probabilistic methods for hydrological predictive models are well developed, although the probabilistic aspects ultimately are subjectively based. The degree of subjectivity can be minimized with sufficient generic data used as a starting point. Any waste-site evaluation must incorporate site-specific data; ways to incorporate such data using inverse methods and geostatistical procedures are available. Whereas methods for porous-media models are developed or undergoing rapid development, fractured-media models are not as well defined.

For porous media, deterministic procedures may be useful if data are extensive. Deterministic porous-media models may especially be useful if inverse methods are used to incorporate site-specific data. Monte Carlo procedures may be used to determine the release probabilities and bounds on estimated probability values. Subjective probability estimates should be clearly and explicitly stated.

Stochastic models are an attractive alternative to deterministic porous-media models, because they include the inherent variability of the media. These may be combined with parameter information of the type discussed above and with geostatistical inverse procedures to perform site-specific evaluations. Monte Carlo analyses may be used both to estimate probabilities and to incorporate data via conditional simulations. These models include the variability inherent in nature more explicitly.

The best fractured-media models are stochastic models of the type studied by Brown (1984), Smith and Schwartz (1984), and Smith et al. (1985). These models require further validation and study.

5

Climatology

Patrick J. Bartlein
Thompson Webb III
and Steven Hostetler

INTRODUCTION

Research on long-term climatic variations has progressed rapidly over the past 15 years. Much of the stimulus for this work came from publication of *Understanding Climate Change—A Program for Action* (U.S. Committee for the Global Atmospheric Research Program, 1975). For variations on time scales of 10^3 to 10^6 years, key studies have revealed the controlling influence of variations in solar radiation determined by variations of Earth's orbital elements (Hays et al., 1976; Imbrie and Imbrie, 1980; Berger et al., 1984). Further research has shown how the land-sea contrast translates seasonal radiation variations into major changes in monsoonal climates (Kutzbach, 1981; Kutzbach and Street-Perrott, 1985; COHMAP Members, 1988). This progress in understanding and modeling long-term climatic variations has led to a few projections of future climates based on factors that influenced past climates (e.g., Imbrie and Imbrie, 1980). No comprehensive effort has been made, however, to predict climates of the next 10^3 to 10^6 years, although theory and techniques are now available that can be applied to the task (Craig, 1987; Goodess et al., 1990). This chapter evaluates (1) current understanding of the climatic system, (2) models of the "slow" and "fast" processes of the climatic system that can be used for climate prediction, (3) the combination of modeling efforts required to estimate future climatic variations within the region of a proposed repository site, and (4) requirements for determining the uncertainties of those predictions.

The need for methods for predicting future climatic variations arises from the great impact these variations can have on other environmental systems (Goudie, 1983). Global-scale variations of the atmosphere, ice sheets, and oceans that typify climatic variations of the Quaternary Period have been accompanied by regional and local changes in hydrology and ecology. Because variations of the hydrologic system are driven by climatic variations (e.g., Street-Perrott et al., 1983), the course of future climatic variations is an important consideration in evaluating the stability of local environments, for example.

Evaluating the predictability of climatic variations that occur on long time scales (>10^3 years) requires consideration of components of the climatic system that vary at those scales, the nature of potential external controls of those variations, and the degree to which observed climatic variations at those scales can be explained.

Components of the Climatic System

Basic Components. The components of the climatic system are the atmosphere, hydrosphere, cryosphere, surface lithosphere, and biomass (Gates, 1981; Figure 5.1). Individual components, such as ice sheets, land surface, and mixed-layer and deep ocean, have different thermal characteristics and response times. The atmosphere responds relatively rapidly to changes in external controls, the mixed layer of the ocean more slowly, and the deep ocean and continental ice sheets extremely slowly (Saltzman, 1985, Table 2). Individual climatic variables may be categorized into (1) those that describe *boundary conditions* or external controls of the climatic system, (2) those that describe *slowly varying* components, and (3) those that describe *fast-response* components.

Individual components are related to one another through transfers of mass and energy. These transfers occur along many pathways and involve many feedback mechanisms. One result of such interconnectedness (Saltzman, 1983, p. 203) is the possibility that variations in a particular component of the climatic system may result from purely internal, "free" variations, e.g., short-term climatic variations like those described by Namias (1975), as opposed to external "forced" variations, like those caused by insolation (cf. Wigley and Raper, 1990). Because they occur in the absence of external changes in the system, these free variations are inherently unpredictable in detail, although their statistical properties can be described (Kominz and Pisias, 1979). In other words, the exact record of a particular component of the system as it participates in such free variations may not be describable, but general characteristics of the record (its temporal

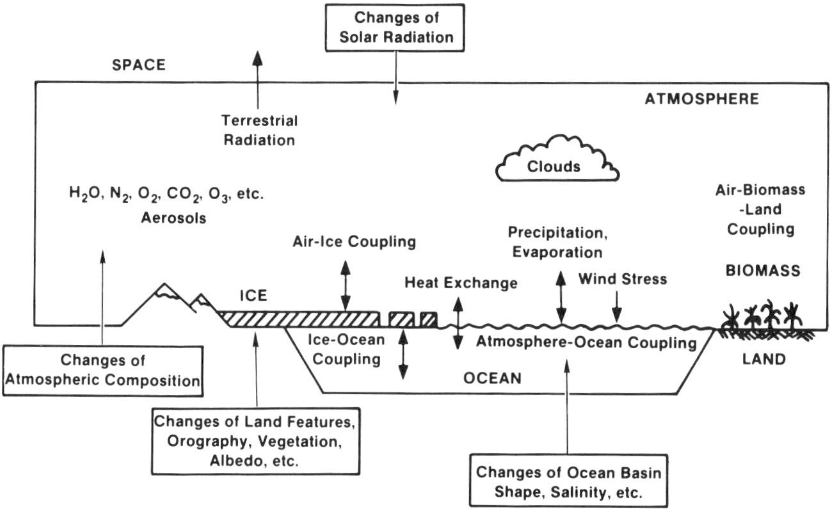

Figure 5.1. The climatic system (after Gates, 1981). Components of the climatic system appearing in boxes represent external controls. Copyright 1981 by Kluwer Academic Publishers.

variability or persistence, for example) may be. The opportunity for such internally driven variations to exist thus ensures that the success of explicit long-term climatic predictions has an upper limit.

Controls of Climatic Variations. The climatic system has several external controls, including incoming solar radiation, composition of the atmosphere, and arrangement and size of the continents. Additional controls exist, but their explicit definition depends on the context within which the climatic system is being considered. At shorter time scales, i.e., shorter than 10^4 years, the size (volume, area, and elevation) of ice sheets and the temperature of the oceans may also be regarded as external controls, because their response times are that long or longer. The relevant set of external controls thus depends on the time scale of climatic variation of interest. For short time scales, controls may include slowly varying components of the climatic system that at longer time scales can be regarded as dependent, internal components of the system.

An additional consideration that complicates the explicit definition of controls of climatic variations is that for some components of the climate system, either their role in the system is imperfectly understood or their future variations are inherently unpredictable. The concentration of CO_2 in the atmosphere is typical of such a component. During the Quaternary Period, CO_2 concentration in the atmosphere has varied in a fashion consistent with global climatic variations (e.g., Shackleton and Pisias, 1985; Barnola et al., 1987). It is impossible to state, however, whether CO_2 variations contribute to global climatic variations, or are driven by them, or both. Other greenhouse gases, such as methane, show similar patterns of variation (Chappellaz et al., 1990). During the past century, the concentration of CO_2 in the atmosphere has increased as a result of industrialization, a process clearly outside of the climatic system, and one whose future course is inherently unpredictable (Kellogg and Schware, 1981).

In a similar fashion, dust in the atmosphere can be considered an externally determined variable, as in the case of volcanic eruptions (e.g., Bryson and Goodman, 1980; Bradley, 1988) or nuclear war (Committee on the Atmospheric Effects of Nuclear Explosions, 1985). In the case of "Dust Bowl"-generated dust, however (e.g., Heathcote, 1983), the dust loading is at least in part an internally determined component of the climatic system. Both dust and CO_2 concentration usually are considered as "prescribed" boundary conditions, because the nature of their variations within the climatic system are imperfectly known.

Climatic variations therefore arise from two sources: (1) external sources, i.e., changes in boundary conditions or changes in the slowly varying components of the system (components with characteristic response times longer than the time scale of variation of interest), and (2) internal sources, i.e., the feedbacks among the various components of the system (Saltzman, 1983; Gates, 1981).

Temporal and Spatial Scales of Climatic Variability

Temporal Variations. A "powers-of-10" approach effectively illustrates the magnitude of climatic variations at different time scales (e.g., Webb et al., 1985, Figure 1; U.S. Committee for the Global Atmospheric Research Program, 1975, Figure A.2; Kutzbach, 1976, Figure 1). Over the past 100 years, the mean temperature of the northern hemisphere land surface has varied by about 0.5°C; over the past 10^3 years by about 1.5°C; and over 10^4 years or longer (i.e., Quaternary Period) by about 10°C. As the length of time considered increases, so does the range of climatic variation.

Spatial Variations. Climatic variations at a particular location are embedded in a hierarchy of larger-scale variations. Precipitation at a point, for example, is generated by passing weather systems with life spans on the order of days. These systems are in turn embedded in hemispheric-scale atmospheric circulation anomalies with life spans on the order of months or seasons (Namias, 1975). Finally, large-scale circulation features in their turn are related to hemispheric or global-scale variations in the energy balance of Earth's atmospheric system. A climatic change in a given region may therefore have a *proximate* cause, such as a change in the duration of different air masses or the relocation of a storm track, that in turn is related to an *ultimate* cause of hemispheric or global extent.

The hierarchical nature of climatic variation implies that the climatic record of a particular region cannot be explained (and, hence, predicted) in isolation (Bartlein, 1988). Local climatic variations are indeterminate with respect to larger-scale variations: the same local record could be generated by a range of larger-scale variations. Climatic prediction must therefore proceed in a top-down fashion: from large-scale, external *controls* to the smaller-scale, dependent *responses*.

Climatic Prediction

Classes of Climatic Prediction. Two different classes of climatic prediction can be envisaged, one in which the time history of the climatic variable of interest is predicted in an *explicit* fashion, and the second in which only *statistical* properties (e.g., average, range, variability, persistence, etc.) over some interval are predicted. Either kind of prediction requires specification both of the controls of the dependent climatic variable of interest and of the model linking controls to the variable. The specific nature of controls and model depends, however, on the context of a particular prediction problem.

When the climatic model can be completely specified, and when controls are known or predictable, the first kind of prediction is possible. Such a situation is not likely to arise for temporal and spatial scales of prediction required to evaluate, for example, the environmental stability of nuclear-waste repositories. Uncertainties exist in both the nature of the controls and the models, and these uncertainties will not be reducible by a simple increase in modeling effort. Some external controls of the climatic system, such as inputs of solar radiation that are determined by orbital variations, are predictable, but other controls that influence long-term climatic variations, such as solar output and atmospheric composition, are not. Similarly, although some climatic models exist that can, for example, relate global ice volume to orbitally determined radiation variations (Imbrie and Imbrie, 1980), models of many other components of the climatic system are not available. Moreover, the opportunity for feedback to occur among components of the climatic system (e.g., Saltzman, 1983; Gates, 1981) makes it unlikely that a comprehensive model of the climatic system could ever be completely specified. Uncertainties in both controls of climatic variations and models of the climatic system thus make possible only the second kind of prediction, prediction of statistical properties of a particular climatic variable.

Climatic Prediction at Different Time Scales. Only at the shortest of time scales ($\leq$10 years) can climatic prediction of the first kind be regarded as "operational" in any sense (Namias, 1975). Short-term prediction of temperature and precipitation patterns over, say, the United States exploits the dependence of surface climatic patterns on characteristics of the large-scale atmospheric circulation. Predictions are made by first extrapolating circulation anomalies (climatic controls) into the forecast period and then using empirically derived relationships (a climatic model) to transform circulation predictions into predictions of surface climatic variables. Explicit prediction of time series of temperature and precipitation is possible only to the extent that the controlling large-scale atmospheric circulation features are predictable. Currently, such predictions have been attempted only up to a season ahead, and considerable uncertainty remains in those predictions.

Medium-term climatic predictions (10 or 100 years) currently are being developed for assessing the impact of rising CO_2 concentrations in the atmosphere (Schlesinger and Mitchell, 1987). These assessments use the full variety of climatic models and are used to predict statistical attributes of altered climatic systems, as opposed to the specific record over this time scale. In a typical example, a three-dimensional model of the climatic system was used to predict the variation of hemispheric mean temperature over the next 100 years in response to variations in such boundary conditions as atmospheric CO_2, dust, and insolation (Hansen et al., 1988).

Atmospheric general circulation models can be used to infer spatial patterns of the response to elevated CO_2. Most models used at this time scale are equilibrium models, which give a snapshot view of conditions after a change in boundary conditions, as opposed to transient models, which attempt to show the course of adjustment to the new equilibrium (Schneider, 1984). Schlesinger and Mitchell (1987) noted that although models tend to show similar responses of globally averaged temperature, they differ on the spatial patterns of the response. In any case, even at a scale that is short relative to the scales of interest here, climatic prediction of the first kind is not possible.

At longer time scales ($>10^3$ years), specific examples of climatic prediction are rare, and these are directed more toward explaining the evolution of climate over the long run than forecasting a specific course of climatic variation. Climatic prediction at long time scales has been concerned mainly with predicting boundary conditions and slowly varying components of the climatic system (e.g., global ice volume). Specific models linking external controls with dependent variables are either simplified or borrowed from studies at shorter time scales.

Potentially Predictable Events and Processes. As time scales increase, temporal and spatial resolution of predictable components of the climatic system decreases. Over the shortest time scales, specific values of climatic variables at individual locations may be predictable. As time scales increase, specific values at a particular location can no longer be predicted, and at the longest time scales, only long-term averages of greatly aggregated climatic variables (e.g., average hemispheric temperature, or global ice volume) have any potential for prediction.

The climatic system offers no intrinsically defined events that are potential targets for prediction in the same way that a volcanic eruption or meteoritic impact is an event. Over long time scales (>10^3 years), even glaciation is not an event; at this scale, global ice volume is a continuously varying component of the climatic system. Although a particular location can be considered to be glacierized or not, it exists under a glacial climate long before ice reaches there. Specific climatic events for which probabilities of occurrence are sought must, therefore, be defined by the user of the predictions.

Climatic Variations on the Quaternary Time Scale (10^3-10^6 Years)

Analyses of climatic variations at the time scale of 10^3 to 10^6 years generally follow one of two interrelated approaches. In a *data-oriented* approach, individual, or networks of individual, records of paleoclimatic evidence are interpreted in paleoclimatic terms (for an extensive review, see Bradley, 1985). In a *modeling* approach, climatic models, initialized using reconstructed boundary conditions, are used to make inferences about the structure of past climates (e.g., Kutzbach, 1981). The data-oriented approach can be thought of as a bottom-up analysis of past climates, in that a large body of individual elements of paleoclimatic evidence is amassed to draw conclusions about the general state of the climatic system at a particular time. The modeling approach, in contrast, can be thought of as a top-down analysis of climate, in that general, or large-scale, boundary conditions are used to initialize a model that then may give a spatially disaggregated simulation of climate.

The two approaches are related in that paleoclimatic data are necessary for specifying boundary conditions of models and for model validation (CLIMAP Project Members, 1976, 1981; Peterson et al., 1979; COHMAP Members, 1988). The models in turn may be used to evaluate the physical consistency of sometimes disparate compilations of paleoclimatic evidence (Kutzbach and Street-Perrott, 1985).

For the purposes of climatic prediction on the long time scale, a top-down, modeling-type approach must be used, and the design of such an approach can make use of experience gained from applications to the paleoclimatic record. Analyses of paleoclimatic data must also be an element of any climatic prediction scheme, because the data can illustrate the range and nature of climatic variations on the long time scale and are necessary for model validation (Webb et al., 1985).

METHODS OF CLIMATIC PREDICTION FOR 10^4 YEARS

Few studies have attempted to predict climate *explicitly* over the long time scale ($>10^3$ years). The range of efforts that have attempted to predict climate *implicitly* is much broader, however, and includes virtually any study that has focused on modeling or explaining Quaternary climatic changes. Methods of climatic prediction for the 10^3-year time scale and longer can be divided into those derived from analyses of paleoclimatic data and those derived from climatic models. In addition to the six explicit attempts, the principal efforts in the broader group will be examined here.

Paleoclimatic-Data-Based Predictions

Prediction of the nature of future climatic variations on the basis of paleoclimatic variations assumes similitude between characteristics of the past record and the desired future one. The nature of past climatic variations can provide information about the nature of future variations only if the operation of the climatic system can be expected to remain unchanged. Study of the past climate can yield only general indications of the future climate; explicit forecasts of the course of future climatic variations are not possible, except as simple extrapolations of past behavior. Because boundary conditions of the climatic system are not likely to undergo a similar evolution over the next 10^4 years as they have in the past 10^4 years, predictions based on analysis of the paleoclimatic record alone will always contain some uncertainty. Climatic predictions based on paleoclimatic evidence can be divided into those that are simply descriptive and those that formally analyze the statistical nature of paleoclimatic variations.

Predictions Based on Descriptive Studies. Two examples of "predictions" of future climatic variations are given by Lamb (1982, Chapter 17) and Budyko (1982, Chapter 6). Both authors examined the record of climatic change. Lamb emphasized the role of climate in history, and Budyko focused on variations through time of the energy balance of Earth and its atmosphere. Both presented predictions in descriptive, qualitative terms, summarizing their own inferences and drawing on the conclusions of other publications. Neither offered an explicit forecast of the future evolution of the climatic system, but instead described in general terms the range of likely possibilities.

Kukla and Matthews's (1972) prediction was based on largely descriptive study of variations during the Quaternary Period. They drew attention to the tendency for past interglacial intervals to last about 10^4 years and noted that the present one had lasted about that long.

A further, more explicit approach uses analogs derived from the paleoclimatic record (Webb and Wigley, 1985). This approach is implemented by identifying those times in the past when the configuration of the controls of the climatic system approximated those expected for the future. The main limitation of this approach arises when exact (or even close) analogs for expected future values cannot be found in the past. For example, the elevated concentration of CO_2 from anthropogenic sources that is expected to occur in the near future has no analog throughout the Quaternary Period.

The largely descriptive approach represented by these studies has utility in formulating climatic predictions over the long time scale, because it focuses attention on the most evident features of the record of the climatic system (such as glacial/interglacial cycles of the Quaternary Period). Such features are likely to be most robust in future variations, and probably the greatest part of the actual predictability of future climates can be attained by such simple means.

Predictions Based on Statistical Analyses. Climatic predictions based on statistical analyses of past climatic variations more directly invoke the assumption that future variations will be like those of the past. By describing the statistical nature of past variations, the nature of future changes can be inferred. In a sense, any analysis of statistical properties of a climatic record can be used for climatic prediction. This discussion will focus on analyses of paleoclimatic records that cover long time scales.

Statistical analyses of time series of paleoclimatic records can be classified into two types: those focusing on the frequency-domain aspects of the series and those focusing on the time-domain characteristics (Imbrie, 1985). Frequency-domain analyses use methods of spectral analysis of time series (Priestley, 1981) to describe variability of a time series as a function of the frequency or period of variation. Time-domain analyses are directed toward building models that explicitly describe the nature of the temporal evolution of a series, i.e., the interval-to-interval persistence or "memory" in the series, or the leading or lagging relationships between series (Box and Jenkins, 1976). The product of a time-domain analysis is a model that explicitly links the current level of a time series with past levels of the series, or additionally, current and past levels of related series. This product is easily adaptable for forecasting. The two types of analyses are related because parameters of the time-domain model can be used to estimate the frequency-domain spectrum and vice versa (Priestley, 1981, p. 211). Both statistical prediction methods probably have their greatest utility in predicting future variations of the slowly varying external controls of the climate system. Although analysis and prediction of local variations are possible, they will suffer from the same indeterminacy described earlier for the spatial hierarchy of climatic variations.

Analyses of the spectrum of paleoclimatic time series include those by

Kutzbach and Bryson (1974), Mitchell (1976), Hays et al. (1976), and Pisias and Moore (1981). Kutzbach and Bryson (1974) assembled a composite spectrum of climatic variations for time scales ranging from 1 to 10^5 years from separate analyses of historical, isotopic, and botanical climatic records. The overall shape of the spectrum indicates that longer-period variations are similar to those of a "red-noise" time series, that is, one in which the range of variation increases with the span of time considered.

The information contained in the spectrum of a time series allows the rates and probabilities of occurrence of climatic changes to be inferred. One such analysis was that by the U.S. Committee for the Global Atmospheric Research Program (1975, Appendix A). That analysis illustrated how probabilistic statements about specific climatic variations can be derived from paleoclimatic data.

Time-domain analyses of paleoclimatic time series include two that explicitly predicted the future course of the series under consideration. Petersen and Larsen (1978) examined a 700,000-year-long record of oxygen-isotope variations, and Kanari et al. (1984) a 500,000-year-long composite record of pollen data. In both analyses, models from the family of autoregressive-moving average models (Box and Jenkins, 1976) were fitted to the data and used to forecast values of the series for future time steps. As is true for all predictions based on statistical models, several sources of uncertainty are attached to these forecasts. In particular, uncertainty arises in the selection of the form of the model and the estimation of the values of its parameters. Variance of forecasts from such a model increases rapidly with the lead time of the forecast (Box and Jenkins, 1976, Chapter 5), and so reasonable predictions are possible for only a limited time into the future relative to the duration of the series.

Climatic-Model-Based Predictions

Climatic models can be classified into those that describe fast-response components of the climatic system, such as the atmosphere, surface-energy and water balances, and mixed layer of the ocean, and those that describe slowly varying components, such as the ice sheets and deep ocean (Saltzman, 1985). At time scales shorter than 10^4 years, slowly varying components may be properly regarded as external, "boundary-condition-like" factors.

Models of the Fast-Response Components. Models of the rapidly varying parts of the climatic system, chiefly the atmospheric ones, can be divided into two groups (Saltzman, 1985). The first group explicitly represents the three-dimensional structure of the atmosphere, and in a sense, simulates day-to-day variations of the weather. The second group instead simulates the longer-term, spatially averaged weather, i.e., the climate. The first group of models includes atmospheric general circulation models (AGCMs or GCMs) and the second contains statistical-dynamical models (SDMs) in general, and energy-balance models (EBMs) in particular.

Atmospheric general circulation models (GCMs) attempt to represent the three-dimensional structure of the atmosphere and simulate day-to-day variations of weather by solving basic equations that govern the motion of the atmosphere over a three-dimensional grid (Meehl, 1984; Washington and Parkinson, 1986). Detailed information on the state of the atmosphere and the underlying surface over that grid is required to initialize the models. Many specific physical processes are explicitly represented in such models, but others are prescribed to remain at set levels and are not free to interact with other elements of the model. Because time constants of the oceans are so long relative to the atmosphere, current computer capabilities make it prohibitively expensive to routinely simulate ocean temperature and circulation in parallel with the atmosphere. In consequence, ocean surface temperatures are often prescribed at climatic-mean values. GCMs incorporating interactive (with the atmosphere) mixed-layer oceans are currently being applied, but models with a fully coupled ocean and atmosphere are still under development.

In practice, GCMs provide snapshot views of atmospheric components of the climatic system under fixed or limited variations (i.e., the seasonal cycle) of boundary conditions, because models are run at most for several hundred model-days at a time. Longer integrations are possible, but then numerical instabilities in the models' equations become important.

Pioneering applications of GCMs for simulating paleoclimates include those by Williams et al. (1974), Gates (1976a, 1976b), and Manabe and Hahn (1977). In these applications, the models were initialized using ice-age characteristics of land and ocean surfaces (e.g., CLIMAP Project Members, 1976). Other applications of a similar nature include those by Manabe and Broccoli (1985) (using a GCM with a mixed-layer ocean), Hansen et al. (1984), Rind et al., (1986), and Broccoli and Manabe (1987).

One of the most extensive applications of GCMs in a paleoclimatic context has been made by Kutzbach and his co-workers using both a low-resolution GCM (Kutzbach, 1981; Kutzbach and Otto-Bliesner, 1982) and the higher resolution NCAR CCM (National Center for Atmospheric Research Community Climate Model) (Kutzbach and Guetter, 1986; Kutzbach and Street-Perrott, 1985). A set of experiments described by Kutzback and Guetter (1986) and COHMAP Members (1988) allowed study of how the model responded to variations of solar radiation input, ocean surface temperature, extent of sea ice, and size of the Laurentide ice sheet during the past 18,000 years. Boundary conditions in these

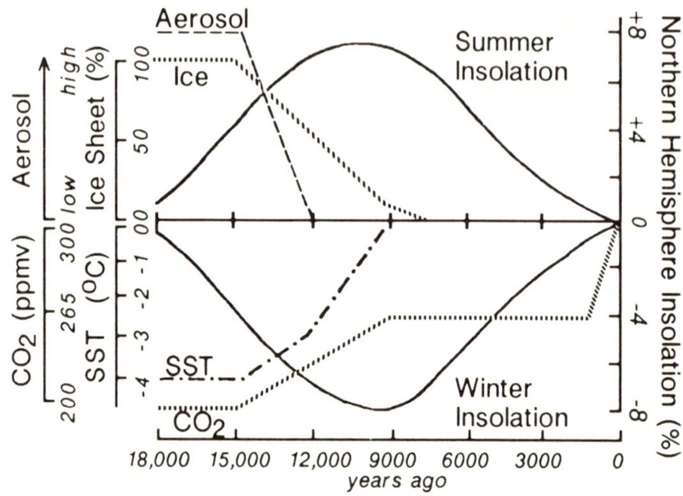

Figure 5.2. Variations in boundary conditions over the past 18,000 years (after COHMAP Members, 1988). The variations of the controls are schematic. Copyright 1988 by the AAAS.

simulations were adjusted to reflect changes in controls of the climatic system that accompanied deglaciation (Figure 5.2), and a set of seven simulations at 3000-year intervals was completed. The COHMAP Members (1988) compared the principal features of the simulated paleoclimates with paleoclimatic evidence (Figure 5.3). This analysis, considered further below, illustrated many elements that a potential scheme for predicting future climatic variations must have.

In all these studies, the models were not used to interpret or reconstruct past climates in a strict sense. Instead, they were used to infer the spatially varying aspects of climate that would be likely under specific boundary conditions and states of the slowly varying components of the climatic system. The potential of these models for climatic prediction lies in their ability to disaggregate the global-scale components of the climate system.

A second group of climatic models that attempts to represent variations of the rapidly varying components of the climatic system has been used in a paleoclimatic context. These include the energy-balance models (EBMs), which are members of the broad class of statistical-dynamical models (Saltzman, 1978). EBMs consider long-term, wide spatial averages of elementary heat-balance-related components of the climatic system. In comparison with GCMs, EBMs are relatively simple in terms of their detail and requirements for initial data and are much less costly to design and use. Introductory reviews of EBMs are given by Meehl (1984) and Ghil (1981), and more comprehensive reviews by Saltzman (1978) and North et al. (1981).

Energy-balance models are used in paleoclimatic contexts by examining the response of highly aggregated climate of the model to changes in external controls. The external control most often considered is solar radiation, as modified by temporal changes in Earth's orbital elements (e.g., North et al., 1984; Adem et al., 1984). The object of such experiments is to examine how such climatic variables as zonally averaged (averaged within a latitude band around the globe) temperature or the latitude of permanent snow cover respond to variations in solar radiation.

Berger (1977) used Sellers's (1969) early model to simulate zonally averaged annual mean air temperature over the past and the next 200,000 years. The model was driven by temporal variations of solar radiation over that interval. Although results of the paleoclimatic simulations by Berger only roughly corresponded to the known record (the magnitude of the climatic response was small, and the large thermal lags of the ice sheets and oceans were not explicitly taken into account), this application illustrates the possible role that such models could play in climatic prediction.

Hyde et al. (1989) recently compared a set of simulations for the past 18,000 years performed with a two-dimensional EBM with those completed by Kutzbach and Guetter (1984) using the NCAR GCM. The hemispherically averaged temperatures simulated by the EBM for the different time periods were in excellent agreement with those simulated by the more elaborate GCM. These results suggest that EBM simulations could play a useful role in evaluating the response of temperature (but not circulation) to changing boundary conditions.

The utility of models of the fast-response components of the climatic system in a comprehensive scheme for predicting future climatic variations is derived from their applicability in providing a "snapshot" view of the climatic system under a particular set of boundary conditions and state of slowly varying components. To portray the behavior of the climatic system as it evolves through time in dynamic equilibrium with changing boundary conditions, models of slowly varying components such as ice sheets and deep ocean that, along with the boundary conditions, generate long-term climatic variations must be considered (Saltzman, 1985).

Models of Slowly Varying Components. Saltzman (1985) grouped models of slowly varying components of the climatic system into two categories. First, a *quasi-deductive* group of models explicitly derives, for example, models of ice sheets, ice shelves, deformable bedrock, and deep ocean from basic physical principles. Second, a more *inductive* group of models seeks a physically reasonable model that produces results consistent with the observational records of paleoclimatic variations. The distinction between these two groups is somewhat blurred. An alternative way to consider models of slowly varying

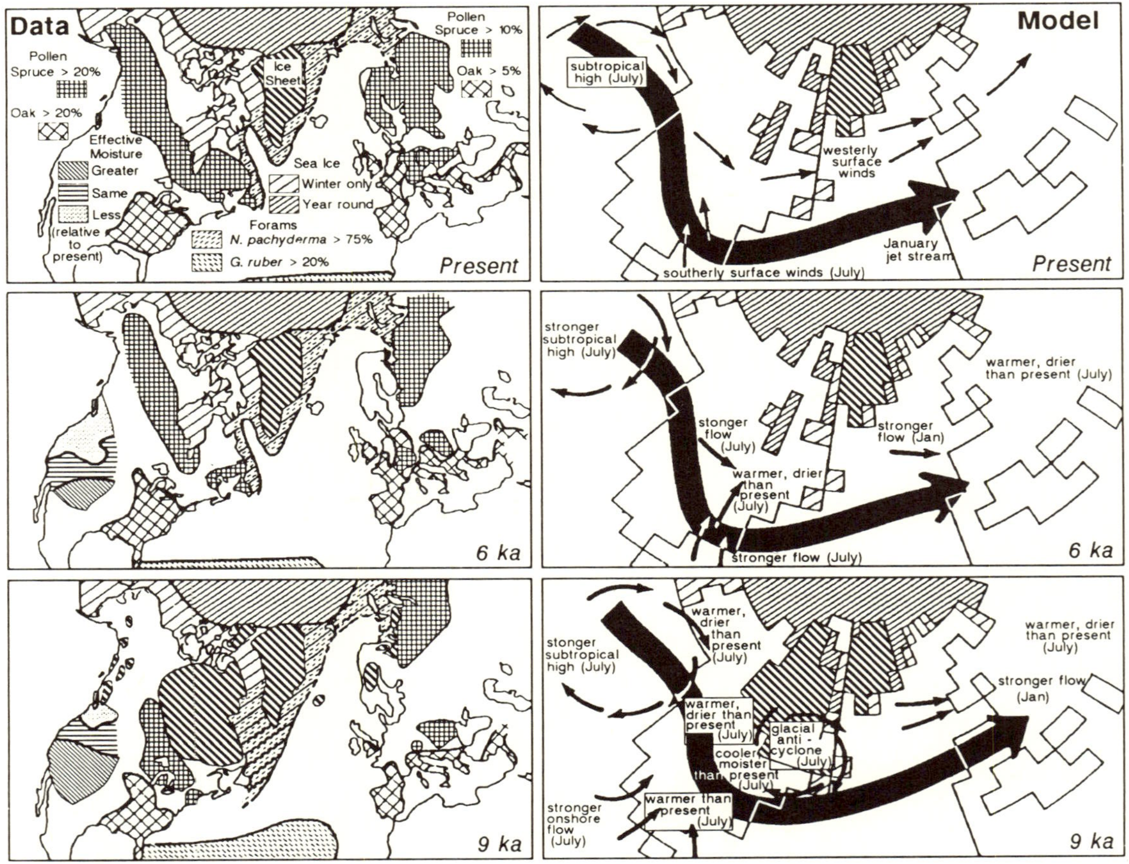
Data
Pollen
Spruce > 20%
Oak > 20%
Effective Moisture
Greater
Same
Less
(relative to present)
Ice Sheet
Sea Ice
Winter only
Year round
Forams
N. pachyderma > 75%
G. ruber > 20%
Pollen
Spruce > 10%
Oak > 5%
Present
6 ka
9 ka
Model
subtropical high (July)
westerly surface winds
January jet stream
southerly surface winds (July)
Present
stronger subtropical high (July)
warmer, drier than present (July)
stonger flow (July)
stronger flow (Jan)
warmer, drier than present (July)
stronger flow (July)
6 ka
stonger subtropical high (July)
warmer, drier than present (July)
warmer, drier than present (July)
stronger flow (Jan)
warmer, drier than present (July)
glacial anti-cyclone (July)
cooler moister than present (July)
warmer than present (July)
stronger onshore flow (July)
9 ka

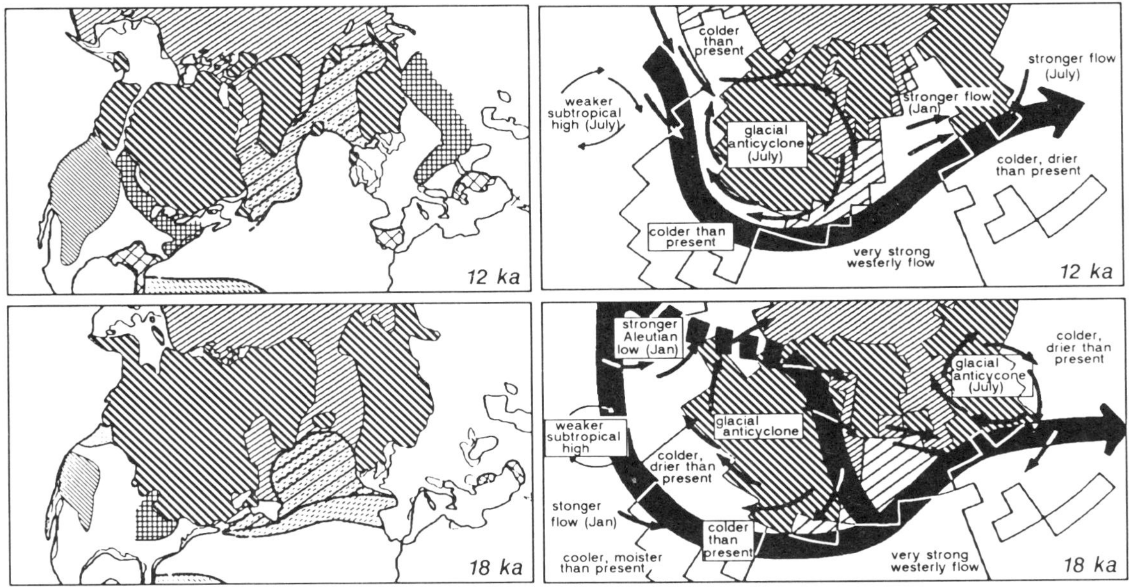

Figure 5.3. Paleoclimatic simulations and data (COHMAP Members, 1988). The left-hand (data) panels illustrate some of the principal paleoclimatic data sources, including terrestrial pollen and marine foraminifera. In western North America, paleoecological and geomorphological records were interpreted in terms of differences in effective moisture from modern conditions. Ice sheets and the inferred limits of sea ice are also shown; these were used as input to the simulations performed with NCAR CCM. The right-hand (model) panels show schematically the principal features of the climates simulated by NCAR CCM. Comparison of the simulations with the observations shows that the simulations are able to explain many features recorded by paleoclimatic data. See COHMAP Members (1988) for details. Copyright 1988 by the AAAS.

components of the climatic system is to arrange them along a gradient from those models that represent the physics of a process as explicitly as possible to those models that are essentially empirically determined "black-box" models of a component of the climate system.

The quasi-deductive group of models is typified by a variety of somewhat related models that represent how ice sheets, ice shelves, sea level, and underlying bedrock respond to variations of solar radiation (for review, see Oerlemans, 1982, 1984). Pollard's (1983) model, for example, simulates Northern Hemisphere ice-sheet thickness and bedrock elevations beneath the ice. It uses solar radiation as input and includes such processes as bedrock deformation beneath and in front of the ice sheet, ice-shelf calving, and other processes of ice-sheet dynamics. The main accomplishment of Pollard's (1983) model was to simulate 100,000-year-long oscillations of the ice sheet, mainly as a result of including the effects of bedrock deformation and ice calving. Although these models contain physically explicit representations of many important processes involved in the growth and decay of ice sheets, they still contain important parameters that are determined only empirically. Saltzman (1984, 1985) argued that the deterministic physics that underlie slow variations of the climatic system may never be known completely.

An alternative approach, therefore, may be to devise models for the slowly varying components of the climatic system that are, in a sense, as geologically reasonable as they are physically explicit. An example of such an inductive model is that given by Imbrie and Imbrie (1980), which attempts to represent variations of global ice volume as forced by variations of Earth's orbital elements. Their model is a simple differential equation governing change in the total volume of ice, expressed as a function of current volume and orbital forcing. The model has five empirically determined parameters, three tuning contributions of the orbital elements to ice volume and two representing response times of the ice sheet that differ between ice growth and ice decay phases, a geologically reasonable development. The ice record simulated by the model agrees well with the observed record over the past 250,000 years, and they used deterministic future values of orbital elements to project ice volume 100,000 years into the future (Figure 5.4).

Models of this type, based on Milankovitch or astronomical climate theory, simulate the past climatic record by modeling effects of variations in solar irradiation brought about by long-term changes in Earth-sun geometry. These orbital elements—eccentricity, precession, and obliquity—are nearly periodic functions with periods of roughly 100,000, 41,000, and 23,000 years, respectively. Recent research using spectral analysis of climate records for the past 730,000 years has confirmed that, in the past, the climatic system has varied at the same time scales as the solar forcing associated with obliquity and precession (Hays et

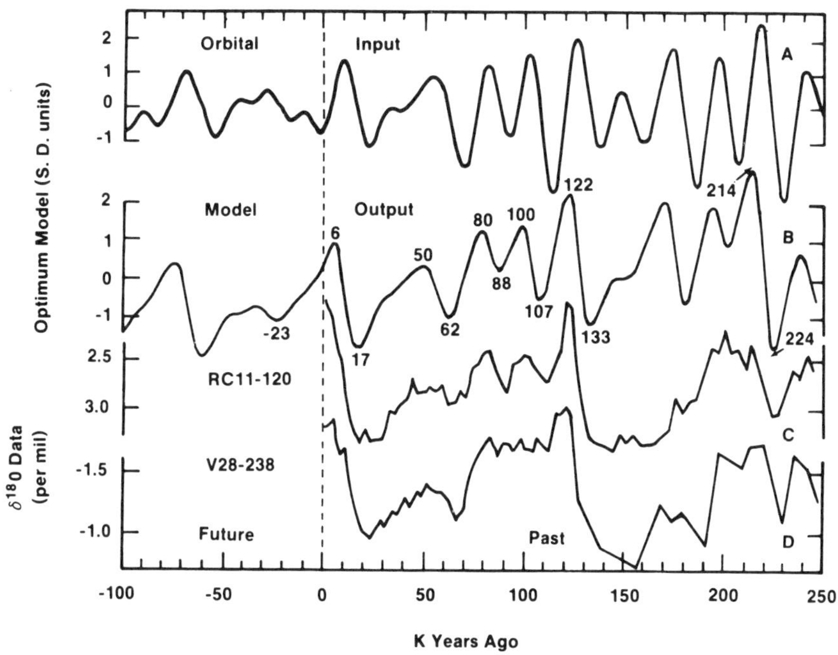

Figure 5.4. Orbital input and past and predicted future ice volumes simulated by the Imbrie and Imbrie (1980) model, with observed ^{18}O records (Imbrie and Imbrie, 1980, Figure 7). Copyright 1980 by the AAAS.

al., 1976). In light of this evidence, Imbrie and Imbrie (1980) argued for a shift in strategy of climatic research, away from searching for evidence of the astronomical theory to using past geologic records for evaluating performance of models based on the theory.

In a similar spirit, Kukla et al. (1981) derived an "astronomical climate index" generated by an empirical function of orbital elements. The function was chosen to yield a time series of the index that matched a composite climatic record consisting of pollen, sea-level, and oxygen-isotope records. Using orbital elements as input, they generated a series of the index for the past and future 10^6 years.

Saltzman and Sutera (1984) and Saltzman et al. (1984) have developed an inductive model that is slightly more elaborate than Imbrie and Imbrie's (1980). This model generates series of both continental and marine ice masses (as opposed to global ice in the Imbrie and Imbrie model) and ocean temperature. A remarkable feature of the model is the ability of nonlinear feedbacks among these components to generate, under random forcing alone, free variations of the model output with a characteristic 100,000-year oscillation. The phase relationships among the output series were incorrect, however. Saltzman et al. (1984) showed that when forced with realistic values of solar radiation, the phase relationships among the series locked in consistently with the observed record. The ice-mass record generated is also in substantial agreement with the observed record. In further developments of this model, Saltzman and Maasch (1988) showed that the global CO_2 system could provide the instability necessary to drive a natural oscillation of the climatic system. Maasch and Saltzman (1990) also showed that under certain circumstances the model could generate longer-term changes in the variability of the output that are qualitatively similar to those noted over the past 3 million years, (e.g., Ruddiman and Raymo, 1988). The role of these inductive models in climatic prediction over the long time scale lies in their ability to account for much of the long-term variations in the global-scale, slowly varying elements of the climatic system.

Summary. Various approaches have been used in the analysis of climatic variations on long time scales, but explicit attempts at climatic prediction are relatively few. Studies in which the specific objective is development of a procedure for prediction of future climatic variations over the long time scale are as yet in a primitive state. Lamb (1982) and Budyko (1982) described in general terms the likely future variations of the climatic system but did not offer explicit predictions for the time scale of interest here. Of six studies among those reviewed above that attempted explicit predictions, most have focused on either the hemispheric or global scales. The U.S. Committee for the Global Atmospheric Research Program (1975) considered generalized climatic records and variations. Petersen and Larsen (1978) and Imbrie and Imbrie (1980) predicted global ice volume, Berger (1977) estimated zonally averaged annual temperature, and Kukla et al. (1981) predicted a global-scale climatic index. Only one study (Kanari et al., 1984) focused on what could be regarded as a local climatic signal. Locally or regionally relevant predictions are currently possible only for fast-response components of the climatic system. Although no specific methods for producing locally or regionally relevant climatic predictions over the long ($>10^4$ year) time scale have so far been developed, the nature of a workable approach can easily be envisioned and is described below.

AN APPROACH TO LONG-TERM CLIMATIC PREDICTION

Prediction Strategy

Consideration of components of the climatic system, controls over their variations through time, and approaches taken in analyzing those variations suggests a potential strategy for developing climatic predictions over the long time scale (Saltzman, 1984). Any prediction scheme for that time scale must necessarily have two parts. The first part consists of models of slowly varying components of the climatic system that generate temporal evolution of those components. The second part consists of models of fast-response components of the system that generate spatial details of a predicted climate that is in equilibrium with boundary conditions and slowly varying components. Although elements of such an overall strategy are individually available (as evidenced by their application to the analysis of *past* climatic variations), their integration into a general scheme for future variations has yet to be accomplished.

Prediction of Slowly Varying Components. A variety of specific models can be used to predict variations of such slowly varying components of the climatic system as ice sheets and deep-ocean temperatures. Although such models currently are not well developed, their ability to match the observed records over relatively long intervals gives them a certain measure of robustness. Models of slowly varying components of the climatic system require as input the external boundary conditions, and for the case of incoming solar radiation, this may be computed for the future with confidence (Berger, 1984).

Several important problems must be overcome before models of slowly varying components can be integrated into a general prediction scheme. Variations in boundary conditions of the climatic system other than solar radiation are not well known. Dust loading and CO_2 concentration in the atmosphere are important controls, but their histories and potential predictability are imperfectly known (Bryson and Goodman, 1980; Hammer et al., 1980; Bradley, 1988; Neftel et al., 1982; Shackleton et al., 1983; Barnola et al., 1987; Kellogg and Schware, 1981). Similarly, the extent of variations in solar output over long time intervals is not known.

Models of the slowly varying components themselves are in need of much elaboration. Models currently generate global or other very large spatial scale averages, and the spatial structure of variations of such components as ice sheets are not immediately forthcoming from the models. Methods for disaggregating the output of or adding greater spatial resolution to the models are therefore needed. Much of the work thus far has focused on ice sheets, leaving models of the deep ocean relatively underdeveloped (Saltzman, 1985).

The relative contribution of forced vs. free variations of the slowly varying components must also be assessed to determine the ultimate predictability of those components. If free variations are indeed important, there may be an as-yet unknown, unacceptably low upper limit to predictability of slowly varying components.

Prediction of Fast-Response Components. Models of fast-response components of the climatic system are needed to resolve spatial details of the equilibrium climate that would prevail under a particular state of the boundary conditions and slowly varying components. Such models are currently available among the family of GCMs and related models. Those models would be used to generate a snapshot of a climate that could be considered to be physically consistent with a particular set of controls. As for the slowly varying components, basic models of the fast-response components are available, but further elaboration of them is necessary before an operational prediction scheme could be developed.

Models incorporating realistic circulation mechanisms and thermal characteristics of the oceans are still in the developmental stage. These could be important in disaggregating a global-scale prediction of ocean temperature, say, into the proper spatial scale of input required for initializing a GCM. Although the character of the ocean surface can be reasonably well specified for the present climate, or for selected times in the past (e.g., 18,000 years B.P., CLIMAP Project Members, 1981), such prescription of ocean conditions currently cannot be made in a satisfactory fashion for predicted future climates. A truly interactive atmosphere-ocean GCM (AOGCM) that could be applied in a routine fashion would help bridge the gap between the primarily global-scale output of models of the slowly varying components and the spatially disaggregated data required as input for models of the fast-response components.

Predictions of global ice volume must be spatially disaggregated to the grid-point resolution of models of fast-response components. Three-dimensional models of glacial dynamics exist (Denton and Hughes, 1981), and these, along with guidance from the geologic record on typical ice-sheet morphology, could be used to supply the needed information.

Even though GCMs are capable of revealing the spatial detail of a particular equilibrium climate, that detail may still be too coarse for the site-specific predictions that may be required for repositories. It should be possible, however, to use the output from a GCM to initialize mesoscale models (e.g., Pielke, 1984) with finer spatial resolution, and work toward this end has begun (Dickinson et al., 1989).

Implementation of a Prediction Strategy

Many of the necessary elements are already available for implementing this essentially hierarchical strategy for predicting local climate at long time scales ($>10^3$ years). Further consideration must be given, however, to selecting individual models for use in an operational method and to validating these models.

Model Selection. Little guidance exists for selecting the best of several models for predicting a particular component of the climatic system. This situation arises mainly because the focus of paleoclimatic modeling and analysis has been on understanding and explaining past variations, and only recently has attention been directed toward climatic prediction. As a result, a somewhat heterogeneous collection of individual data analyses and models has been assembled, each focusing on a specific aspect of the overall problem of explaining past climates. Because prediction has not been an issue, particular aspects of the performance of a model (such as an estimate of the variance of its prediction errors) have not received the same attention they have in the development of models for daily weather forecasting. Also, little thought has been given to the possible integration of individual models into an overall prediction scheme.

Some existing models are likely candidates for inclusion in a prediction scheme, although because no integration attempt has yet been made, too little is known to either select with assurance or discount completely any particular model. A typical configuration of models might consist of Imbrie and Imbrie's (1980) and Saltzman and Sutura's (1984) models to predict long-term variations of global ice volume and ocean temperature, the NCAR Community Climate Model (e.g., Kutzbach and Guetter, 1986) to provide snapshots of the fast-response components of the climatic system, and a mesoscale meteorological model (Pielke, 1984) to provide sufficient spatial resolution of the resulting configurations. Specific models or prediction methods do not exist for all components of the climatic system; until their performance is actually measured, there is no reason to presume this particular group of models is optimal in any way.

A task that may prove as important as the continued refinement of existing models is the articulation of the overall objectives of climate prediction on the long time scale. From this information, the general form of an overall strategy could be developed, and selection and refinement of specific models can be directed toward implementing this strategy.

Model Validation. An important consideration in comparing models is their relative performance (Schlesinger, 1984). A standard benchmark used is how well a model fits the data with which it is calibrated, or how well it reproduces a particular feature of the paleoclimatic record. In the case of climatic prediction, the issue is how well a model will perform given substantially different input than that with which it was calibrated. In essence, most models have been designed with reference to only a single climatic state (the present "modern climate") or to a single series (for example, the observed ^{18}O record). The danger will always exist, even in the case of a model with elaborate physics, that it will be tuned to work well with the modern climate at the expense of its overall performance through time.

The appropriate way to measure the performance of climatic models is to verify their ability to simulate climates different from today's. The paleoclimatic record is the obvious provider of such alternative climates (Gates, 1976a; Manabe and Hahn, 1977; Webb et al., 1987). The way model validation would work in practice is that the boundary conditions and state of slowly varying components would be input to a hierarchy of prediction models to simulate the climate of a particular time, say, the full-glacial climate of 18,000 years B.P. (e.g., CLIMAP Project Members, 1976). The simulated full-glacial climate would be compared with the observed climate as interpreted from paleoclimatic evidence (Peterson et al., 1979).

A number of promising exercises in model verification (e.g., Hansen et al., 1984; Manabe and Broccoli, 1985; Kutzbach and Street-Perrott, 1985; Webb et al., 1987) have been performed, but no systematic program of validation has yet been established. One necessary element of a systematic program would be development of a comprehensive paleoclimatic data base.

Data Bases. Three general data sets are needed for developing a climate-prediction scheme for the long time scale. The first set consists of past records of boundary conditions and slowly varying components of the climatic system. These records include the input of solar radiation to the climatic system, ice-sheet volume and extent, horizontal and vertical temperature structure of the oceans, CO_2 concentration, dust loading, and so on. These data are required for developing and testing prediction methods for slowly varying components of the climatic system.

The second data set consists of comprehensive data on the spatial variability of the present climate. These data are required to initialize models of fast-response components of the climatic system. Included in this data set are grid-point values of the standard meteorological variables, as well as data such as sea surface temperature, land surface albedo, surface roughness, soil characteristics, and so on.

To the extent that models of both slowly varying and fast-response components of the climatic system have been developed, the basic data included in the first two sets could be said to already exist. As indicated earlier, models of slowly varying components presently do not provide much spatial resolution, and further development of the first data set may be required.

The third data set consists of data necessary for validating model simulations. The scope of the data that could potentially be included in this set is large, including virtually any kind of paleoclimatic evidence.

The set of all paleoclimatic data is enormous and varies greatly in spatial and temporal coverage and in its sensitivity to paleoclimatic variations (Bradley, 1985; Hecht, 1985). Comprehensive data sets for specific times in the past are needed for model validation studies. Methods are also needed for quantitatively comparing paleoclimatic evidence with model simulations. Although development of a global-scale data base for the past 18,000 years (COHMAP

Members, 1988) has begun, much work is still needed to increase the spatial resolution of the data and to develop procedures for data/model comparisons.

An Example. The set of GCM experiments and data syntheses for the past 18,000 years analyzed by COHMAP Members (1988) provides an example of how information on the state of the slowly varying components of the climatic system can be coupled with simulation models of the fast-response components to yield regionally specific predictions. In the COHMAP study, the objective was paleoclimatic analysis, as opposed to the prediction of future climates, but basic stages in the analysis would be similar in either case. The particular sequence of simulations and syntheses that were examined encapsulates much of the amplitude of climatic variations during the Quaternary by including the range from full-glacial to full-interglacial conditions.

In the COHMAP study, the sequence of variations in the boundary conditions over the past 18,000 years (Figure 5.2) was determined from the geologic record; in a climate-prediction exercise the boundary conditions would be derived from the output of a model of the slowly varying components. The spatial patterns of the fast-response components that are consistent with these boundary conditions were illustrated by a sequence of GCM simulations (Figure 5.3). The simulations show the nature of the large-scale features of the climatic system (e.g., location of the jet stream) and also provide hemispheric and regional averages of key climatic variables (e.g., Kutzback and Guetter, 1986). In a fully developed prediction scheme, the spatially coarse simulations of fast-response components provided by the GCM would be disaggregated further using mesoscale simulation models.

The comparisons between data syntheses and model simulations that were a key part of the COHMAP study also illustrate the use of the paleoclimatic record for model validation (Figure 5.5). Many of the principal features of the climatic variations over the past 18,000 years that are recorded in the data (e.g., glacial climates of North America and Europe 18,000 years ago, or the greater effective moisture in the monsoon region of Africa and Asia 9000 years ago) were simulated by the model. Such agreement, if replicated in other studies, would provide strong validation of the overall approach of using a GCM to simulate fast-response components of the climatic system under a particular configuration of the slowly varying components.

SUMMARY

At present, no operational method exists for climatic prediction at the long time scale (10^3-10^4 years and longer). Many of the necessary components of an overall strategy exist, however, including a variety of models, methods for transforming predictions into probabilistic statements, and the basic data required

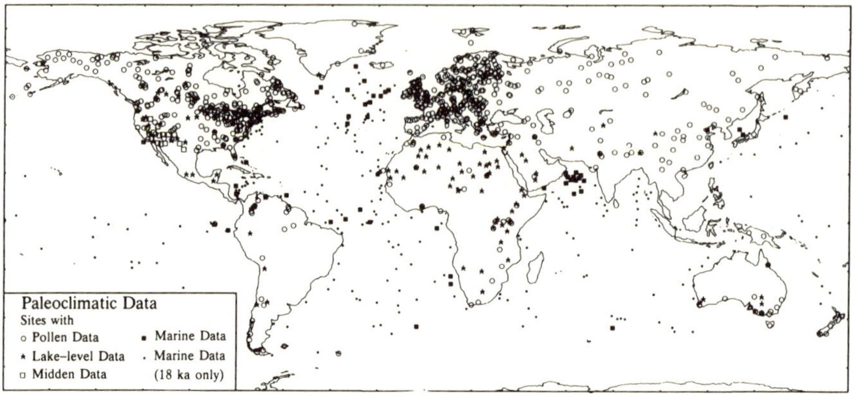

Figure 5.5. **Paleoclimatic data sources (COHMAP Members, 1988). The map shows the distribution of sites with paleoclimatic data spanning the past 18,000 years. Copyright 1988 by the AAAS.**

for model validation. A program to adapt existing data and methods and integrate them into a general method of long-term climatic prediction would include three elements. First, the overall goals of an effort to predict climatic variations over the long time scale would be clearly identified by the potential users of those predictions. Identifying specific objectives of climatic prediction would aid development of an overall scheme by revealing where gaps exist. In addition, the influence of climatic variations on the surface-water system, and in turn on the ground-water system, would be reviewed to further refine these objectives. Second, a comprehensive research program would be initiated to adapt existing and develop new methods of climatic prediction. Individual tasks in such a program would include the development of methods for

- predicting the boundary conditions and external controls of the climatic system
- predicting the slowly varying components of the climatic system
- predicting the fast-response components of the climatic system
- developing site-specific predictions of important climatic variables
- assessing the influence of climatic variations on other environmental systems

For each of these tasks, model-validation studies would be required, which would include the development of the appropriate paleoclimatic data and the methods for comparing model simulations with paleoclimatic observations. Finally, the scientific basis of climatic prediction would be periodically reviewed, as new understandings of the causes of climatic variations develop and as new targets for prediction arise.

6

Tectonics and Seismicity

Jonathan F. Callender

Determining the probability of tectonic activity and the probability of impaired repository performance are to some extent separate activities. Techniques for probabilistic prediction of tectonic motion and seismicity are discussed here, and probabilistic prediction of seismic effects on a site is discussed in Chapter 7.

The geology of a repository site is the fundamental source of data for quantitatively evaluating tectonic and seismic hazards that could affect a nuclear-waste repository. Site geology depends on processes acting on various scales, including the whole Earth (tectonic plates), regional (the tectonic province), and local (within a 100-km radius of the site).

In general, quantitative models at the whole-Earth scale predict plate movements over millions of years, identify broad areas of Quaternary activity or potential activity, predict variations in plate positions and velocity that may greatly change intraplate strain, and generally identify areas of little orogenic or epeirogenic activity (cf. Chapple and Tullis, 1977; Richardson and Solomon, 1979). Identifying areas of little orogenic or epeirogenic activity is imprecise. For example, continental interiors affected by glacial loading in the 10^4- to 10^5-year time span, and intraplate seismic belts, in which the recurrence interval is long but energy release may be significant (e.g., New Madrid, Charleston, Rio Grande rift), are areas that might be broadly identified without much quantitative prediction allowed by the geologic data.

Regional analyses evaluate broad patterns of faulting, tectonic sedimentation, epeirogenic movements, changes in fluvial and marine regimes, crustal stress environments, etc. Defining regional boundaries and obtaining reasonable temporal and stratigraphic precision may be difficult. If regional areas are seismic, classification of tectonic phenomena is reasonably straightforward (Electric Power Research Institute [EPRI], 1985). In aseismic areas, i.e., areas in which historic earthquakes have not been recorded, tectonic provinces are largely determined by basement boundaries, and basement effects are poorly understood and largely unquantified. To adequately define regional geologic-province boundaries for waste-repository studies and to identify potential sources of tectonic hazards, the time scale of relevant deformation must be expanded to include the late Tertiary Period; the Quaternary Period is far too brief.

The most quantitative and precise tectonic data available to predict future behavior at a given site come from local areas that have been active in the Holocene or late Quaternary ($<10^5$ years). Areas of older tectonic activity yield information of significantly less precision ($\pm 10^4$-10^6 years or more).

Four kinds of geologic data are most useful for predicting potential tectonic activity: surface ruptures along fault zones, stratigraphic sequences related to tectonic activity, landforms related to surface rupture and regional uplift or subsidence, and relationships between seismicity and tectonic features. As a rule, as the size and geologic age of the area considered in a specific investigation decrease, the accuracy, precision, and validity of probabilistic estimates increase.

Earthquake prediction and detailed seismic analysis are not extensively discussed here, even though seismic data play a critical role in evaluating young tectonic phenomena. Agnew and Ellsworth (1991) have provided an excellent review of the recent literature on earthquake prediction for periods roughly equivalent to the operational lifetimes of geologic repositories. The short instrumental record and society-dependent historic earthquake record can be an inadequate statistical base for evaluating even late Quaternary tectonic phenomena, however. For example, whether earthquake processes are episodic on the scale of 10^5 years is important (cf. Allen, 1975). If they are, seismic analysis based on the historic record may be meaningless for long-term assessments, and an understanding of the relationship between seismic and tectonic events is more tenuous than assumed.

This chapter discusses the major quantitative geomorphic or geologic techniques for estimating tectonic hazards in waste-disposal site evaluation and reviews the empirical relationships between tectonics and seismicity. The basic analytical methods are presented in three categories: rupture of land surface (paleoseismicity); recent vertical crustal movements, emphasizing regional tectonic phenomena; and correlation of tectonics and seismicity.

METHODS

Rupture of Land Surface

Fault Recurrence. Recurrence methods use reasonably well-dated Quaternary deposits to infer the number, ages, and style of fault offsets. Ages of the most recent or multiple earlier ruptures are derived from the relationship between a fault trace and Quaternary deposits of known or estimated age.

Recurrence methods examine the presence and amounts of offsets of a given fault trace in one or more dated units exposed naturally or artificially. Most stratigraphic units either predate or postdate a given rupture event; thus ages of the deposits that are and are not faulted by a given rupture bracket its age (Sieh, 1978a; Weldon and Sieh, 1985). Some deposits are inferred to have occurred during the rupture event [e.g., colluvial wedges along the buried trace of a fault (Swan et al., 1980)]. Some faulted stratigraphic units may be correlated with offset landforms like stream channels or terraces or with topographic fault scarps (Sieh, 1984; Sieh and Jahns, 1984; Weldon and Sieh, 1985).

Recurrence methods can estimate the age of the most recent rupture and the number and frequency of prior rupture events. The latter data can be converted to estimates of either recurrence intervals or mean slip rates. The resolution varies widely, depending on the quality and availability of accurately dated units; the number of such units relative to faulting events; the repeat time of the fault relative to the age span of the units; and the clarity of geologic relationships between fault and deposits. Repeat times range from 10^2 to $>10^5$ years/event, depending on the fault zone; the precision of slip rates varies accordingly. Slip rates are usually given in millimeters per year or millimeters per 10^3 to 10^4 years. Statistical treatment of repeat intervals may be used where a successive series of multiply offset, dated deposits occurs (Sieh, 1978a; Weldon and Sieh, 1985). Data reduction leads to a probabilistic repeat time, based on past faulting history.

The use of recurrence methods requires a *well-exposed* sequence of deposits and fault; adequate age control of deposits and fault; a clear relationship between dated deposits and fault trace(s), especially with respect to presence or absence of offsets; and a geologic setting suitable for continued deposition and preservation of deposits near the fault trace.

Several basic assumptions are significant in this method. Assumptions about the magnitude of the rupture are especially critical where only vertical exposures are available (cf. Sieh, 1978a, 1978b, 1984); this problem can be mitigated by examining lateral offset of landforms, if present. In some cases, relatively uniform repeat times between events are assumed; this is especially critical when uniform slip rates are inferred from a few faulted deposits of drastically different ages. Finally, past fault behavior is assumed to accurately reflect, and even predict, future rupture patterns.

This method has several limitations. Good age control on deposits is frequently lacking. The fault surface may be poorly exposed, or fault behavior may be nonuniform in time and space. Long recurrence times, relative to the age span of deposits, e.g., 10^5 years in late Quaternary, or long repeat times relative to cited slip rates, e.g., 10^4 or 10^5 years converted to millimeters per year, cast doubt on the real slip rate. Poor bracketing relationships, such as a long time span between youngest faulted and oldest unfaulted deposits, relative to a given rupture, critically affect the resolution of rupture-age estimates. The presence of relatively few ruptures at a given site reduces the statistical data base and hence reliability of any repeat-time projections. The relationship between faults and sedimentary deposits assumed to be related to faulting can be ambiguous. These methods generally are costly, because they require trenching and accurate dating of units.

In ideal situations, recurrence methods provide the best data for probabilistic prediction of long-term behavior of a given fault, especially for faults active in the Holocene Epoch (Sieh, 1978a, 1978b, 1984; Sieh and Jahns, 1984; Weldon and Sieh, 1985; Swan et al., 1980). Many, if not most, faults lack one or more requirements for high-resolution studies of this type, however. The most common problem is poor age control, usually because of a lack of dateable materials (Bonilla, Villalobos, and Wallace, 1984; Pearthree and Calvo, 1982; McCalpin, 1982).

Dating Techniques for Quaternary Deposits. Recurrence methods require some means of dating deposits ruptured by a fault. Techniques for dating Quaternary deposits vary in their reliability and precision. Bradley (1985), Goudie (1981), and Colman and Pierce (1979) have given good general discussions.

Radiometric techniques can be used to date a wide variety of materials, but the usefulness of any given technique depends on the availability of materials of a certain type and age. Carbon-14 dating is the method most commonly used for late Quaternary deposits from certain environments, such as bogs. Radiocarbon is not generally useful in oxidizing environments. Potassium-argon dating is useful in volcanic rocks that are older than 50,000 to 100,000 years and younger than 10^6 years. Uranium-series dating uses the relative properties of various decay products of ^{238}U or ^{235}U, especially ^{230}Th, ^{231}Pa, and Pb isotopes. Material with ages of 10,000 to 350,000 years can be dated. The ratio of $^{235}U/^{238}U$, usually in shells, can yield dates in the 40,000- to 1,000,000-year range (Muhs and Szabo, 1982). Closed system decay of ^{238}U to ^{230}Th yields ages in the range of 10,000 to 300,000 years, for pedogenic carbonate drawn from the densest inner core of carbonate-coated gravels, for example. Fission track dating is mainly limited to minerals (e.g., zircon, apatite) with fissionable U in volcanic rocks, but has no inherent limitations on age range. Thermoluminescence dating is best for stratigraphic correlation and relative ages, rather than absolute ages. Electrospin resonance (ESR) has been used for dating fault gouge, but has only been tested once in Japan, using ESR in quartz (Ikeya et al., 1982).

A number of nonradiometric dating techniques can also yield absolute ages for specific materials. Dating using paleomagnetism requires material with sufficient stratigraphic and time ranges to expose an adequate number of reversals, criteria not commonly found in fault trenches. The thickness of weathering rind developed on clast margins, such as that developed by obsidian hydration, can be used to estimate the age of some deposits (Colman and Pierce, 1981). Cation-ratio dating of desert varnish (Dorn, 1983) is limited to arid regions. Tephrochronology, the identification and correlation of undated and dated volcanic ashes, requires suitable ash deposits and detailed geochemical and petrographic analyses (Sheets and Grayson, 1979; Self and Sparks, 1981). Dating of organic material using amino-acid racemization is based on the assumption of time-dependent diagenetic conversion of one form of amino-acid polymer structure to another (Bada, 1985; Bada and Protsch, 1973). Lichenometry estimates ages from the size of lichens on gravel or boulders in certain specialized environments, such as young glacial deposits (Locke et al., 1979). Soil-profile ages can be estimated from measured amounts of accumulated pedogenic materials (e.g., Machette, 1978). Dendrochronology can be used to determine ages of trees disrupted by faulting or fault-related phenomena such as landsliding (Page, 1970; Sieh, 1978b).

Relative age dating, i.e., relative age assignment to suites of deposits, is the least reliable and precise method of dating. In some cases, crude age estimates may be inferred by correlation with dated deposits or with sequences with better age control. Relative age dating is generally based on the relative degree of soil-profile development or on the relative degree of weathering.

In many cases, the relative degree of soil-profile development (chronosequence) in either buried soils or relict surface soils on surficial deposits (Machette, 1982) is the only available age indicator. Techniques range from careful laboratory measurements (McFadden and Tinsley, 1982; Machette, 1982) to field description of soil profiles. Field descriptions are most useful when data are converted to a numerical index that allows semi-quantitative comparisons (Harden, 1982). The estimated resolution and accuracy vary with degree of quantification of soil-profile characteristics and age of the soil. Resolution decreases with increasing soil age. Resolution also depends on the degree of age calibration in the soil chronosequence. Adequate age control for soil chronosequences commonly is difficult to achieve; however, at least order-of-magnitude estimates of the ages of Quaternary deposits can be inferred in most settings (Matti et al., 1982; Pearthree and Calvo, 1982; Pearthree et al., 1983; Keller et al., 1982; Chadwick et al., 1984).

Ages may also be estimated by evaluating the relative degree of weathering of surface and subsurface clasts in sedimentary deposits such as glacial moraines. The method requires some independent means of age calibration for a given suite of deposits. It is useful in conjunction with soil-development or weathering-rind data.

The use of absolute age determinations is limited by their resolution and the age interval that may be dated. For young deposits, ^{14}C dating is the standard technique. For older deposits, various techniques, many of which contain serious limitations, are used; none serves as a standard. Relative age determinations are crude and depend strongly on environment. One of the major shortcomings of quantitative probabilistic studies on the effects of tectonics is the imprecision of age determinations.

Quaternary Features Used for Recurrence Studies. Fault-related or -controlled deposits may occur within or adjacent to a fault zone, such as sag-pond fill (Sieh, 1978a, 1984), antithetic graben fill (Weldon and Sieh, 1985; Swan et al., 1980), and colluvial wedges derived from paleoscarp degradation (Swan et al., 1980; Weldon and Sieh, 1985; McCalpin, 1982). Colluvial wedges often are difficult to relate to specific rupture events. Alluvial fans, debris cones, fluvial terrace deposits, and lacustrine shoreline deposits may be ruptured by or extend across a fault zone (Pearthree and Calvo, 1982; Matti et al., 1982; McCalpin, 1982). They may be deposited in response to nontectonic factors (i.e., climatic change) of potentially much different age; thus rupture ages may only be estimated by bracketing between ages of faulted and unfaulted deposits (with potentially poorer resolution).

Stratigraphic indicators of earthquake shaking in unconsolidated sediments include sand boils, mud volcanoes, or other liquefaction-related features. The presence of such features associated with a rupture event has been used to infer minimum earthquake magnitude (Sieh, 1978a; J. E. King, 1978; Youd and Perkins, 1978; Russ, 1982; Cox and Talwani, 1984; Obermeier et al., 1985).

Earthquake-induced landslides, slumps, rockfalls, disrupted lake sediments and soil horizons, and earth or soil flows are useful if a clear relationship to earthquake intensity and timing along a fault zone can be established (Keefer, 1984; Youd and Perkins, 1978; Youd and Keefer, 1985; Adams, 1981). The earthquake must be dated. Shear zones, cracks, and fissuring may also be related to earthquakes or tectonic strain (Zellmer et al., 1985). Some means of dating, or of relating these features to dated deposits if a crack occurs in stratigraphic sequence, must be available.

Amounts and ages of fault ruptures can be estimated from offset or displaced landforms such as stream channels, fluvial terraces, and marine terraces. Stream-channel offsets are especially useful in unraveling lateral displacements on transcurrent fault zones, which cannot be directly observed in vertical trench walls (Keller et al., 1982; Sieh, 1978b; Sieh and Jahns, 1984). Abrupt thickening of alluvial channel deposits across a fault zone may record recent faulting as well (Bull et al., 1979). Stream-channel offsets can be used to infer vertical uplift rates (Hamblin et al., 1981). Active and potentially inactive paleochannels must be correctly correlated across the fault zone. Good preservation of older paleochannels and good age control (usually obtained by determining age of a deposit or surface underlying the paleochannel, or by dating units above paleochannel gravels) are also needed. The method works best in combination with stratigraphic offsets.

One or more offset fluvial terraces or alluvial fan surfaces along a stream that crosses a fault zone may record vertical or lateral displacements (Suggate, 1960; Lensen, 1964, 1968). Terraces that record multiple successive offsets along a fault are especially valuable; in this case, a detailed time-progressive history of vertical, horizontal, and oblique displacements can be developed. Data requirements include the ability to correlate and recognize terraces across a fault zone; terraces that are traceable for some distance; and a well-understood terrace form and longitudinal profile (particularly to determine vertical offsets) (Bull et al., 1979; Hamblin et al., 1981; Rockwell et al., 1984). If tectonic in origin, terrace formation occurs at the time of rupture; if of climatic or other origin, terrace ages bracket faulting. Thus the origin of the terrace must be determined, most commonly by comparing the longitudinal profile of terrace to modern channels. Because of base-level fall and stream incision along the fault where the stream crosses it, profiles of tectonic terraces diverge downstream toward the fault zone. Some means of dating terraces is required, either by dating terrace deposits or by estimating the age of the surfaces by the degree of soil development. In general, paired strath terraces are least ambiguous for this type of analysis. A well-preserved flight of successively offset or warped, laterally traceable terraces, with good age control, can provide a fairly detailed history of a recurrent fault. The history may be treated statistically and converted to fault slip rates; however, these situations are rare.

Uplifted flights of marine terraces can be used to estimate the ages of a series of earthquake or fault events (Matsuda et al., 1978). Problems include correlation of individual terrace levels with an earthquake origin, i.e., determining that the uplift of terraces is co-seismic and not related to larger-scale coastal uplift or sea-level changes.

Stratigraphic sequences are probably the best single data source for fault recurrence studies, especially when deposits are explored in trenches across the fault zone and when they contain good radiometric age control (e.g., Sieh, 1978a,

1984). A good stratigraphic sequence, with multiple deposits of varying ages, is especially useful, if not required, for resolution studies of active faults. For other stratigraphic indicators, features of nontectonic origin must be excluded. Offset, displaced, or anomalous landforms must be well preserved and bear a clear relationship to faulting events. The ages of these features usually are obtained by dating underlying or associated deposits, by using soil development to obtain the age of a geomorphic surface, or by extrapolation from a few dated geomorphic surfaces in a sequence, assuming some constant slip rate (Weldon and Sieh, 1985). Some assumptions as to constancy of slip rates are required; thus slip rates must be independently checked. This method works best if it is combined with soil-profile data.

Stream-Profile Analysis. Stream-profile analysis estimates the presence, age, and possible amounts of fault movement by determining the size and position of anomalous inflection points (knickpoints) in the longitudinal profile of a stream above a fault zone.

Stream-profile analysis requires detailed profiling of stream channels and terraces and isolation of any knickpoints (Reed, 1981). The analysis models rates of knickpoint retreat up the stream channel from the original point of offset where the stream crosses a fault. Modeling uses either an exponential decay law (Hamblin et al., 1981) or a diffusion equation (Begin, 1983; Begin et al., 1981; Mayer, 1979, 1982; Mayer et al., 1981). This method requires some important assumptions about the origin and type of modification of observed knickpoints, e.g., that knickpoints migrate upstream after their formation instead of remaining stationary while declining in slope (although knickpoints probably decline in slope as they migrate). Diffusion modeling requires the additional assumption that knickpoint retreat mimics a heat/mass flux, a process mathematically described by the standard two-dimensional continuous diffusion equation for heat conduction.

Fault-Scarp Morphology. The age of fault displacements (usually the most recent, although in some cases, one or more prior events) can be estimated directly from the morphology of the topographic fault scarp produced by the associated surface rupture. Several techniques are currently available for morphologic age estimation of fault scarps. All use the same basic data: field-based topographic profiles oriented perpendicular to the scarp trace. The key data are scarp height and maximum slope angle and identification of steeper versus more gently inclined slope segments if the scarp is composite (reactivated scarp with more than one rupture episode). Wallace (1977) produced initial deterministic age estimates, based on a graph of maximum slope angle vs. age. His results were poorly constrained by independently dated scarps and had poor age resolution (order of magnitude only).

Bucknam and Anderson (1979) introduced a more rigorous method. They noted systematic variation of maximum slope angle with height for a given fault scarp, definable by linear regression of maximum angle versus the log of scarp height. The regression line changes position and angle of slope with increasing age of fault scarps. Thus the age of an undated scarp can be estimated by comparing its height-angle regression with those of dated scarps. Machette and McGimsey (1983) and Machette and Personius (1984) illustrated this approach. Mayer (1982, 1984) improved on the method by adding 95% confidence limits to the reference scarp-regression lines and by using discriminant function, statistical analyses to assign the height/slope data of undated scarps to one of several dated reference scarp data sets. Nash (1980) first applied the diffusion equation as a means for estimating rates of fault-scarp degradation. Mayer (1982) computerized the technique; he emphasized the use of multiple profiles of a given scarp to allow more stochastic analysis of natural variations in topographic form. Hanks et al. (1984) and Colman and Watson (1983) offered means to analytically solve the diffusion equation based on a given topographic profile. Nash (1984) also offered alternative graphical means for solving the diffusion equation for a given profile. These methods offer the first really promising means of directly estimating the age of the topographic expression of a surface fault rupture. The height/slope-angle regression methods generally yield at least an order-of-magnitude age estimate and generally are reliable to about 10^3 years for scarps $<10^4$ years old. Resolution decreases on older scarps ($\pm 10^4$ years). The diffusion-equation solutions provide more precise numerical ages, with estimated accuracies of $\pm$30% to 60% (Nash, 1984).

Dating using fault-scarp morphology requires several important assumptions and restrictions. The fault scarp is assumed to be the surface manifestation of faulting at depth. This assumption is supported by recent seismic-reflection profiling across fault scarps (Crone and Harding, 1984) and observations of recent earthquake traces (Crone and Machette, 1984) and associated shear zones. The scarp must be of tectonic origin, however, and not related to other geomorphic processes such as stream erosion or land subsidence due to human intrusion. Fault-scarp height is assumed to be a reasonable proxy for tectonic displacement; usually tectonic displacement represents about 50% to 70% of topographic height. Machette and Personius (1984), Wallace (1980), and McCalpin (1982) offered means of calculating displacements from scarp height. More important, all morphological age estimates are based on the assumption of a consistent sequence of time-progressive scarp degradation (first outlined by Wallace, 1977), essentially a slope-decline model with a well-defined time of initial rupture.

Morphological variations between fault scarps are considered primarily dependent on age differences; thus other factors, such as climate, slope aspect, scarp parent material, scarp gullying, and so forth, must be constant, especially between dated reference scarps and undated scarps. Fortunately, age estimates do not seem overly sensitive to small variations in these factors, given the broad resolution of the method. Ideally the upper and lower original surfaces are exposed above and below the scarp. On many higher scarps, however, the lower surface is buried beneath slope-wash material derived from scarp degradation or from adjacent alluvial fans. This yields a minimum estimate of scarp height, which usually affects offset estimates more than age estimates. The position of a scarp at or near the base of a larger escarpment or mountain front will influence the rates and processes of scarp degradation. Linear-regression analyses require at least 5 to 10 profiles of varying height along the same scarp for reliable age estimates; obtaining so many profiles is not always possible.

Diffusion-equation solutions require determination of the critical rate constant (k), which is assumed to incorporate the combined effects of the other non-time-related factors. This usually requires solution for k on a dated fault scarp of similar parent material, climate, and so on; finding a similar, dated scarp is sometimes difficult.

Uniform past rates of scarp degradation must be assumed, but late Pleistocene scarps and Holocene scarps in areas where climatic changes have significantly affected slope vegetation and erosion processes may not have similar rates. This affects diffusion more than the regression analyses, as the rate constant k may change with climatic regimes.

Morphological age estimates are most suitable for single rupture scarps, where the entire scarp forms at once. There is some controversy as to applicability of these techniques to composite scarps, in which the effects of successive ruptures are superimposed on a previously degraded scarp. Sometimes the latest event can be isolated as a steeper, short segment if recurrence times are long. With shorter repeat times, single events tend to merge. Nash (1980, 1984) restricted analyses to single-event scarps; Mayer (1982, 1984) thought that the precision of the method depends on recurrence time and age of last rupture. Hanks et al. (1984) applied diffusion-equation modeling to both composite and single-event scarps. All things considered, it is probably best to work with single-event scarps, using progressive offsets of older surfaces or deposits to infer prior ruptures and repeat times, and to restrict morphologic age estimates to clearly segmented composite profiles (Mayer, 1984; Machette and Personius, 1984). Fault rupture is also assumed to occur along a single, spatially restricted zone, creating an initial free-face vertical step, and not to be distributed over a complex rupture zone, because a complex rupture alters the assumed initial scarp configuration (Mayer, 1982, 1984; Nash, 1984).

This method is mainly sensitive to scarps with at least some vertical separation. Other methods are required for primarily lateral-separation (transcurrent) faults. Despite the limitations of this technique, results from a number of studies suggest that morphologic age estimates are surprisingly robust and are applicable in some form to most scarps. However, results should be stated in broad age ranges, more than one profile or scarp should be incorporated in analyses, and results should be calibrated against age constraints provided by offset stratigraphy. The final result of these analyses is an estimated age of one or more ruptures, recurrence intervals, or slip rates. Statistical analyses and probabilistic reduction of future fault-rupture behavior are subject to the same interpretational problems discussed earlier with respect to offset stratigraphy.

Vertical Crustal Movements

Analysis of vertical crustal movements requires identification of recent long-term movements of the crust. This review emphasizes vertical displacements distributed over a large region (up to epeirogenic scales). The discussion focuses on tectonic zones of uplift or subsidence, although other, nontectonic processes can induce large-scale movement (e.g., isostatic adjustments, salt or mud diapirs, ground-water dissolution or withdrawal).

Two major types of analytical methods that differ in scale and resolution are used to quantify vertical crustal movements. Geophysical methods, primarily geodetic leveling surveys, are directed primarily at small vertical displacements over short (10- to 100-year) intervals (Reilinger et al., 1984; Riecker, 1979; Brown, 1978b; Brown and Oliver, 1976). The validity of leveling data for interpreting tectonic strain is controversial. Geologic methods are sensitive to larger movements over longer intervals (10^3 to 10^6+ years). Geologic methods fall into several categories: analysis of fluvial adjustments, such as changes in longitudinal profiles and terraces, regional dissection patterns, and map-view channel patterns; evaluation of marine and lacustrine terraces, shorelines, or deposits; use of other geologic indicators such as lava flows; and quantitative models of geologic processes.

Fluvial Adjustments. Fluvial systems are sensitive to vertical changes within their drainage basins. The methods discussed here identify patterns of vertical or lateral fluvial adjustments and attempt to isolate those of tectonic origin from those arising from such perturbations as regional climatic change, complex responses of fluvial systems, normal stream evolution, and so forth.

Longitudinal profiles or fluvial terrace profiles identify amounts and rates of uplift on the basis of the forms of stream profiles and terraces (e.g., Huber, 1981). There are two methods: analysis of profile knickpoints and of fluvial terrace profiles. Analysis of *profile knickpoints* is an attempt to isolate knickpoints in stream profiles and qualitatively or quantitatively (using diffusion-equation modeling) evaluate knickpoint data in a fashion similar to knickpoint analysis for fault rupture. Knickpoints associated with vertical movements are commonly larger than those associated with faults; are found on larger, more regional river systems; and include composite effects of multiple offsets (if tied to one specific structural zone), with consequent decrease in potential resolution of uplift effects.

Fluvial terrace profiles can be used as indicators of paleostream channels to isolate older intervals of uplift. The age of a terrace is inferred from the relative degree of soil-profile development or from dated deposits. The anomalous pattern sought is downstream divergence of terrace longitudinal profiles relative to each other or to modern channels. Downstream profile divergence suggests downstream base-level fall at or following the time of terrace formation. Downstream divergence, followed by convergence, isolates the probable zone of relative uplift. Downstream profile convergence is less diagnostic, as nontectonic processes, especially climatic change in the catchment basin, can produce this pattern. Fluvial terrace profiles can indicate amounts, rates, and possible locations of uplift at either local or regional scales. The possible resolution varies, ranging from meters/10^2 to 10^6 yr. These methods are not particularly useful for quantitative studies.

For stream-channel profiles, either topographic maps or detailed field surveys can provide the required data. For terrace longitudinal profiles, relative heights of the terrace above the modern channel must be measured, and terrace remnants downstream must be correlated. Extensive field studies are usually required, including a series of elevation transects up flights of terraces, possibly augmented by topographic maps. Mayer's (1982) longitudinal analysis of the Colorado River and its tributaries in the western Grand Canyon established cumulative uplift due to faulting along Grand Wash Cliffs. Begin et al. (1981) developed an experimental and theoretical derivation of this type of analysis.

Good topographic maps and well-preserved or at least correlative terrace remnants must be available for these studies. Terraces that are traceable for large sections of a river profile, with minimal large gaps requiring long-range correlations, are most reliable. Knickpoints and terrace-profile anomalies must be tectonic in origin and not controlled by bedrock variations or normal convergence of climatic fill-terraces. Either a diffusion constant for knickpoint analysis or good control on the ages of terraces must be available.

Regional Patterns of Dissection or Degradation. Regional dissection or degradation can be used to infer presence, amounts, and minimum rates of regional uplift based, first, on comparison of modern and paleochannel profiles (without specific knickpoints or terrace profile divergence) (Hamblin et al., 1981); and second, on uplift rates inferred from regional denudation rates (Ahnert, 1970). Identification of regional dissection (long-term valley downcutting) is based on reconstruction of paleochannels by inverted topography, where a dated unit such as a lava flow or geomorphic surface overlies channel deposits adjacent to and above the modern channel (Hamblin et al., 1981). This technique includes plotting and comparing modern and paleolongitudinal profiles, calculation of minimum mean downcutting rates (from the age of overlying dated deposits and the amount of subsequent downcutting), and direct conversion into maximum uplift rates.

Mean uplift rates are obtained in meters/10^3 to 10^6 yr, with widely variable resolution. Uplift estimates are mean rates, averaged over the entire interval of downcutting since the formation of an upper time datum (such as a lava flow). Estimates are maxima, because downcutting may have begun some time after the datum or have ended before modern channel development. For best results, an inverted topography or dissected landscape with some dateable upper horizon is necessary, preferably with more than one dated horizon. Without time control, the presence of probable uplift can be estimated, but no limits on rates or timing can be established. For minimum ambiguity, the longitudinal profile of an older inverted channel should be compared with the modern profile. Similarity in profile form (but not in elevation) suggests reestablishment of an equilibrium profile following or during uplift (Hamblin et al., 1981).

Inverted or well-dissected topography and well-dated deposits must be present for this method to provide quantitative results. This combination of circumstances is rare in North America outside the western United States. The window for rate estimates is relatively broad.

Regional Denudation. Regional denudation can be used to estimate ages. Ahnert (1970) correlated measurements of mean denudation rates with the amount of relief in large denudations. From relief data, he provided a means of estimating order-of-magnitude uplift rates, using an equation derived from regression of denudation-relief data from 20 control basins, assuming certain rates of summit denudation and relative rates of stream incisions (Ahnert, 1970). Uplift rates are estimated in meters/10^3 yr. Regional denudation is an indicator rather than a quantitative measure, because of assumptions in the method. It is probably best used to estimate general long-term uplift rates.

Anomalous Map-View Channel Patterns. In some cases, anomalous map-view channel patterns can be related to regional tectonic uplift or subsidence. Adams (1980) and Ouchi (1985) related observed changes in channel sinuosity to zones of uplift, subsidence, or tilting. Ouchi (1985) also conducted flume experiments to document channel changes under controlled base-level changes. Nansen (1980) measured anomalous preferred directions of meander migration, which he attributed to regional tilting. An anomalous channel pattern mainly indicates distribution, presence, and amount of uplift, tilting, and so on, during fairly recent times. Estimates are difficult to convert to tectonic rates unless some longer-term time control is available. Adams (1980) estimated a "numerical age" for tilting, assuming a sudden start and a uniform rate of deformation. Age bounds can also be established using the age range of flood-plain deposits. Identifying and quantifying sinuosity and other channel patterns requires good aerial photographs and topographic maps. These techniques provide some measure of potentially subtle tectonic processes over time spans between geodetic and most geologic data (10^2 to 10^3 years).

Marine and Lacustrine Terraces and Shorelines. Anomalously high or distorted elevations of dated marine or lacustrine shoreline landforms or deposits may indicate uplift or warping (Ward and Chappell, 1975; Bloom et al., 1974; Chappell and Veeh, 1978; Pillans, 1983; McLaughlin et al., 1983; Williams, 1982; Lucchitta, 1979). Altitudes of marine or lacustrine features are identified and measured, and ages of features are established as precisely as possible. For marine terraces, mean sea-level fluctuations are removed using sea-level curves derived from either tectonically stable control areas (e.g., Bahamas) or areas with probable uniform uplift rates (e.g., Barbados, parts of Papua New Guinea). Rates of uplift are inferred from anomalous elevations versus age of feature. Studies with abundant time control can yield fairly accurate estimates of uplift rates, in meters/10^3 to 10^4 yr or mm/yr.

Typically, field surveys normal to coast shorelines or terrace flights, sometimes augmented with topographic maps, are needed to establish elevations. Terraces are usually dated by U-series or U-Th of shells on terrace levels (the best control), amino-acid racemization, or ^{14}C. In some studies, dating has been restricted to several control points, and ages have been extrapolated to the rest. Some studies have fitted best-fit regression lines to height-age data to estimate uplift rates.

Correlation of Tectonics and Seismicity

A seismic hazard analysis, which calculates the probability that some level of earthquake ground motion will be exceeded at a repository site, depends on three critical sources of data. These are identification of the earthquake source, an estimate of the rate and magnitudes of earthquakes that might occur within a given source, and an estimate of the ground motion that might result at the site from the earthquakes (Housner, 1969; Electric Power Research Institute, 1985). The relationship between an earthquake source, its recurrence rate, and earthquake magnitude (and thus a prediction of potential maximum earthquake magnitude) is the primary subject of this section. Studies of earthquake precursors and associated deformation play an essential role in understanding and predicting the magnitude of a future event. Therefore, a brief review of some important references in the tectonic and seismic literature on earthquake prediction leads this discussion.

Earthquake Prediction. The literature on earthquake prediction is extensive (cf. Rikitake, 1976, 1982; Brown, 1978a; Ward, 1978; Simpson and Richards, 1981; Agnew and Ellsworth, 1991). Prediction is based on the presence and interpretation of long- and short-term precursors, from which models and the assignment of probabilities are derived. As yet, efforts to predict times and places of earthquake occurrence have met with little success, even though long- and short-term precursors of moderate to large seismic events have been extensively studied and quantified. One critical view of earthquake prediction is that great earthquakes are not random occurrences, that the historic record generally provides an inadequate data base for proper statistical and probabilistic analysis, and thus that deterministic models may be more useful. A second view suggests that earthquake risk cannot be expressed in deterministic terms because of the enormous uncertainties about the nature and location of future earthquakes. The most significant problem with earthquake prediction is probably the lack of acceptable models that explain why precursors work.

A critical data base for earthquake prediction is the historical record of earthquakes at local, regional, and plate scales (Ambraseys, 1971; Coffman and von Hake, 1973; Rikitake, 1982; Shimazaki and Nakata, 1980; Veneziano et al., 1983). Recognizing the shortcomings of historical data (Allen, 1975; Veneziano and Coppersmith, 1985), various workers have attempted either to relate historical data to phenomenological or statistical relationships such as Weibull distributions (Hagiwara, 1974), seismic gaps (Mogi, 1968; Sykes, 1971; McCann et al., 1978), and migration of seismicity (Gelfand et al., 1972), or to ascertain and rectify historical incompleteness (Stepp, 1972; Veneziano and Coppersmith, 1985). Measurements of crustal strain comprise a second important source of data for identifying long-term precursors. This type of analysis describes the accumulation rate of strain (Langbein et al., 1983; Rikitake, 1974) and ultimate crustal strain (Rikitake, 1975b), as well as site-specific strain and geodetic measurements (Castle et al., 1976; Jackson et al., 1980; Prescott et al., 1979; Mark et al., 1981). The usefulness of long-term precursors largely depends on historical records. The more complete the record and the larger the experience for a specific tectonic feature, the greater is the confidence in predictions. As a long-term earthquake predictor, the monitoring of crustal strain probably has more potential than historical records, especially in North America where earthquake records date back only a few hundred years. Crustal strain studies require repeated geodetic surveys and experiments to calculate the strain required for crustal failure that results in large earthquakes.

In active seismic areas, careful analysis of anomalous uplift (Rikitake, 1976); changes in sea level (Ota, 1975; Plafker and Rubin, 1978; Sato, 1977; Rikitake, 1975a; Wyss, 1975, 1976a, 1976b, 1977); crustal tilt, strain, or stress (Rikitake, 1984; Sauber et al., 1983; Jamison and Cook, 1980; McGarr and Gay, 1978; McGarr et al., 1982; Abou-Sayed et al., 1978; Keys et al., 1979); and even rainfall (Huang et al., 1979) can often yield interesting if not always significant information that may serve as a basis for earthquake prediction.

Seismologic precursors are extensively used (Ohtake et al., 1978; Acharya, 1984; Ebel, 1984; Kagan and Knopoff, 1976; Simpson and Richards, 1981; Kanamori, 1981; Von Seggern, 1982; Electric Power Research Institute, 1985). They include foreshock (Rikitake, 1982); anomalous seismic activity (Evison, 1977a, 1977b; Ishida and Kanamori, 1977; Sekiya, 1977); seismic gaps of the second kind (Ohtake, 1976; Ohtake et al., 1977); growth and decay of seismic activity (Rikitake, 1976); b-values (Hasegawa et al., 1975; Kier, 1984); source mechanisms (Gupta, 1975; Ishida and Kanamori, 1978; Lindh et al., 1978); hypocentral migration of microearthquakes (Engdahl and Kisslinger, 1977; Gupta, 1975); and changes in seismic wave velocity (Aggarwal et al., 1973; Rikitake, 1979b; Semenov, 1969; Wyss, 1981).

In addition, other geophysical data offer potential for prediction. These include Earth-tide measurements (Latynina and Rizaeva, 1976; Mikumo et al., 1977); geomagnetic, geoelectric, and other potential-field values (Raleigh et al., 1977; Barsukov, 1972; Honkura, 1978; Rikitake, 1979a; Smith and Johnston, 1976; Rikitake, 1982; Corwin and Morrison, 1977; Noritomi, 1978; Varotsos and Alexopoulos, 1984; Dmowska, 1977; Morrison et al., 1979; Yamazaki, 1975, 1977; Hagiwara, 1977; Jachens et al., 1983); and ground-water data (Kovach et al., 1975; Merifield and Lamar, 1981; Moyle, 1980; Sato, 1977; Shimamura, 1982; Tagutschi et al., 1983; Wesson, 1981).

Short-term precursors are usually carefully monitored only in seismically active areas. One of the most reliable short-term precursors to large earthquakes is land uplift, although not all uplifts are followed by earthquakes. Changes in sea level, tilt, strain, and crustal deviatoric stress have all had mixed success in prediction studies. Seismologic precursors are widely used for prediction and will continue to be important as the meanings of characteristic seismic and microseismic patterns are clarified. Premonitory Earth tides are not widely accepted as precursors. Geomagnetic and geoelectric precursors are being intensely studied, and although some methods are subject to experimental error and their results are often extensively interpreted, the potential exists for significant forecasting using these geophysical tools. No integrated theory that explains why changes in the potential fields occurs prior to large earthquakes has been developed. Geochemical precursors, such as anomalous radon concentration in ground water (C. Y. King, 1978; Aric et al., 1984; Hauksson, 1981), have been useful in earthquake prediction. Macroscopic phenomena, such as animal behavior, lights, and noises (Rikitake, 1982), are excellent very-short-term precursors that are not considered temporally relevant here.

Various workers have attempted to quantify precursor information (Aki, 1981; Coats and Murray, 1984; Fedotov, 1982; Kagan and Knopoff, 1977; Rhoades and Evison, 1979; Vere-Jones, 1978) or to develop predictive models (Ward, 1978; Rikitake, 1982; Mallio and Peck, 1981; Anagnos and Kiremidjian, 1984; Bufe et al., 1977; Niazi and Mortgat, 1984). The results of such models are often the subject of debate (Kerr, 1979), although consensus about their practicality is sometimes reached (Reddy, 1983; Electric Power Research Institute, 1985).

In summary, interpreting earthquake precursors reliably requires models of the fundamental tectonic processes that create the earthquake source. Adequate and acceptable models are not yet available. Furthermore, scientific understanding of intraplate tectonic processes and available data (for example, in North America those areas east of the Rocky Mountains and Rio Grande rift) are currently not adequate to assess causative models or describe sources of potentially large earthquakes with a good degree of certainty (Electric Power Research Institute, 1985). Similar problems are evident even in the more tectonically active western part of North America, exclusive of the intensely studied late Quaternary-Holocene San Andreas fault system.

Fault Rupture and Seismicity. Fault models that relate rupture geometry to earthquake magnitude are based on empirical and theoretical relations between the area of fault rupture, average fault slip, and earthquake magnitude. With this type of model, large-scale deformation patterns may be predictable (Brown, 1978a).

The relationship between fault geometry and earthquake magnitude has been extensively studied (Tocher, 1958; Aki, 1966; Brune, 1968; Kanamori and Anderson, 1975; Cluff et al., 1980; Swan et al., 1980; Wallace, 1981; Schwartz and Coppersmith, 1984; Shaw and Gartner, 1984). Generally these studies are of two types. One type relates rupture *length* to magnitude (Mark, 1977; Tocher, 1958; Bonilla and Buchanan, 1970). A second type considers seismic moment, the relationship between rupture *area* and displacement on the rupture surface (Aki, 1966; Brune, 1968; Kanamori and Anderson, 1975; Hanks and Kanamori, 1979). Seismic moment may also be related to stress drop (Geller, 1976; Purcarn and Berkhemer, 1982).

In seismic-moment studies, the approximately linear relation between magnitude and rupture area (Kanamori and Anderson, 1975) allows one to predict the extent of rupture along a fault, and by inverting the data, to calculate an earthquake body-wave magnitude (Nuttli, 1983a; Veneziano and Coppersmith, 1985). Without inversion, the application of elastic dislocation models (Savage, 1972; Bischke, 1974; Savage and Hastie, 1966; Turcotte and Schubert, 1982) to a specific crustal volume can theoretically predict deformation patterns and the effective radius of deformation (Rikitake, 1976).

Length, shape, depth, orientation, area, amount of slip on the earthquake rupture surface, and recurrence interval and history of the source feature are critical parameters in this type of analysis. Length, shape, and surface orientation are measured topographically and bathymetrically. Depth and subsurface orientation are determined by geophysical means, generally by earthquake and reflection seismology. Area of slip is commonly measured after rupture and is based on length of surface offset, the seismologically active zone during rupture, and geophysically determined depth. This process is much more difficult, if not impossible, in aseismic areas. To obtain potential rupture area in aseismic zones, the geometry of the inactive rupture often is used to segment the fault zone into discrete lengths along which it is assumed to displace coherently. Alternatively, a percentage factor may be used to estimate rupture length (cf. Slemmons, 1982). The amount of slip is determined by surface offset on exposures of rupture surface, from focal mechanism data, and from seismic slip calculations. Recurrence interval is calculated from the historical earthquake record (cf. Nuttli, 1974, 1983b), with prehistoric geologic data sometimes added in active fault zones (cf. Sieh, 1984; Sieh and Jahns, 1984).

The exposure of surface-rupture geometry may be limited by sediment, water, or erosion. Seismic activity may not completely delineate the areal extent of rupture in active zones; in inactive areas, the resolution of geophysical methods for exploring the crust may be severely limited. The historical earthquake record is generally not sufficient to constrain the maximum earthquake in a region (McGuire, 1977b; Chinnery, 1979). The best constrained seismic-moment data come from interplate boundaries; continental interiors are largely unstudied, and empirical moment relationships may not be appropriate to intraplate tectonic settings (Nuttli, 1983b), although a scaling factor may be derived in some areas (Nuttli, 1983a).

Prediction of the maximum magnitude of an earthquake (and thus future deformation of a repository site) from the relationship between geologic features and seismicity associated with an earthquake source is a corollary problem to earthquake prediction, with the same inherent difficulties. In seismic areas, satisfactory resolution depends on a combination of detailed seismologic information, experience with previous earthquakes in similar settings, and reliable geologic mapping of potentially active faults (Allen, 1975). For aseismic settings, these methods largely depend on geologic characteristics of the region, along with intuitive perceptions and poor-resolution geophysical data about the composition and fabric of the crustal lithosphere beneath the specific site and regionally, as well as comparisons with "like" features that are active elsewhere. In both seismic and aseismic areas, prediction of the magnitude of future events is largely based on the qualitative perception of the possible extent of rupture; few quantitative data exist (cf. Veneziano and Coppersmith, 1985), although the state of stress of a specific region provides quantitative limits on effective stress release (Zoback, 1983).

CRITICAL COMPARISON OF METHODS

In aseismic areas, geologic methods best ascertain the potential of a tectonic hazard. The best possible data set is obtained when fault recurrence can be established using well-dated stratigraphic sequences to infer the number and ages of fault offsets. Because age is a critical parameter in recurrence studies, appropriate materials must be available for age dating to ensure reasonable precision. Relative age determinations are crude. In every case, the best data are obtained when late Quaternary or Holocene tectonism and related sedimentation have occurred. Older Quaternary activity is imprecisely documented, and late Tertiary tectonism is understood only within a precision of $\pm 10^6$ years. Analysis of stream profiles and fault-scarp morphology offers significant potential for quantifying Quaternary tectonism, but the methods are not completely tested or acceptable. Regional studies on vertical crustal movements and their relation to tectonism are too broad and are usually imprecise geographically or numerically.

In seismic areas, fault recurrence can be quantitatively established using a combination of historical earthquake records, detailed seismic information along the tectonic feature, geologic mapping, and experience with similar seismic features elsewhere. The prediction of the maximum magnitude of an earthquake is possible along interplate boundaries. Inside continental plates, the lack of models describing processes that potentially could act on the plates severely limits the certainty with which deformation can be predicted.

At present, no tectonic or seismologic method is completely adequate to quantitatively assess, with a good degree of certainty, the probability of tectonic activity at a repository site during the long periods for which performance must be predicted.

AVAILABILITY OF DATA BASES AND VARIABLES OF EACH

A number of good data bases are readily available for evaluating tectonic and seismic phenomena; a few of the more significant are listed here by issuing organization. No attempt has been made to evaluate these data except to note their nature.

Tectonic Data

The major sources of tectonic data are the publications of the U.S. Geological Survey (USGS), state surveys of the United States, Geological Society of America (GSA), American Association of Petroleum Geologists (AAPG), and National Aeronautics and Space Administration (NASA). Most of these data are in the form of geologic and tectonic maps, photoimages, cross sections, isopach and structural contour maps, and fence diagrams at various scales and by various authors. All are subject to technical review before publication, including photoimages from NASA.

For example, the USGS publishes a series of field maps (e.g., Ziony et al., 1974; Hadley and Devine, 1974; Howard et al., 1978; Machette and Personius, 1984), open-file reports (e.g., Bucknam et al., 1980; Gable and Hatton, 1980; Clark et al., 1984; open-file reports like these normally contain extensive references), and professional papers (e.g., Poppe, 1979; Wentworth and Mergner-Keefer, 1983). Similar publications are published by almost every state geological survey in the United States. The GSA and AAPG also publish many maps, as well as their respective bulletins and reports (e.g., Sexton et al., 1982). NASA provides space photography and interpretation (e.g., NASA, 1979). The National Earthquake Hazards Reduction Program has generated major new sources of interpreted data for tectonic and seismic analysis (e.g., U.S. Geological Survey, 1980).

Seismic Data

The major sources of data for seismic analysis come from the World Wide Standardized Seismograph Network (e.g., Oliver and Murphy, 1971), National Geophysical Data Center (NOAA, no date), National Oceanic and Atmospheric Administration (e.g., Coffman and von Hake, 1973), U.S. Coast and Geodetic Survey (e.g., Stepp et al., 1965; Barazangi and Dorman, 1969), UNESCO (e.g., Rothe, 1969), and the National Academy of Sciences (e.g., DeBra, 1979; Brown, 1978a). Some of these data are raw seismic records on computer tapes; other data have been plotted and interpreted. Not all data are technically reviewed or formally published.

In addition, data bases are found in the journals of major geophysical publishers, including the Bulletin of the Seismological Society of America (e.g., Doser and Smith, 1982; Stevens, 1980; Ryall et al., 1966), the American Geophysical Union (e.g., Zoback and Zoback, 1980; Wallace, 1981; Coppersmith and Schwartz, 1984), Elsevier (e.g., Lisitzin, 1974; Holdahl and Morrison, 1974; Smith, 1977; Brown, 1978b; Reilinger et al., 1984), and Birkhauser (e.g., Swanberg and Morgan, 1978). This material is all subject to technical review.

Large numbers of unpublished data exist in reports and manuscripts submitted to the Nuclear Regulatory Commission (e.g., Chiburis, 1981; Shannon & Wilson, 1976), DOE consultants (e.g., Fugro, Inc., 1980), and national laboratories (e.g., Powers et al., 1978; Aldrich and Laughlin, 1982). Much of this material is subject to at least in-house technical review.

PHENOMENA NOT PRESENTLY PREDICTABLE BY PREVIOUSLY DISCUSSED METHODS

Perhaps the most significant uncertainty in tectonic analysis of waste repository sites is the determination of hazard in areas where little or no seismicity exists and where Quaternary-late Tertiary sedimentation and evidence for tectonism are minimal or nonexistent. The eastern United States is an example of this type of environment. There the main strategy for hazard analysis is quantifying the regional state of stress; understanding the processes that cause this stress regime; and identifying and characterizing tectonic features that may become hazards (Coppersmith and Kulkarni, 1985). To a large extent, this approach entails evaluating inactive basement features with the potential to be reactivated. (In seismically active areas, basement reactivation can be an important component of seismicity and might seem to introduce significant uncertainty into probabilistic calculations. In these cases, questions of basement reactivation could be resolved by studies using, for example, high-resolution seismograph networks, surface and borehole strain measurements, and resurveying.)

Regional State of Stress and Strain

The first step in the tectonic analysis of aseismic or weakly seismic areas is determining the regional stress-strain regime, mainly by using in-situ stress determinations (e.g., Carr, 1974; Abou-Sayed et al., 1978; McGarr and Gay, 1978; Keys et al., 1979; Zoback and Zoback, 1980) within the context of the worldwide plate environment (Richardson and Solomon, 1979). Any earthquake data from the region must be integrated into this data set, because there is often a spatial relationship between zones of weakness (and thus local stress-strain regimes) and earthquakes (Bollinger, 1973; Barstow et al., 1981), although this is not always the case (Smith and Bruhn, 1984). The strength of crustal materials can sometimes be predicted by this analysis, combining laboratory (Kirby, 1980) and seismic (McGarr, 1984) data. Geodetically (e.g., Mark et al., 1981; Reilinger et al., 1984) and geologically (e.g., Schwartz and Coppersmith, 1984; Bull, 1978a, 1978b) observed strain is also critical.

Although in-situ stress can be measured with precision, the determination of strain mainly depends on interpreted geological observations (e.g., Prescott et al., 1979). Theoretical strain analysis can yield a broader regional picture (Reches,

1983). In any case, the final product requires time-consuming, site-specific data gathering and analysis.

Tectonic Processes

A second, more theoretical step in this analysis is the evaluation of tectonic processes that may be acting on the region to yield the observed regional stress-strain patterns (see Weldon, 1991, for a recent review). A number of important regional tectonic processes have been suggested. The first is the stress regime created by plate-tectonic processes. Most modern authors consider trench-pull and associated traction on the subducted slab to be the critical force that generates stress in the plates (Chapple and Tullis, 1977; Richardson and Solomon, 1979; Turcotte and Schubert, 1982; Turcotte, *in* Coppersmith and Kulkarni, 1985, p. 3-12). Additional significant stresses are created by thermal perturbations (e.g., Liu, 1980), ridge-push, variations in crustal and lithospheric thickness, membrane stresses from variations in Earth curvature, and erosion and sedimentation (Turcotte, op. cit.).

In addition to plate stresses, the regional picture is complicated by more local phenomena such as stress corrosion and chemical effects within the lithosphere (Anderson and Grew, 1977; Etheridge et al., 1984); localized movement along zones of different mechanical properties (e.g., Negmatullayev et al., 1984; Sibson, 1984; McGarr, 1984); large- and small-scale lithospheric compositional (e.g., Campbell, 1978; Kane, 1977) or rheological (e.g., Byerlee, 1968; Kirby, 1980; Sibson, 1977) inhomogeneities; stress amplification (e.g., Segall and Pollard, 1980); enhanced fluid pore pressure (e.g., Bulau et al., 1984; Etheridge et al., 1984) and associated hydrolytic weakening of minerals such as quartz in the upper lithosphere (Sibson, 1981; Blacic and Christie, 1984; Kronenberg and Tullis, 1984); stress-induced crack growth in stressed regions (Crampin et al., 1984; Swanson, 1984; Segall, 1984); and brittle reactivation of previously ductile zones due to later uplift (e.g., Sibson, 1977; Watts and Williams, 1979; Cranwell and Donath, 1980; Passchier, 1982; Grocott, 1977).

The influence of each of the above processes has not been thoroughly tested in any regional context. Intuition demands that each may be important locally, but only study of specific regions will ultimately resolve the relative importance of each factor. Undoubtedly additional relevant processes will appear in the course of analysis.

Tectonic Features

Finally, a map of the major tectonic features of the region must be created. This map depends on two critical sources of data: surface and near-surface geology, which can be obtained with some confidence; and subsurface crustal and lithospheric features, which depend on generally poor-resolution geophysical mapping and significant interpretation for their identification. Some of the more important tectonic features on which exploration should be focused are ductile shear zones in all regions of the lithosphere, particularly the upper and middle parts (e.g., Sibson, 1983, 1984; Sykes, 1978; Ratcliffe, 1983; Passchier, 1982); plutons (e.g., Campbell, 1978; Sykes, 1978); water at depth (e.g., O'Neill and Hanks, 1980); other lithospheric inhomogeneities (e.g., Kane, 1977); and brittle fracture zones (Sykes, 1978; Sibson, 1983; Wentworth and Mergner-Keefer, 1983; Zoback, 1983).

Assessment

Predicting basement reactivation along preexisting zones of weakness requires that these zones be recognized (generally by geophysical mapping); that stresses and strains be measured around the structure; that critical strain accumulation and mechanisms for failure be modeled and predicted; and that tectonic processes that may cause seismic activity along the structure be delineated and quantified. The critical assumption for earthquake hazard analysis in this situation is that once the stress regime acting on a known tectonic feature with predictable failure mechanism is understood, the magnitude and location of the seismic hazard may be predicted. This assumption has yet to be tested, and not all data necessary for such a test have been accumulated for any specific site.

EVALUATION OF BEST METHOD FOR PERFORMANCE ASSESSMENT

A number of studies have attempted to outline procedures for performance assessment in tectonics and seismology (e.g., Geotechnical Engineers, 1980; D'Alessandro et al., 1980; Reddy, 1983). The major problems associated with such assessments are inadequate data collection; disagreement among professionals about the quality and significance of the data; significant uncertainty in the level of current scientific understanding and in the available data; and the inability to assign meaningful and generally accepted probabilities.

A recent report by Dames & Moore (Electric Power Research Institute, 1985) is probably the best effort now available to develop methods for evaluating seismic hazard. In order to understand the tectonic/seismic hazards of a region, three data bases were established in the EPRI study: the causes of crustal stress in the region; the present state of the regional stress regime; and the identity and characteristics of the tectonic features in the region that might produce moderate to large earthquakes. A matrix of the physical characteristics of a region was interpreted independently by six earth science teams.

SUMMARY

The geology of the repository site is the fundamental source of data for quantitatively evaluating tectonic and seismic hazards that could affect a nuclear-waste repository. Four kinds of geologic data are most useful for predicting potential tectonic activity near a waste repository: surface ruptures along fault zones; stratigraphic sequences related to tectonic activity; landforms related to surface rupture and regional uplift or subsidence; and relationships between seismicity and tectonic features.

Recurrence methods provide the best data set for probabilistic prediction of long-term behavior of a fault zone, but these methods often are severely limited by proper age control. Stratigraphic sequences are probably the best single data source for fault recurrence studies, but only relatively young (late Quaternary or younger) sequences offer significant resolution. Landform analysis, especially fault-scarp morphology, offers potentially significant information, but the precision is poor, particularly from regional analysis. In every case, the best data are obtained when late Quaternary or Holocene tectonism and related sedimentation have occurred.

Earthquake prediction depends on the availability of long- and short-term precursors, a well-established historical seismic record, and detailed monitoring networks for precursors of all kinds. The greatest weakness of earthquake prediction is the lack of understanding of the fundamental tectonic processes that create the earthquake source, particularly in intraplate settings. The prediction of the maximum magnitude of an earthquake from the relationship between geologic features and seismicity associated with an earthquake source is a corollary problem to the prediction of earthquakes. In both seismic and aseismic areas, prediction of the magnitude of future events is largely based on the qualitative perception of the possible extent of rupture; few quantitative data exist.

Earthquake hazard analysis in aseismic or weakly seismic regions requires that the process of reactivation of basement structures be understood. If the stress regime acting on a known tectonic feature with predictable failure mechanisms is known, then it may be possible to predict the magnitude and location of the seismic hazard. This approach has yet to be tested quantitatively.

Methods that establish the causes and present state of crustal stress and that characterize the tectonic features of a region offer significant potential for creating more realistic performance assessments. At present, no tectonic or seismologic method is completely adequate to quantitatively assess the probability of tectonic activity at a repository site with certainty.

7

Seismic Hazard Assessment

C. Allin Cornell
and Gabriel Toro

SEISMIC HAZARD INTEGRATION

Probabilistic analysis of the threat of seismological effects at a repository site requires an integration of two basic elements. The first is a general earthquake recurrence model: a probabilistic prediction of future earthquake dates, locations, and source characteristics (magnitude, rupture length, etc.). The second element is a probabilistic prediction of effects at the site given the location and source characteristics of an event. The site effects of interest may include one or more measures of strong ground motion and faulting.

Integration of recurrence and site-effects models into a site hazard assessment is straightforward and does not deserve extensive treatment here. Although differences in computational details may exist, all such integration procedures apply the total probability theorem, which states that the probability of event A is the sum of a set of products. Each term of the sum is a product of two factors: (1) the conditional probability of A given the occurrence of some event B_i and (2) the probability of B_i. The sum is over all possible events B_i (the B_i's are a mutually exclusive, exhaustive set). In the current application, event A is the site-effect hazard statement of interest (e.g., peak ground acceleration exceeds 0.5 g at least once in the next T years). Its conditional probability can be obtained from the probabilistic site-effects model. The B_i's are possible future earthquake occurrences (e.g., a particular sequence of dates, locations, and magnitudes). For the small probabilities of engineering interest, normally the second factor above may be replaced by the expected number of events in T years of each size originating from each unit volume of Earth's crust in the surrounding region.

Probabilities of the B_i's are outputs of the stochastic earthquake recurrence model. In general, probabilities or expectations related to the occurrences of future events should depend on the current "state," e.g., times, locations, and sizes of the more recent events in the region (Cornell and Winterstein, 1986). These are examples of spatial and temporal dependence.

Examples of descriptions of seismic-hazard integration schemes have been given by Cornell (1968), McGuire (1978b), der Kiureghian and Ang (1977), Savy et al. (1980), Bernreuter et al. (1985), Esteva (1976), and Cornell and Winterstein (1986). All but the latter two take advantage of the strong independence assumptions of the Poisson model.

Assessment and analysis of uncertainties in such probabilistic hazard studies has become routine in practice. Uncertainty, as defined here, reflects the degree of statistical and physical information upon which the probability estimates are based. In seismic hazard analysis, "logic trees," i.e., discrete representations of alternative, uncertain parameter values and/or alternative model elements (e.g., different functional forms of ground-motion prediction equations or different tectonic models) have become popular (Yankee Atomic Electric Company, 1985; Bernreuter et al., 1985; Electric Power Research Institute, 1985). This discrete tree-structure format has the advantage both of facilitating the treatment of conditional uncertainty assignments (e.g., the mean rate of activity may depend on the tectonic model) and of encouraging as explicit a treatment of model uncertainty as is normally given parametric uncertainty. Here model uncertainty might include the consideration of one tectonic model that implies stationary seismicity and another, alternative model that favors evolutionary or even abrupt changes in seismicity over the next 10,000 years.

Any proper uncertainty methodology will mechanically propagate uncertainty in the model and on the parameters through the hazard analysis to produce uncertainty or confidence bands on the probability assessments.

RECURRENCE MODELS

Standard Poisson Exponential Model

The standard assumption in most probabilistic analyses of seismic shaking (e.g., Cornell, 1968; Bernreuter et al., 1985; Electric Power Research Institute, 1985) is that earthquakes occur as a Poisson process (Parzen, 1962b). A key property of the Poisson process is lack of memory; the probability distribution of the time to the next earthquake is independent of the time since the previous earthquake. This unique property manifests itself in two characteristics of the Poisson process: exponentially distributed interarrival times, with a coefficient of variation (COV) equal to 1 and constant hazard function. The hazard function h(t) indicates the probability that an earthquake will occur in the near future given that the last earthquake occurred t years ago. In addition to the Poisson assumption, the magnitude and location of the next event are routinely assumed to be independent of the time since the previous earthquake and of its magnitude and location.

Finally, magnitudes generally are assumed to follow a doubly truncated exponential distribution and locations to follow a piecewise-uniform distribution; each region of uniformity is called a seismic source and is usually associated with a geologic or tectonic feature or an area of much historic seismicity. The characteristic earthquake magnitude model studied by Youngs and Coppersmith (1985) does not have an exponential distribution of magnitude, but it is still a Poisson model.

This is the simplest probabilistic model of earthquake occurrence that can be formulated; it has the greatest possible degree of probabilistic independence. In essence, the only parameters of this model are the expected number of earthquakes per year for each magnitude class and each seismic source.

The Poisson model requires as data a catalog of time, magnitude, and location of past earthquakes. Traditionally, most studies have used the historic record of seismicity. Even if the earlier period of noninstrumental reporting is included, this data base in the United States spans less than 400 years. If only the later period of instrumental recording is included, the data base spans less than a century. The entire process of estimating parameters for a Poisson occurrence model (including analysis of dependent events, catalog incompleteness, common magnitude scale conversion, determination of spatial rates of occurrence, and estimation of magnitude distribution) is documented by Veneziano and Van Dyck (1985), who have recently developed the most rigorous, systematic approaches to these various problems.

In recent years, investigators have begun to combine historical seismic records with paleoseismic information on earthquakes (e.g., Schwartz and Coppersmith, 1984; Youngs and Coppersmith, 1985). Paleoseismic methods have been widely applied in the western United States; application to the eastern United States has met with more limited success.

Another critical parameter in these models is the maximum possible magnitude that a feature can generate. The typical historical record of earthquakes during a few hundred years is usually of limited use in evaluating maximum magnitude (McGuire, 1977b) because most maximum-magnitude events have mean inter-arrival times much longer than the historic record. Estimates based on fault dimensions (Slemmons, 1977; Wyss, 1979) have been widely used in the western United States. More subjective methods have been used in the eastern United States (Electric Power Research Institute, 1985).

The assumptions of Poisson arrivals and full independence among magnitudes, locations, and interarrival times are physically unrealistic; they are not consistent with widely accepted theories on accumulation and release of tectonic strains, which suggest a complex pattern of dependencies between occurrence time, magnitude, and location of earthquakes.

Poisson models do not reproduce the observed pattern of smaller earthquakes that may occur before (foreshocks) and after (aftershocks) a large earthquake (Kagan and Knopoff, 1976; Kanamori, 1981). Studies have shown, however, that secondary shocks contribute only modestly to the seismic hazard (Merz and Cornell, 1973; Veneziano and Van Dyck, 1985). Thus, inability of the Poisson models to predict secondary shocks is not a major drawback.

More significantly, recent work on historical and prehistorical patterns of occurrence of large earthquakes in selected seismic features indicates that the pattern of large earthquakes exhibits more regularity than the Poisson model would predict (e.g., Sykes and Quittmeyer, 1981; Sykes and Nishenko, 1984). This regularity manifests itself in a distribution of interarrival times with a COV smaller than unity and a hazard function that increases with time (as time since the last large earthquake increases, probability of a large earthquake in the near future increases). The Poisson assumption works better on an aggregate of seismic sources than on an individual feature, because the sum of several independent non-Poisson processes approaches a Poisson process (Brillinger, 1982). Thus, even if the Poisson assumption is not realistic for a single feature, it may produce realistic hazard estimates for a site that is threatened by earthquakes from several features.

A recent study (Cornell and Winterstein, 1986) suggests that deviations from Poisson may not have a large effect on the hazard calculations even for nuclear plants threatened by a single feature, mostly because these plants have design lives of only some 50 years. Effects on hazard of longer-term deviations from Poisson have also been studied by McGuire (1979) and McGuire and Barnhard (1981), who used the Chinese earthquake catalog. They concluded that the effect was small because of the short design life of typical facilities. The effect on hazard of deviations from Poisson has also been studied by Brillinger (1982). Deviations from Poisson behavior are likely to be more important for waste repositories than for nuclear plants for two main reasons: the long design life and potential for cumulative effects. This issue requires considerably more study.

A problem in estimating seismicity for the evaluation of seismic hazard to nuclear repositories, which is related to the design life of waste repositories and which is not limited to Poisson models, is the possibility of changes in current tectonic regimes. Even when good paleoseismic information is available, estimating steady-state seismicity levels may be difficult. Estimating trends, let alone sudden changes (Bender, 1984), is clearly difficult and requires additional research. Some evidence suggests that such changes occur slowly. For instance, Zoback and Zoback (1980, 1981) note that directions of principal stresses and patterns of deformation in some regions vary on a time scale of tens of millions of years. The possibility of such a change could be examined using discrete model alternatives (logic trees).

Models with Temporal Memory

A number of magnitude-recurrence models that contain more memory than the Poisson model have been proposed. Some of these models try to incorporate certain dependencies suggested by a physical understanding of tectonic processes. Renewal-process models (e.g., Esteva, 1976; Cornell and Winterstein, 1986) differ from the Poisson model in that the distribution of interarrival times is not necessarily exponential. Thus, an interarrival time distribution could be chosen with the desired COV and with an increasing hazard function.

Dependence between interarrival time and magnitude is introduced in the so-called time-predictable and slip-predictable models. In time-predictable models (Shimazaki and Nakata, 1980; Anagnos and Kiremidjian, 1984), the distribution of the time to the next earthquake depends on the size of the previous earthquake: the larger the previous earthquake, the longer the wait to the next earthquake. In slip-predictable models (Shimazaki and Nakata, 1980; Kiremidjian and Anagnos, 1984), the distribution of earthquake magnitude depends on the time since the previous earthquake. A combined time-slip predictable model (with backward and forward memory) has been analyzed by Cornell and Winterstein (1986). The general semi-Markov model (Patwardhan et al., 1980) can also include all these possible dependencies.

A "characteristic-process" model, in which a feature produces two kinds of earthquakes, has been postulated by Cornell and Winterstein (1986). Large (characteristic) earthquakes follow a renewal-process model with a small COV of interarrival times, and smaller (background) earthquakes follow a Poisson-exponential model.

Recurrence models contain more parameters than the Poisson-process model and require more data. The simplest renewal-process model has one more parameter that must be estimated: the COV of the interarrival time (Cornell and Winterstein, 1986). The time-predictable and slip-predictable models require at least one more parameter to characterize dependence between interarrival times and magnitudes.

In addition to model parameters, the current state of the process should be estimated. This might be the date and size of last major event on the fault. The interarrival times of characteristic earthquakes may be large, however, larger in fact than the period of observed seismicity. Thus, the time and magnitude of the previous earthquake may be unobserved. Such state descriptions can also be analyzed, however.

These models introduce some physical realism, which is lacking in the Poisson-process model. They are still crude, ignoring, for example, spatial relationships among earthquakes (Singh et al., 1981). These models have not been tested in practice and the methods for parameter estimation and treatment of parameter/state uncertainty are not fully developed. A few studies have attempted to apply these models to practical hazard analysis (Esteva, 1976; Savy et al., 1980; Cornell and Winterstein, 1986).

The effects of departures from Poisson on the seismic hazard at a nuclear repository, considering the long time frame and the possibility of cumulative effects, have not been investigated. In principle, the magnitude of these effects depends on the extent and type of departure shown by the data and on the resulting uncertainty in parameters and states. The larger the uncertainty on parameters and states, the closer the marginal hazard (integrated over all uncertain variables) will be to the corresponding result for a Poisson model. Uncertainty in the hazard (as a result of uncertain parameters) may not be the same for the Poisson and non-Poisson models.

Models with Temporal and Spatial Dependence

Some models attempt to predict the temporal and spatial dependence of seismicity. An example of this dependence is the enhanced likelihood of earthquakes occurring on seismic gaps (Singh et al., 1981). Veneziano and Cornell (1974) developed a hybrid stochastic-mechanical model of earthquake occurrences in space and time. Kiremidjian and Anagnos (1980) proposed a Markov model in space and time. See also Anagnos (1985).

Physical models that include stress accumulation, rupture initiation, and static or dynamic rupture propagation have been investigated (e.g., Burridge and Knopoff, 1967; Dieterich, 1972, 1979; Andrews, 1978; Stuart, 1985). Combined with a stochastic-process description of strength along the fault, these models have a potential for engineering applications.

Estimating parameters and current state for these models is difficult, especially for the mechanical and hybrid models. The state variables (current state of stress on different fault segments) are many and cannot be observed directly (Veneziano and Cornell, 1974). As a result, estimation is complicated, and the estimates obtained have little precision.

Models with temporal and spatial dependence may produce more realistic seismic-hazard estimates. Further understanding of the mechanics of faulting and direct measurements of stresses will increase the applicability of these models. In application, these models might be used to improve the specification and estimation of parameters in simpler temporal-memory models.

These models are still research tools and have not yet reached the level of development needed for engineering applications. In particular, model uncertainty is still large, and parameter/state uncertainty will be large if past earthquakes are the only available data.

GROUND SHAKING

The possible effects of ground shaking on the behavior of a repository are important. Various methods for predicting ground shaking are described, evaluated, and compared here. State-of-the-art reviews on the prediction of earthquake shaking have been written by Swanger et al. (1980, 1981), Hays (1980), Aki (1982), Boore (1983a), and Scholl and King (1985).

Earthquake-induced shaking may affect the performance of repositories during the period of waste emplacement and after closure. Mechanisms of damage during the emplacement would be similar to those experienced in a deep mining operation: damage to above-ground facilities or the vertical conveyance system, rock spalling, and collapse of tunnels. These mechanisms may lead to releases of radioactivity from waste temporarily stored above ground, in transit, or in permanent storage. The mechanisms of damage after backfilling and closure of the repository may compromise the integrity of the rock mass in the vicinity of the opening, the tunnel liner, the closure system (backfill and plugs), or the waste package. Mechanical damage to the waste package would likely be caused by unsatisfactory performance of the backfill or the rock mass, not by shaking.

A number of analytical studies and limited empirical evidence from earthquakes and explosions suggest that tunnels are less susceptible to shaking-induced damage than conventional, above-ground structures (Newmark, 1963; Dowding and Rozen, 1978; Dowding, 1979; Dowding et al., 1983; Hendron and Fernandez, 1983, Wu et al., 1982). The presence of backfill would further improve the strength of the tunnels and reduce the spatial extent of damage compared with empty tunnels. The magnitude of this effect depends on the mechanical properties of the backfill material.

Site ground shaking during an earthquake is most commonly characterized in terms of peak acceleration, peak velocity, and peak displacement. Another common measure, more directly related to the shaking of structures, is spectral displacement SD(f, d), which is defined as the peak displacement of a lincar mechanical oscillator of natural frequency f and damping (a measure of energy dissipation) equal to d percent of critical. A plot of SD(f, d) as a function of frequency f is called the *response spectrum*; approximate methods exist for constructing response spectra from peak acceleration, velocity, and displacement. Less common in practice, yet useful (at least on a conceptual basis), are the Fourier amplitude spectrum of ground acceleration and the duration of significant shaking (Trifunac and Brady, 1975; McGuire and Barnhard, 1979; Vanmarcke and Lai, 1980). These quantities are discussed in detail in the earthquake engineering literature (e.g., Newmark and Rosenblueth, 1979; Clough and Penzien, 1975; Newmark and Hall, 1982). Peak acceleration is the most commonly used measure of shaking, but it has numerous drawbacks (Kennedy et al., 1984). For instance, two shaking episodes with the same peak acceleration, one from a large-magnitude, distant earthquake and the other from a moderate-magnitude, close-in earthquake, produce substantially different shaking in a typical structure. This is particularly true if the close-in earthquake occurs in the eastern United States, where ground motions may contain a substantial fraction of their energy at high frequencies (>10 Hz); such energy does not affect the vibration of typical structures with low and moderate fundamental frequencies.

The amplitude, frequency content, and duration of shaking at a site on the surface is the combined effect of three factors: seismic source characteristics, propagation of seismic waves in the crust, and local effects due to soil and subsoil conditions. Most measurements of shaking are made at sites on the surface, but earthquake ground shaking in deep tunnels differs from that on the surface, because amplification caused by a free surface and filtering of high-frequency waves by weathered bedrock are absent.

The free-surface effect on the different wave types is well understood and documented (Anderson, 1976; Aki and Richards, 1980, Chapter 5). For a simplified model of the crust (i.e., a homogeneous half-space), theory predicts amplification at the surface by a factor of two for S_H waves and by a factor that depends on the angle of incidence (but remains approximately equal to two for a wide range of angles) for S_V waves. Empirical studies cited by Wight (1979) indicate that ground-motion amplitude at depth is typically 50% to 70% of that experienced at the surface, in agreement with theory. A recent review of downhole recordings by Chang et al. (1986) indicates amplitudes at depth equal to about 20% of those experienced at soil sites on the surface; soil is probably responsible for the difference from the value given by Wight.

Hanks (1982) argued that the seismic source generates energy at relatively high frequencies and that the observed sharp cutoff at about 10 to 15 Hz (in California) results from filtering of high-frequency ground motions by weathered rock. In fact, measurements of ground motions in deep mines (McGarr et al., 1981) during small earthquakes at small hypocentral distances indicates uniform frequency content up to frequencies of several hundred hertz. (Papageorgiou and Aki

[1983a, 1983b] argued that the frequency cutoff is a source effect; furthermore, Aki [1987] argued that the source cutoff frequency is greater for smaller earthquakes.) If Hanks's (1982) argument is correct, ground motions with more energy at high frequencies should be expected in deep tunnels during close-in earthquakes. For conventional structures, frequencies above 50 Hz are much greater than the structure's fundamental frequency and the structure does not respond to energy at these frequencies. For tunnels, on the other hand, the transverse fundamental frequency calculated from a diameter of 4 m is roughly 400 Hz (McGarr et al., 1981) and the tunnel responds to but does not amplify energy at frequencies below 200 Hz. This high-frequency energy may or may not be damaging to the tunnel and lining depending on their rheological properties. For example, some materials, such as wood, are much more resistant to short-term loading than to long-term loading.

Direct Empirical Methods

Empirical methods for the prediction of ground shaking originated in the 1960s and have been used extensively in probabilistic seismic-hazard analysis. These methods use regression analysis (usually linear) to fit an equation to a data base on the peak amplitude of ground motion, earthquake magnitude, and distance to the earthquake. The functional form most commonly used in these regressions is

$$\ln Y = a_0 + a_1 M + a_2 \ln R + a_3 R + e \tag{7.1}$$

where Y is peak amplitude of ground motion, M is earthquake magnitude, R is distance between the recording site and the earthquake, and e is random error. Equations of this type are often called *attenuation equations*. The functional form has some theoretical justification: the dependence of ln Y on M is based on the definition of earthquake magnitude (Richter, 1958), and the distance terms are based on the theoretical decay of body waves in homogeneous viscous media. Coefficient a_2 represents decay due to geometric spreading of energy, and coefficient a_3 represents energy dissipation (often called *anelastic attenuation*).

A logarithmic form of the above equation is used to make it linear on the unknown coefficients a_0 through a_3. The error term e has been observed to have a normal (Gaussian) distribution. Many variations on the above equation have been used. Some include a term for soil effects; others include magnitude-distance interaction terms (Campbell, 1981b); a few others have functional forms other than logarithmic (Bolt and Abrahamson, 1982; Brillinger and Preisler, 1984a, 1984b).

Y may be any measure of ground shaking discussed earlier, although peak acceleration is the most common. Root-mean-square (rms) acceleration was used by McCann (1980); response spectral ordinates were used by McGuire (1977a) and by Boore and Joyner (1982); Fourier spectral ordinates were used by McGuire (1978a), Trifunac (1976), and Cornell et al. (1979). M is the earthquake magnitude, which is a measure of earthquake size; many magnitude scales exist (Nuttli and Herrmann, 1982). The most commonly used measure of distance, R, is the closest distance to the fault trace. As large earthquakes are usually associated with long ruptures (Slemmons, 1977; Bonilla, 1967; Joyner, 1984), the predicted isoseismals (i.e, lines of equal shaking) from large earthquakes are elongated rather than circular.

Idriss (1978) summarized the early work on empirical prediction of ground motion. Among the more recent empirical studies on California data, the work of Joyner and Boore (1981), Campbell (1981b), and Boore and Joyner (1982) is in common use. These studies include data from the 1971 San Fernando and 1979 Imperial Valley earthquakes, which produced more recordings than previous earthquakes. To prevent these two earthquakes from excessively influencing the regression analysis, Joyner and Boore (1981) and Boore and Joyner (1982) used a two-stage regression scheme, whereas Campbell (1981b) applied a weighting scheme. A more elegant, and statistically more efficient, treatment of this problem is given by Brillinger and Preisler (1984a, 1984b), who used a "random effects" model. A similar approach was used by Veneziano and Heidari (1986) in their analysis of ground shaking in eastern North America.

An important issue in empirical ground-motion prediction is whether, at small distances, the marginal increase in amplitude of shaking is smaller for large magnitudes than for small magnitudes (a phenomenon known as saturation). Campbell (1981b) predicted saturation, but Joyner and Boore (1981) showed that the data cannot distinguish between models with and without saturation.

Prior distributions on one or more of the parameters a_0 through a_3 may be introduced when data are limited; simple regressions may lead to unacceptable estimates or large standard errors on some of the parameters (Veneziano and Heidari, 1986).

A large number of processed earthquake recordings (e.g., Hudson, 1976; McGuire and Barnhard, 1977) are needed to obtain a large number of points in the space of (ln Y, M, ln R, and R). California data can be considered adequate for the prediction of shaking at distances greater than 20 km for magnitudes smaller than 6. For shorter distances and larger magnitudes, data are scarce, and predictions depend strongly on the assumed functional form (Askins, 1980). For other regions of the United States, where the seismic activity and density of strong-motion seismographs are substantially less, the available strong-motion data are much more limited.

Toro and McGuire (1986) compiled data on shaking from North American earthquakes east of the 105th meridian. The largest earthquake in this data set has $m_{b_{Lg}}$ 5.6 ($m_{b_{Lg}}$ is a magnitude scale commonly used for earthquakes in the eastern and central United States; see Nuttli and Herrmann, 1982); data for distances smaller than 100 km are available only for $m_{b_{Lg}}$ smaller than 5.

Data from recordings of smaller earthquakes and/or large distances are available for eastern North America (Atkinson, 1984; Gupta et al., 1985). In spite of their small amplitudes, these data often are used in the absence of better data. A small number of underground recordings were described by Chang et al. (1986).

Direct empirical methods are adequate if sufficient data over the range of magnitudes and distances that dominate hazard at the site are available. Insufficient data cause two problems: predictions may be extremely sensitive to the assumed functional form of the regression model (model uncertainty), and the parameter estimates may have large standard errors (statistical uncertainty). In the worst, but not unrealistic, case, some parameter estimates may not be statistically significant or may take values that violate physical principles. For all regions except California, the available ground-level earthquake recordings are insufficient for the fitting of empirical relations to be used in the prediction of ground shaking for the magnitude and distance ranges of engineering interest. (Relationships derived from California data are presumably applicable to the entire Pacific Northwest, as long as differences in focal depth are properly considered.)

The situation is much worse for underground recordings. Calculation of "pseudo"-underground recordings by removing surface effects from ground-level recordings is possible (Joyner et al., 1981; Chang et al., 1986) but requires knowledge of the local soil profile and introduces additional uncertainties.

In summary, given the limitations in the available data, empirical methods, although perhaps adequate for conventional surface facilities, are not well suited for predicting ground shaking in underground repositories.

Indirect Empirical Methods

The available instrumental strong-motion recordings are few for most areas outside California. Noninstrumental information, from reports of perceived intensity of shaking and the observed extent of structural damage, is often more abundant. This is the case in the eastern and central United States. Noninstrumental information is measured on the Modified Mercalli Intensity scale (MMI; Richter, 1958).

Attenuation relations for intensity have been obtained (e.g., Bernreuter et al., 1985; Veneziano, 1985) in terms of magnitude and distance, of epicentral intensity and distance, and of distance (for specific individual earthquakes).

Several authors have substituted these attenuation relations into empirically obtained relationships between instrumental measures of shaking as a function of intensity and possibly other variables such as magnitude and distance. The net result is an attenuation function of the standard type; namely, amplitude of shaking as a function of magnitude and distance.

McGuire (1984) and Bernreuter et al. (1985, and references therein) illustrated the use of these methods. Cornell et al. (1979) applied these methods to California data and concluded that most substitution procedures were biased; i.e., they did not yield the same coefficients as the direct regression of magnitude and distance on amplitude of ground motion. Veneziano and Heidari (1986) have developed an unbiased procedure and used it to estimate peak accelerations and response spectral ordinates for the eastern and central United States. These authors have developed statistical diagnostics that indicate whether the regressions being combined are compatible and procedures to treat incompatibilities.

Sources of data on Modified Mercalli Intensity from earthquakes in the eastern United States and Canada were listed by Bernreuter et al. (1985) and Eckert and Atkinson (1985). Intensity data are generally more abundant than instrumental data, but their quality is poorer. To avoid biases in using intensity data, the analyst should be aware of how intensity is assigned to a given locale and how intensity information is compiled (Veneziano and Heidari, 1986). Abundant data from California, where both instrumental and MMI information are available, can be used to fit relationships for amplitude of shaking as a function of intensity, distance, and possibly magnitude (e.g., Bernreuter et al., 1985).

For regions where MMI data are abundant, but few or no instrumental data exist, unbiased formulations of the indirect method (as developed by Veneziano and Heidari, 1986) are superior to direct methods because they incorporate additional information. Like direct empirical methods, these methods are hampered by the virtual absence of instrumental recordings at depth, even in California.

Semiempirical Methods

In trying to overcome the dearth of instrumental ground-motion data in regions outside California, investigators have devised a number of procedures that combine data with predictions from theory. Nuttli (1979) fixed the coefficients a_1 through a_3 in Equation 7.1 using theoretical arguments. Coefficient a_0 was calibrated using data from the San Fernando, California, earthquake. Campbell (1981a, 1982) used a variant of Equation 7.1, with a_3 fixed to a value consistent with anelastic attenuation in California. He fit that equation to the California data set of Campbell (1981b) and modified the resulting coefficients to convert from the magnitude scales used in his regression to m_b and to reflect anelastic attenuation in the central United States. Nuttli (1983c) fixed the values of coefficients a_1 through a_3 based on theory (the value of a_1 is based on the scaling of source spectra proposed by Nuttli, 1983a) and calibrated a_0 to data from the central United States. Bernreuter et al. (1985) described a similar method.

Semiempirical methods are at least partially successful in overcoming the limitations of sparse data in areas other than California. These methods are not without their problems. The magnitude coefficients (i.e., a_1 in Equation 7.1) used by Nuttli (1979, 1983c) are based on the spectral-scaling model of Nuttli (1983a; Street et al., 1975). A spectral-scaling model determines how the Fourier amplitude spectrum of acceleration near the source varies with earthquake size. This spectral model has been criticized (Boore and Atkinson, 1987). The eastern U.S. data used by Nuttli (1983c) come mostly from recordings on soft soils, which may have amplified ground shaking.

One possible problem with Campbell's (1981a, 1982) relationship is the magnitude conversion, which is effectively performed by substituting one regression into another. As documented by Cornell et al. (1979) and Veneziano and Heidari (1986), substitution of one regression into another may lead to biased results. The biases should be smaller, however, for magnitude conversions, where correlation is stronger, than for MMI-to-ground-shaking conversions.

As with direct and indirect empirical methods, removal of local site effects (from the input data or from the model's result) to obtain peak motions at depth is difficult and introduces additional uncertainty.

Stochastic Methods

Stochastic methods use a probabilistic representation of the energy radiated by the seismic source or of the slip on the fault. They use simple concepts of wave-propagation theory to model the effect of propagation path and free-surface amplification. Stochastic methods may be divided into two broad categories: those that start from a more abstract representation of the seismic source and those that model the kinematics or dynamics of slip on the fault. Aki (1985a, 1985b) reviewed methods in these two categories and their relationship to deterministic methods.

Most methods in the first category use the source spectral representation of Brune (1970, 1971), which relates the amplitude and shape of the Fourier spectrum of acceleration to seismic moment (a measure of earthquake size; Aki and Richards, 1980; Hanks and Kanamori, 1979) and stress drop on the fault. (Some authors prefer to consider stress drop an empirical, rather than physical, parameter.) For given seismic moment, stress drop, and distance, these methods yield the power spectral density and duration of ground acceleration. Peak accelerations, velocities, spectral displacements, and other measures of ground motion are then calculated using random-vibration theory (Crandall and Mark, 1963; Boore, 1983b). In essence, these methods are not too different from the semiempirical scaling methods of Nuttli (1979, 1983c); they differ only in their assumptions about source scaling and in their mathematical rigor.

Stochastic methods satisfactorily predict ground motion at some locations in California (Hanks and McGuire, 1981; McGuire and Hanks, 1980; Boore, 1983b; McGuire et al., 1984). In these applications, stress drop has been assumed constant and equal to 100 bars. The same value of stress drop has been used to predict ground motions in eastern Canada (Atkinson, 1984) and eastern North America (Boore and Atkinson, 1987; Toro and McGuire, 1986). Results show reasonable, albeit not perfect, agreement with the limited data available.

Herrmann (1985) extended these methods to predict eastern U.S. ground motions at distances of several hundred kilometers by introducing, in an approximate manner, the dispersive nature of Lg waves, which dominate shaking at those distances. This result allows comparison of model predictions with the more distant recordings that constitute most of the instrumental data in those regions (Eckert and Atkinson, 1985; Gupta et al., 1985; Toro and McGuire, 1986).

Methods in the second category date back to the work of Aki (1967) and Haskell (1966). More recently, Papageorgiou and Aki (1983a, 1983b, 1985) have proposed a model that represents earthquake rupture as an aggregate of circular cracks of equal diameter, which represent co-seismic slip. As the rupture front sweeps the fault plane, a stress drop takes place in each crack. The diameter of the cracks is a model parameter. Savy (1980) and Savy et al. (1980) have proposed a model in which slip along the fault and propagation of the rupture front are random functions. Other models have been proposed (Boatwright, 1982; Gusev, 1983; Faccioli, 1986). These models have been used mostly in an inverse mode; i.e., given a well-recorded and documented earthquake, which parameters will best reproduce the observations?

Stochastic methods require data for calibrating parameters and verifying assumptions. Recordings from seismograph instruments (which measure shaking from more distant events) often are useful for these purposes (e.g., Sommerville et al., 1985), thus increasing the availability of data. Other diverse sources of data that can be used to calibrate and verify the assumptions of stochastic models include geological observations and theoretical studies in seismology.

Stochastic methods depend less on scarce strong-motion data than empirical methods, which makes them more attractive for areas like the eastern and central United States. Stated differently, if the data force extrapolation from distant and low-magnitude recordings, extrapolation with the model that has the strongest theoretical justification is best.

Some assumptions in the first category of methods are controversial. For the eastern United States, debate continues on whether stress drops are constant or increase with seismic moment (Nuttli, 1983a; Sommerville et al., 1985; Boore and Atkinson, 1987). Methods in the second category have more parameters, which are sometimes hypothesized to vary from earthquake to earthquake and presumably from region to region. These parameters must be specified, either as one deterministic value or as an uncertain (fixed but unknown) or random quantity that must be treated accordingly in the analysis.

In principle, these methods are not sensitive to the lack of data on ground motion at depth, i.e., all that is needed to predict ground motion at depth is to remove the factor of two that accounts for free-surface amplification and the low-pass filter that accounts for the effect of weathered rock near the surface. In practice, this is not entirely the case. For instance, Boore (1986) revised the stress-drop estimate for California on the argument that surface recordings on rock were amplified, because the rocks typical of California rock sites are less rigid than Earth's average crust. Two conflicting views on the origin of the high-frequency cutoff observed on ground-level recordings are held.

Deterministic Methods

Deterministic methods generate synthetic "seismograms" from a (usually kinematic) representation of slip at the source and detailed modeling of wave propagation through detailed models of the crust. These methods arise from the work of Aki (1968) and Haskell (1969), in which a smooth temporal and spatial distribution of slip along the fault plane is assumed. More recently, investigators have refined the models by adding the effect of free-surface and layered-crustal models (e.g., Archuleta and Frazier, 1978; Heaton and Helmberger, 1978; Archuleta, 1985). Most of these methods use nonuniform spatial distributions of slip; usually the distribution of slip is modified until agreement is obtained with the observed suite of seismograms being modeled.

To model the frequencies of engineering interest properly, observed records from smaller events are summed to yield the ground motion from a larger earthquake (Kanamori, 1979; Hadley and Helmberger, 1980; Hartzell, 1985). Munguia and Brune (1985) add a random component to the smooth slip distribution in order to represent the observed ground motions.

Deterministic models require detailed knowledge of crustal properties. The popular layered-earth models require knowledge of the depth, density, wave velocity, and damping for each layer. A priori specification of the distribution of fault slip in space and time is difficult.

In their present state of development, simulation methods have a limited range of applicability for predicting high-frequency ground shaking of engineering interest.

NEW FAULTING

New faulting near an underground repository may seriously affect repository performance. Faulting may create a path of reduced or increased flow resistance for contaminated ground water. In addition, if the fault rupture intersects a repository, the integrity of the tunnel liner, backfill and plugs, and waste package may be compromised. The effects of faulting also may compound the effects of failure due to shaking of the liner, backfill, or waste package.

Current estimates of the probability of new faulting near a repository vary widely. For instance, Wight (1979) suggested that the probability be taken as unity, whereas Trask (1982) suggested a small probability.

Two classes of models are available for evaluating probabilities of faulting. First, the same models used to estimate probabilities of ground shaking can be modified to predict probabilities of faulting at the site. These are called magnitude-occurrence models. Second, a probabilistic characterization of fault generation can be developed without using earthquake magnitude as an intermediate variable. These are called faulting-occurrence models.

Magnitude-Occurrence Models of Faulting

In a magnitude-occurrence model of faulting, earthquake occurrences are characterized as a marked point process in space and time; magnitude is used to characterize earthquake size. Instead of an attenuation function, an influence function characterizes the extent of faulting. The influence function takes values of zero (no faulting) or one (faulting), may be defined in three dimensions (i.e., it includes rupture depth), and may be anisotropic (i.e., rupture is likely to follow some preferred directions of faulting). The extent of the influence function depends on magnitude, but it has a random component. Kiremidjian (1984) used a model of this type to investigate the probability of damage of aqueducts by faulting. This model uses anisotropic influence functions in two dimensions and assumes that ruptures will follow the orientation of existing faults. All that is needed is the characterization of fault size (and possibly the extent of fault slip) as a function of magnitude. Kiremidjian's (1984) method can be modified to consider the three-dimensional geometry of rupture and repository and for regions where no active faults have been identified and earthquakes are hypothesized to occur on broad seismic sources. Magnitude-occurrence models have also been used by Otsuka (1964) and Logan and Berbano (1977).

In addition to the data needed to characterize the spatial and temporal distribution of earthquakes, a magnitude-occurrence model of faulting requires a probabilistic description of rupture geometry (length, width, depth, dip, and complexity) as a function of magnitude. Much observational evidence is available on rupture length, width, and depth as functions of magnitude, collected in California and other regions where seismic activity is common and earthquakes rupture the surface.

Rupture length as a function of magnitude has been investigated by Bonilla (1970, 1979, 1982), Bonilla and Buchanan (1970), Bonilla, Mark, and Lienkaemper (1984), and Slemmons (1977, 1982). Observational information on rupture width may be obtained from aftershock maps, but no such compilation of widths as a function of magnitude is known. Physical considerations provide some guidance as to width (Sibson, 1977, 1982). Information on rupture width may also be obtained from the relationship for rupture area as a function of magnitude proposed by Wyss (1979). Focal depth was studied by Sibson (1982) and by Chen and Molnar (1983). Some of these regressions predict magnitude as a function of rupture dimensions and should not be used in reverse. Instead of inverting these equations, the original data should be used to develop regressions for rupture dimensions as functions of magnitude.

Fault complexity, subsidiary faulting, and fault-zone thickness affect probabilistic calculations (in the sense that the rupture may have a non-infinitesimal thickness). Subsidiary faulting will affect calculations of ground-water flow given that faulting near the repository has occurred. These aspects of faulting have been studied by Bonilla (1979), Sharp (1979), Wallace and Morris (1979), and Speed (1979). There is no agreement on whether fault complexity on the surface is representative of fault complexity at depth. According to Wallace and Morris (1979), fault-zone thickness is related to cumulative fault displacement. No quantitative (empirical or physical) studies have been performed on predicting fault complexity as a function of magnitude and surface geology, however; a notable exception is a compilation by Wight (1979) of fault-zone thickness from a number of large earthquakes.

Most observational data mentioned above come from earthquakes that occurred on existing faults (Bonilla, 1979). Predicting rupture geometry when the earthquake ruptures intact rock or healed faults is also of interest. Rupture dimensions (inferred from aftershock or geodetic studies) from earthquakes in the eastern United States and similar regions can provide some data for developing empirical relationships for fault dimensions. Physical considerations, such as relationships between seismic moment, stress drop, and source dimensions (Brune, 1970, 1971) are also useful in this context.

The critical requirement of the magnitude-occurrence model is the probabilistic characterization of rupture geometry as a function of magnitude. For regions like California (and, perhaps, everywhere west of the Rocky Mountains), observational data and understanding of the tectonic processes are sufficient for the development of such a probabilistic model; a systematic statistical reevaluation of the empirical data is probably in order. For other regions, data are scarce, and seismological models of faulting must be used. In consequence, model uncertainty is likely to be greater.

The probabilistic characterization of earthquake occurrences used in these models is the same one used in probabilistic models of ground shaking. Thus, experience gained in seismic shaking studies during the last 15 years can be transferred to this problem. For the eastern and central United States, methodological developments for treating uncertainty and estimating recurrence parameters from historic seismicity and tectonic regime are available (Electric Power Research Institute, 1985).

Faulting-Occurrence Models

Faulting-occurrence models directly attempt to characterize the occurrence of faulting as a random process in space and time. A probabilistic model of this type has been proposed by Cranwell and Donath (1980), who modeled fault generation as a Poisson process. They suggested a mechanical model that calculates the rates of faulting from the distributions of fault orientation and strength and from the tectonic stress regime. Brooke (1978) has studied the spatial distribution of faults in selected regions.

Trask (1982) suggested that the neotectonic history of broad regions be used to estimate the rates of recurrence of fault movements in these regions. He proposed to use slip rates determined from paleoseismic studies; slip rates can be transformed into recurrence intervals using theoretical (J. G. Anderson, 1979) or empirical (Slemmons, 1977) approaches. Trask (1982) summarized information on state of stress and slip rates in seven regions and estimated the annual rate of faulting events.

Several authors (e.g., Bonilla, 1982; Zoback and Thompson, 1978) have noted that most earthquake rupturing occurs in preexisting faults or other zones of weakness. Studies that focus on existing faults should allow for the possibility of undetected faults (Wight, 1979; Logan and Berbano, 1977; Bertozzi et al., 1977; Cranwell and Donath, 1980). A few other models that are specific to particular sites illustrate alternative approaches.

Engineering Decision Analysis Company (1979) and Benjamin and Associates (1980) have estimated the probability of faulting beneath a building located between two faults. The basic assumptions of the former model are (1) faulting events constitute a Poisson process (Parzen, 1962b); (2) given that one slip event occurs, there is a probability q that the slip occurs between the existing faults instead of occurring in the currently existing faults, independently of where previous slips occurred; (3) given one occurrence of slip between the faults that currently exist, the location of that fault follows a certain probability distribution.

A Bayesian analysis (Benjamin and Cornell, 1970) was used to estimate the annual rate of slip events and q, using as data the information on historic offsets at the fault during a period of 100,000 to 200,000 years and the evidence of no faulting between the two known faults during that period.

Howland et al. (1983) used a mechanical (finite element) model of the local geologic structure to investigate the probability that the offset beneath a building exceeds a certain value. The critical parameters of the mechanical model were assumed to be random variables. The force and displacement boundary conditions driving the mechanical model were not described. In principle, these boundary conditions may be either deterministic (e.g., constant slip rates) or probabilistic.

Fault-occurrence models require information on the spatial and temporal occurrence of faulting events and on the distribution of fault size and orientation. Information on the spatial pattern and density of faulting is obtained from geologic reconnaissance studies, remote imagery, aerial photography, geophysical measurements, and microseismicity. Methods based on the surface expression of faulting are difficult to apply in the eastern United States.

Information on the yearly number of events on a fault is obtained from slip rates. Slip rates are determined using paleoseismic methods, which provide approximate dates and displacements of past slip events. In the western United States, slip rates have been estimated for a large number of faults (Trask, 1982; Wallace, 1977). In the eastern United States, there are relatively few faults on which Quaternary movement can be demonstrated and for which slip rates can be estimated (Trask, 1982; Electric Power Research Institute, 1985). There is also more uncertainty in how to convert slip rates to recurrence rates in the eastern United States. The regional approach proposed by Trask (1982) partially alleviates the problems created by the dearth of data. If all faults in a region are assumed to have identical behavior, more conclusions can be drawn from slip rates determined on a few faults.

Conceptually, fault-occurrence methods are attractive because they deal directly with the phenomenon of interest, namely faulting. Most magnitude scales measure the moderate-to-high frequency content of earthquake-induced motion, whereas the extent of faulting is better measured by the zero-frequency component of motion. As a consequence, regressions between rupture size and magnitude show large scatter (e.g., Bonilla, 1982). Thus, direct collection of data in terms of rupture size, as opposed to magnitude, should lead to greater precision in the probabilistic estimates of faulting.

The limited success of paleoseismic methods in the eastern United States restricts the applicability there of fault-recurrence methods. Also, there is considerably less experience on the application of fault-occurrence methods than on magnitude-occurrence methods for the prediction of shaking.

SUMMARY*

The quantitative assessment of seismic hazard at a site includes both probabilistic predictions and uncertainty statements. The former result from an integration of a stochastic earthquake-recurrence model and a probabilistic site-effects prediction model. Hazard uncertainty statements result from explicit propagation of parameter and model uncertainty through the hazard calculations. Methods for these analyses are straightforward and widely applied in current practice.

Current hazard assessment, which is applicable at least to the operational phase of a repository, is based almost solely on the memoryless Poissonian earthquake-recurrence model. Its data needs are relatively limited, although low-hazard, long-time-window estimates will be subject to significant, if quantifiable, uncertainty. Adequacy of this Poisson model for long-window hazard predictions has not yet been well investigated. Questions of seismicity stationarity become more dominant in the long-window case; nonstationary and stochastically changing parameter values have not been well studied or applied.

Non-Poissonian models of earthquake recurrence having characteristic magnitudes, characteristic interevent times, earthquake "cycles," stochastic dependencies between interevent times and magnitudes, and temporal and spatial memory have all been modeled. They parallel the scientific developments in the geosciences (Chapter 6). Their data needs are more intense than those of Poisson models. Practical experience is negligible.

Site-effect prediction models are separated into two major categories: ground shaking and faulting. Ground-motion prediction models probabilistically estimate important characteristics of the motion at a site some distance from an event of specified source characteristics. Current methods include direct and indirect empirical methods as well as more recent semiempirical, stochastic, and deterministic methods. This sequence of models involves an increasingly stronger physical basis, increasing instrumental data requirements, and often increasingly more parameters or independent (source and site) variables. All these estimators display large prediction variability. Current development, however, is rapid.

Site-faulting prediction models are categorized as "magnitude-occurrence" and "faulting-occurrence" models. The former are more closely analogous to ground-motion prediction models; the latter do not use earthquake magnitude as an intermediate variable. These prediction models are all probabilistic and strongly empirical.

*This manuscript was originally prepared in 1986. Since then, development of models and data referred to in this section has been substantial. See, for example, recent volumes of *Bulletin of the Seismological Society of America.*

8

Volcanology

Alexander R. McBirney

The primary basis for any estimate of volcanic hazards that could affect a nuclear-waste repository is the geologic context of the site and the scale on which forecasts of probabilities are attempted. Igneous processes can be considered on three scales: broad continent-wide regimes, regional settings of a single geologic province, and specific igneous centers. Each has its own characteristic features and must be evaluated according to different criteria. As a general rule, as the scale of investigation is restricted to smaller regions and shorter time periods, the data needed to assess future activity are better defined, and the validity of probability estimates tends to improve. The special problems associated with each scale of igneous activity will be outlined before considering the basis of probabilistic calculations that can be made for specific sites.

GEOLOGIC BACKGROUND

Broad Continent-Wide Regimes

The potential for igneous events in broad tectonic regimes is governed by global plate motion and large-scale crust-mantle relations. On this scale, methods of calculating probabilities depend almost entirely on interpretations of theoretical concepts, many of which are difficult to test by rigorous standards. The validity of these concepts must be evaluated from a fragmentary geologic record, and in many instances this involves judgments with a large element of subjectivity. Data are drawn from geophysical observations on a continental scale and from a geologic record extending back millions of years. Two approaches can be used.

Tectonic Analyses of Structural Evolution. The first approach focuses on the distribution of volcanism in a lithospheric plate for periods extending back into the Cenozoic for 15 or 20 million years. The data consist mainly of inferred plate motions, ages of volcanic episodes, compositional variations of magmas, and relations of activity to identifiable structural features on a continental scale. The approach seeks to establish patterns in the distribution of igneous centers that can be related to plate motions and intraplate stress orientations. The approach is empirical in that it uses the past record of activity and extrapolates trends on the assumption of no major changes in the tectonic regime on a continental or global scale.

For example, the temporal and spatial distribution of volcanism along the western margin of the North American continent has been explained in terms of the interaction of the North American Plate with the East Pacific Rise. As the continent has overrun the spreading axis, subduction has ceased along the coast of North America, and after a delay of a few million years, the andesitic volcanism that dominated the region for several hundred million years has been progressively extinguished. This decline of activity has proceeded southward and northward at a nearly linear rate from the time of the initial intersection of the ridge with the American plate on the coast of California. Today it has reduced the zone of andesitic volcanoes to the Cascade Range and the Mexican Volcanic Belt. The region between the southern extremity of the Cascade Range in northern California and the mouth of the Gulf of California is essentially devoid of this type of volcanism.

Dating the rate of extinction of volcanism in a setting of this kind indicates that activity may continue for as long as 10 million years after the end of subduction, but most of the decline takes place in the first 2 or 3 million years. Thus, one could conclude that a region such as central California would be free of volcanism if at least 10 million years had passed since subduction ended and that volcanism would not resume unless another major change in the tectonic regime occurred.

Analyses of Stress Distribution in Continental Plates. A second approach has been applied to intraplate regions where volcanism is related to large-scale deformation and tectonic stress patterns. Smith and Luedke (1984) studied the temporal and spatial distribution of igneous activity in the continental United States and provided a basic source of information on potential volcanism on a regional scale. They examined the distribution of volcanism for the past 16 million years and found activity in the western states to be related to linear fault systems reflecting the orientation of stresses in the North American plate. As in the first approach, structural patterns identified in the field have been dated to measure rates of propagation. Trends in time and space obtained in this way can be projected into the future, and so long as the stress patterns remain the same, location of future activity can be predicted. Smith and Luedke emphasized that the data are imprecise and that more study is needed.

Many studies of volcanism in different types of tectonic regimes have demonstrated that igneous activity seldom if ever reaches near-surface levels in regions undergoing strong compressive stresses (Williams and McBirney, 1979). Regions of intense folding and thrust faulting, such as parts of the Rocky Mountains, therefore can be ruled out as potential sites for most types of volcanism.

Potential Value for Probability Estimates. Most geologists would agree that both these methods are sound and are based on widely accepted tectonic interpretations. Although no quantitative estimates of probabilities have been attempted, large regions of the continent have been classified as tectonically active or stable (Scott, 1978; Smith and Luedke, 1984; Tilling and Bailey, 1984). The entire eastern half of the continent, for example, could be assigned a very small probability of igneous activity, because it meets the criteria provided by this type of analysis.

The possibility of volcanism cannot be ruled out entirely, however. Most intraplate volcanoes in stable continental regions are thought to be related to persistent thermal anomalies or "hotspots" in the underlying mantle. These are readily identifiable as regional features and will be considered subsequently. Other eruptions come from isolated eruptive centers that seem to break out at random. Volcanoes of this latter type have appeared infrequently in intraplate regions, such as Kansas, Missouri, and Kentucky, that are otherwise stable (Williams and McBirney, 1979). Most are small and so rare that if probabilities of such events were calculated, they would be extremely small. The effect of such events on a repository could be important, however, because even though the eruptions may be small, they tend to be extremely explosive. Lacking historic examples, how well such an eruption could be forecast by geophysical means is unknown. The rate of rise of such magmas is thought to be rapid, and premonitory phenomena may provide only a few hours, or at most a few days, advance warning.

Global Distribution of Volcanism in Geologic Time. Increasing evidence has indicated that volcanism is not distributed randomly in geologic time but tends to be concentrated in episodes that appear to be in phase on a global scale. Kennett et al. (1977) distinguished periods of brief but intense activity at intervals of roughly 5 million years since early Miocene. No clear relation has been identified between these episodes and tectonic events. Strong concentrations of volcanic activity in relatively short intervals have not been explained by current knowledge. Such an episode of unusually intense volcanism may be occurring now. Kennett and Thunell (1975) analyzed the distribution of ash layers in marine sediments to demonstrate the strongly episodic nature of the most recent volcanism in the Circum-Pacific region. They stressed the sudden increase of activity in the past 2 million years and suggested a possible causal relationship to Pleistocene climatic changes. Judging from the duration of previous Cenozoic episodes, which has been of the order of 1 or 2 million years, the current episode may be approaching an end. Johnpeer et al. (1981) also addressed the problem of estimating rates of eruptive activity in provinces where volcanism has been episodic and showed how this factor can affect probability estimates.

Any attempt to estimate rates of prehistoric volcanism must take this episodic distribution into account. Extrapolation of data on eruptive events that are compiled for a period falling between two such episodes would grossly underestimate future rates, for example.

Regional Settings of a Single Geologic Province

The potential for igneous events on a regional scale must be assessed in terms of the record of prior activity and the thermal and stress regimes in specific regions. Investigations required to assess probabilities at this scale would focus on a region within the effective radius of any igneous event that would have an impact on the specific site under investigation, but the geographic limits of the area to be examined will differ according to the type of activity being considered. Most studies have concluded that the most serious possible event on this scale would be a large-scale fissure eruption capable of blanketing thousands of square kilometers with flood lavas or pyroclastic flows (Bailey et al., 1983; Simkin et al., 1981).

Data will come from regional geologic and geophysical studies designed to identify potential centers and the types of activity that could be associated with them. Studies of this type seek to identify either of two types of conditions, *intraplate hotspots* or *dilational zones of faulting* of the kind known to have been associated with volcanism in the past. The particular setting need not, however, have a history of past activity: potential volcanism may consist either of renewed eruptions from an established system or initial outbreaks from new vents.

An example of the hotspot setting would be the Yellowstone region, which, although it has no record of historic volcanism, has produced great sheets of rhyolitic ignimbrites covering tens of thousands of square kilometers in Wyoming, Utah, and Idaho. These outpourings have come in three distinct cycles, about 2.0, 1.3, and 0.6 million years ago. The probability of a future eruption on this time scale might be estimated in two ways. First, one can date previous events and find the average interval of repose between large-scale eruptions. These are normally on the scale of thousands to millions of years, but the regularity of spacing differs widely from one place to another. Second, no body of magma of this size could be mobilized without being preceded by conspicuous geophysical phenomena. At Yellowstone, uplift of the central part of the caldera complex during the last decade has been interpreted as the effect of a large mass of siliceous magma. A similar condition has been identified in Long Valley, California. No attempt has yet been made to estimate probabilities of such premonitory phenomena leading to an igneous event, because no historical examples are known and, therefore, no records of their associated geophysical phenomena exist.

Nevertheless, quantitative estimates of the time scale of previous activity may be derived and projected forward in time and space. For example, the San Francisco volcanic center in central Arizona has been shown to be migrating in a northeasterly direction at a rate of about 25 km/10^6 yr. Predicting the general area and time in which the next outbreak is likely to appear is a simple matter. Other examples of migrating centers of volcanism are found in central California and Oregon (Scott, 1978; Smith and Luedke, 1984; Tilling and Bailey, 1984) and the Snake River Plain of Idaho (Niccum et al., 1980).

The effective radius of large siliceous eruptions from centers related to these hotspots may extend for hundreds or even thousands of kilometers, if the activity produces large pyroclastic flows. The area affected by such an eruption would be so large compared with the dimensions of the vent complex that the exact location of the outbreak would be less important than the timing of the eruptive cycles. A clear relation has been demonstrated (Figure 8.1) between the length of the period of repose of these large eruptive centers and the magnitudes of ensuing explosive eruptions. The longer the interval between eruptions, the stronger the discharge tends to be (Simkin et al., 1981; Simkin and Siebert, 1982, 1984). The strongest explosive eruptions are most likely to come from old volcanic centers that have been inactive for periods of thousands or tens of thousands of years. This relation presents a problem in evaluating potential hazards. The longer the period of repose, the less complete will be the geologic evidence of past eruptions, but at the same time, the consequences of renewed activity can be correspondingly greater. The consequences of this relationship will be considered when discussing attempts to arrive at probabilistic estimates for such eruptions.

An initial outbreak of volcanism from a hotspot with no record of previous eruptive activity is such a rare event that it is impossible to evaluate its probability. If such an eruption were to occur, it would probably be preceded by centuries of identifiable geophysical phenomena, such as increased seismic activity or heat flow. For example, a large thermal anomaly under southern

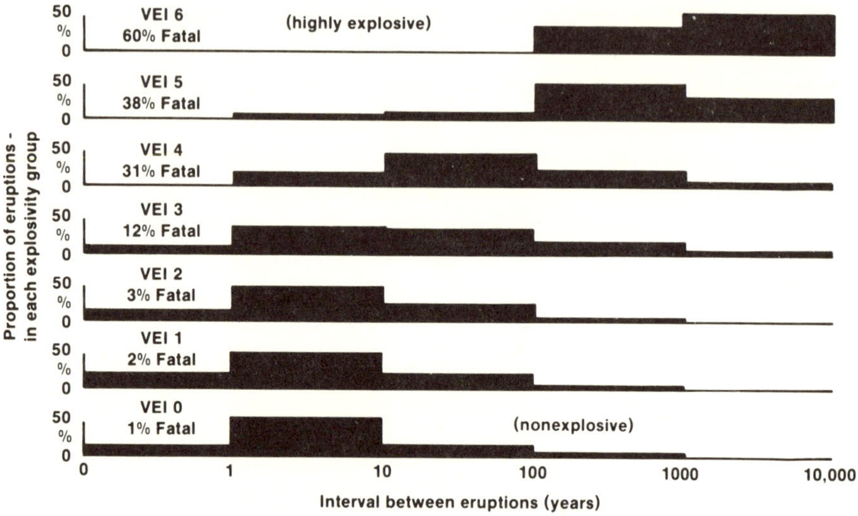

Figure 8.1. Relations between explosivity and time intervals between eruptions (Simkin et al., 1981). Events are groups in each class of "volcanic explosivity" index (VEI) according to the time interval elapsed after the last eruption. The numbers of eruptions in VEI groups 0 to 6 are, respectively, 354, 338, 2882, 617, 102, 19, and 8. For each group, a value is given for the percentage of historical eruptions that have resulted in at least one fatality. Copyright 1981 by Hutchinson Ross.

California has been identified by seismic tomography. Techniques for identifying large thermal anomalies and bodies of magma have been developed to the extent that they provide a means of surveying any region and detecting potentially dangerous conditions. The method has been described by Iyer (1984). Although Iyer dealt mainly with centers of recent volcanism, he described geophysical methods by which bodies of magma could be located in other regions of potential activity. The significance of these techniques lies in their potential for studying the crust under a particular site and detecting potentially important bodies of magma. Even though interpretation of such anomalies needs refining, the means of detecting them have been developed to the degree that their presence may safely be assumed to be known, and a site over such a feature could be avoided.

A second major type of large-scale volcanism is related to dilational faulting and continental rifting. During at least seven episodes, widely spaced between Precambrian and Miocene, huge outpourings of flood lavas have been discharged from multiple fissures to cover wide areas of North and South America, Siberia, Greenland, India, and South Africa. An example is the series of Miocene basalts of the Columbia Plateau, which cover an area of 220,000 km^2 in Oregon, Washington, and Idaho. Individual flows with volumes as great as 1500 km^3 erupted from fissures tens of kilometers in length and, in periods of a week or so, spread for distances of up to 300 km (Williams and McBirney, 1979).

Most flood lavas are associated with continental rifting and result from such large-scale tectonic events that they could not occur without many years of conspicuous premonitory phenomena. Fissure eruptions from the central rift of Iceland, though much smaller, are the only modern example that could provide an analog for the past. Icelandic geologists have developed reliable methods of forecasting eruptions of this sort and can recognize sites of coming eruptions months or even years in advance (e.g., Tarantola et al., 1982). Although a few regions, notably the Rio Grande Rift and the Salton Sea area at the head of the Gulf of California, have been interpreted as zones of recent dilational faulting, they show few signs at this time of evolving into systems capable of generating large-scale eruptions of flood basalts. The problem of evaluating probabilities of eruptions in such regions of more moderate tensional faulting is discussed in the following section.

Recent studies of the Deccan flood basalts of India have shown that their outpouring came, not as a response to crustal rifting, but from initiation of a hotspot. The ages of volcanic islands and seamounts extending northward across the western Indian Ocean have been shown to increase linearly with distance between a presently active volcano, Piton de la Fournaise on Reunion, to the Eocene centers from which flood basalts erupted on the Indian continent. After a million years or so, the scale of volcanism declined abruptly to intermittent outpourings that have generated a chain of more-or-less evenly spaced volcanoes as the lithosphere moved northward over the hotspot (Duncan et al., 1989).

Sources of Information on Regional Patterns of Volcanism. The information required for probabilistic calculations of volcanic hazards in regions with no historical record of activity comes mainly from regional geologic and geophysical surveys. Field studies are undertaken to find evidence of past activity in the stratigraphic record. With sufficient data on the ages and volcanic character of the rocks, the frequency of past activity may be established and the nature of future eruptions may be predicted. In using data obtained in this way, certain limitations that arise from the basic nature of volcanic activity must be considered. The most serious of these arise from the following well-recognized relationships.

First, the most powerful volcanic eruptions tend to leave the most widely dispersed and hence the thinnest deposits (Walker, 1982). Because they tend to be light and unconsolidated, products of strong explosive eruptions may be quickly eroded and lost. Even deposits of great historical eruptions are difficult to find only a few decades after they were laid down. Second, large eruptions tend to be much less frequent than small ones. Hence, in trying to assess the probability of a major eruption, one is dealing with a small number of widely spaced events in the history of a region. Missing even one large eruption results in a seriously distorted statistical record. Finally, the structural edifice of a large explosive volcanic center tends to have inconspicuous relief. Many are low-rimmed calderas buried in their own debris. For this reason, they are less likely to be identified than smaller structures of lesser importance. Thus the available record is not only incomplete, but also biased toward small events, such as eruptions of lava, that leave a more enduring record.

Hence the problem of assessing hazards on a regional scale is basically one of finding the sources and products of past eruptions, determining their ages, and measuring their extent and volume. Although many major eruptive centers have been identified in the western states, it is difficult to say with assurance that all are known. The eruptive history is even less certain.

Specific Igneous Centers

Because repositories are not likely to be sited in close proximity to volcanic centers with a record of historical activity, methods used to monitor active volcanoes and to forecast consequences of eruptions (e.g., Newhall, 1981, 1982) are largely irrelevant to the question of nuclear-waste disposal and need not be considered. Nevertheless, some methods of calculating probabilities could have application to regions of less frequent activity. Dubois and Cheminee (1991) have recently applied fractal analysis to the distribution of repose times for four large volcanoes with long historic records of activity. They found a clustering of events over both long and short intervals and concluded that the patterns of activity of each volcano are similar on different time scales. If this observation proves to have general validity, long-term probabilities possibly could be estimated from records of more limited duration.

The efficiency of monitoring techniques in forecasting eruptions may be an important factor in estimating probabilities, because they will determine the lead time that may be available before an igneous event could disrupt a repository. These techniques could be applied to a variety of types of regions, including those with no record of previous activity, and could be important tools in the future. For example, Adamchuk et al. (1984) described a potentially useful means of forecasting volcanism in regions with no recent activity, but the techniques are untested and their reliability in a given region cannot be assessed. Archambault et al. (1982) described modern techniques for detecting thermal anomalies associated with potential volcanic activity. Burton (1982) showed how the present state of a large but long-inactive volcanic center can be evaluated by geological and geophysical methods. Tarantola et al. (1982) provided an excellent description of the use of seismic methods and strain measurements to forecast volcanic activity, mainly in regions of rifting. Tazieff and Sabroux (1983) collected papers on all aspects of monitoring volcanoes and anticipating the nature of their activity. The main conclusion is that no major igneous event is likely to occur, even in a region with no record of recent activity, without premonitory phenomena; as geophysical techniques are further refined, the lead time provided by these methods will become longer.

Other factors that will enter probabilistic analyses on a local scale are geologic conditions (such as rock properties, ground-water levels, fracture patterns, and geothermal gradient) that can affect the manner in which magma could rise to intrude the repository or erupt close enough to the facility to affect its performance. Carr (1974) provided an example of the techniques used to determine the nature of regional stresses in the crust and their relations to past and present centers of volcanic activity. Vaniman and Crowe (1981) showed how studies of geologic structures and compositions of igneous rocks can be used to interpret the current stage of tectonic and magmatic evolution of a region. In the more refined studies, probabilities can be zoned and delineated on maps, as shown by Booth (1979) and Westercamp and Traineau (1983). The latter, in a study of a recently active volcano, outlined a method of zoning the magnitudes of different types of volcanic hazards on maps. Westercamp and Traineau also illustrated techniques that can be used to determine the prehistoric eruptive record of a large volcanic complex.

Translating local geologic and geophysical conditions into quantitative probabilistic estimates entails theoretical analyses of the factors governing the rise of magma from deep levels and is somewhat speculative. With the current level of understanding of magmatic processes, the validity of these interpretations is impossible to evaluate.

METHODS OF ESTIMATING PROBABILITIES OF VOLCANIC HAZARDS

Probabilities of an igneous event occurring within a given distance of a repository site have been calculated by various means. In most instances, the critical factor in the calculation is the frequency of past igneous events within an effective range of the site. The simplest approach is to determine the number and total time span of all igneous manifestations that can be assigned ages and to divide one by the other to obtain an average frequency. The probability can then be estimated by one of two methods that are illustrated by the following examples.

Crowe and his co-workers (Crowe, 1980, 1981; Crowe and Carr, 1980; Crowe, Amos, et al., 1982; Crowe, Johnson, and Beckman, 1982; Crowe et al., 1983) have dealt with a region in and around the Nevada Test Site where volcanism has occurred at 14 localities within the past 5 million years. Radiometric dating and paleomagnetic measurements of lavas suggested that all 14 events occurred between 0.7 and 2.5 million years ago. The probability of an igneous event affecting the site was calculated from the relation

$$P_{\text{volcanic disruption}} = R \times A \qquad (8.1)$$

where R is the rate of volcanism and A is an area ratio defined as the area of the repository or the area of an appropriate zone of volcanic disruption, whichever is larger, divided by the area for which the rate of volcanism applies. Several possible assumptions can be used in this approach. First, the rate of events obtained from dating eruption products depends on the number of sites found and the area examined. In this case, both number of events and area were small. Geologic reasoning could be used to define the area within which an igneous event was capable of affecting the site, but, because data may be incomplete, it is difficult to know whether the rate represented by the volcanic rocks found in that area is representative. A second, more serious problem involves the nature of the volcanic episode and whether the distribution of events is linked to some other factor that may not have a uniform distribution in time and space.

Smith and Luedke (1984) addressed this latter question in a study of patterns of distribution of igneous activity throughout the western United States. They concluded that the rate has been increasing for the past million years (Figure 8.2). This increase, detected from broad regional studies, is not taken into account in probability estimates derived from past rates at restricted localities.

Johnpeer et al. (1981), in another study dealing with a much larger province, assumed the probability of a future event to have a Poisson-type relation to the number of past events per unit time, and that this relation is given by an equation with the form

$$P = 1 - \exp(-\lambda t) \quad (8.2)$$

where P is the probability that at least one eruption will occur in a given time, t, and λ is the average annual rate of igneous events. The assumption of a Poisson probability relation was arbitrary and probably invalid. Probabilities of a particular disruptive event were calculated by taking a series of possible consequences of eruptions, assigning probabilities to each, and calculating the final probability as the product of the individual values for the probability of an igneous event and the probabilities of all consequences that are conditional on that event:

$$P_{E/A} = P_{E/D} \bullet P_{D/A} \quad (8.3a)$$

$$P_{D/A} = P_{D/C} \bullet P_{C/A} \quad (8.3b)$$

$$P_{C/A} = P_{C/B} \bullet P_{B/A} \quad (8.3c)$$

where $P_{E/A}$ is the probability of an event dependent on prior events D, C, and B, all of which can follow from A.

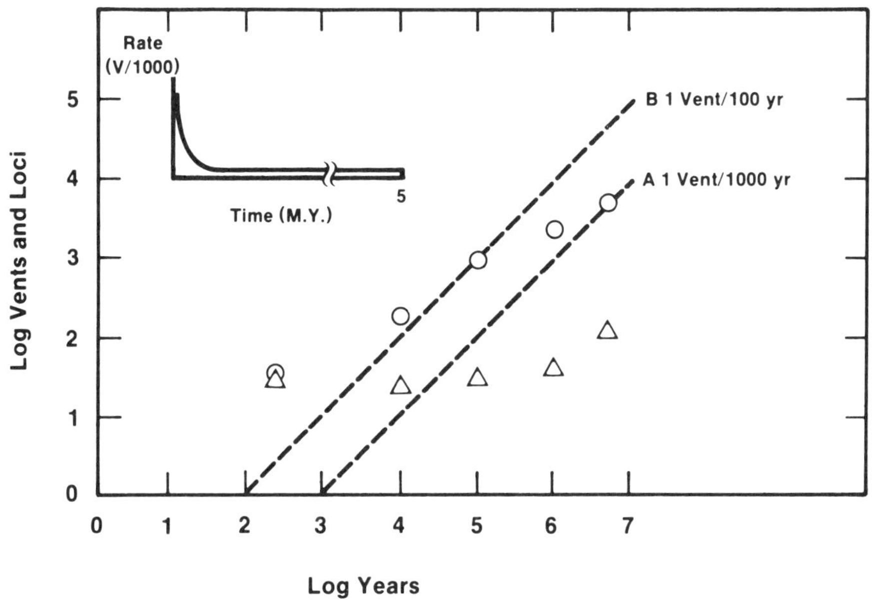

Figure 8.2. Cumulative curves showing volcanic vents and loci active during specific time periods over the past 5 million years, exclusive of the Cascade zone (Smith and Luedke, 1984). The data show an increase in the rate of vent formation within the past million years, particularly over the past 100,000 years. The circles indicate vents, the triangles loci. Copyright 1984 by the National Academy of Sciences.

When Johnpeer et al. used this method for the Hanford Site in Washington, they estimated frequencies of eruptions for three potentially active areas: the central part of the Cascade Range, the Columbia Plateau, and an adjacent part of the Basin and Range province. For example, the frequency for the Columbia Plateau was taken as one event every 2×10^6 years ($\lambda = 5 \times 10^{-7}$/yr). Substituting in Equation 8.2 gives the probability of an eruption within 10,000 years as

$$
\begin{aligned}
P_{(10{,}000\ \mathrm{yr})} &= 1 - \exp(-5 \times 10^{-7} \times 10^4) \\
&= 1 - \exp(-5 \times 10^{-3}) \\
&= 0.005 \text{ or } 0.5\%
\end{aligned}
$$

Having estimated the probability of an eruption, they estimated probabilities for each of a series of subsequent conditions, such as the nature and magnitude of the eruption, the probability that it would reach the site or dam the Columbia River, and so on. These "conditional probabilities," each of which is contingent on a preceding probability, were calculated in the same way as the initial probability.

Finally, the probability of a disruptive event affecting the site was obtained by multiplying the eruption probability by all the conditional probabilities. Probabilities of these consequences were estimated from a variety of geologic factors.

For example, the range of volumes and areal extents of possible lava flows can be estimated from previous events. Then if one of these types of eruptions were to occur, the further possibility of damming a major drainage system could be postulated. A probability for that conditional event would be needed to estimate the probability of flooding at the elevation of the repository. Petrie et al. (1981) showed how this exercise can be developed as an interactive computer code.

As Johnpeer and his co-workers (1981) recognized, the most serious deficiency of this approach lies in the temporal distribution of events used to estimate a rate of eruption. Work on well-documented volcanoes with long records of historical activity (Chester, 1986; Mulargia et al., 1986) has led to differing assessments of the assumption that events have a Poisson probability relation. Although eruptions may be random on a short time scale, they are not random on the scale of centuries or more, and the probability of a given event is not independent of previous events. On the Columbia Plateau, activity was strongly concentrated in a period of about 10^6 years. In fact, geologists working in the region estimate that at least 99% of the erupted material was discharged during a brief period 15 million years ago: radiometric age determinations of products of the main episode are indistinguishable within the precision of the analytical methods (Watkins and Baksi, 1974). In other words, 99% of the eruptive events could have occurred during a period of less than 10^6 years, and the rest were distributed at irregular intervals over the next 14 million years. The "eruption rate" obtained from these data will differ dramatically depending on the time interval used and the way in which events are distributed within that interval. Because volcanism is so episodic, both locally and globally, this problem will be encountered in almost every province and on every scale of measure. Unless one can define the temporal pattern of events and where the present is in such a cycle, probabilistic calculations have little meaning.

There may be ways of dealing with the question of episodicity. The temporal distribution of volcanic events has been studied on several scales, ranging from individual volcanoes to Earth as a whole. The treatment of individual eruptive centers is the most rigorous, because more data are available and dating of events is more precise. Even though a repository is not likely to be sited in the vicinity of a volcanic center with such a record of historical activity, the method of analysis may be useful, because it could conceivably be applied on time and distance scales relevant to a site in a region of little or no Holocene activity. As noted above, recent studies (Dubois and Cheminee, 1991) indicate that clustering of events tends to have a similar pattern on different time scales.

Wickman (1966) provided an important statistical analysis of the frequency distribution of eruptions and methods of using such data to estimate probabilities of activity after differing periods of repose. In some instances, he found a linear relationship between the log of the number of repose periods (or intervals without recorded eruptions) and the durations of these intervals (Figure 8.3). Differences in patterns from one volcano to another or even in the record of a single volcano are probably related to local geologic conditions and are poorly understood, but with sufficient data this method could be extended to large-scale episodic activity. Studies of Etna are the best example of what can be done with large numbers of reliable data (Mulargia et al., 1985, and references therein).

The number of data that can be obtained for a given region will depend, of course, on local geologic conditions. In many places, previous studies will already have revealed the general level of earlier volcanism, and little additional investigation may be needed. But if much of the geologic record is missing, owing to extensive erosion, for example, no amount of geologic work will improve the data base needed for probability estimates.

Assuming that a realistic rate could be deduced from detailed geologic and geochemical data, probabilities could be calculated for volcanic hazards at a particular locality using a variant of Booth's (1979) method for delineating zones of risk around a known eruptive center (Figure 8.4). Booth calculated probabilities using a Poisson model, but, as mentioned above, the Poisson distribution may not in all cases be appropriate.

Another type of probabilistic calculation treats the problem of magma rising through existing or newly established conduits. The rise of magma from a source in the mantle or deep crust can be analyzed in terms of the physics of flow of a viscous fluid through a colder medium. The probability of magma rising a given distance depends mainly on its viscosity and rate of heat loss to its surroundings. Several persons have analyzed the problem in detail (e.g., Wilson and Head, 1981; Marsh, 1981, 1984; Hardee, 1986). The data needed for such an analysis include measurements of thermal gradient, estimates of probable width and length of fissures, probable magma compositions, and ground-water levels and permeabilities. In general, a region with a steep geothermal gradient, particularly where the crust has been heated by previous intrusions, is more vulnerable to magmatic intrusions, because of reduced heat loss from the rising magma to the country rock. Silica-rich magmas have greater viscosity and rise more slowly than silica-poor magmas and are less likely to rise to shallow levels without being frozen. Statistical studies confirm these relations, and enough data are available to permit a sophisticated analysis of the probability that a given type of magma will reach near-surface levels. Marsh (1981, 1984) has provided graphical solutions for estimating the probability that a magma of any common composition will erupt (Figure 8.5), and Hardee (1986) has used the same approach to estimate

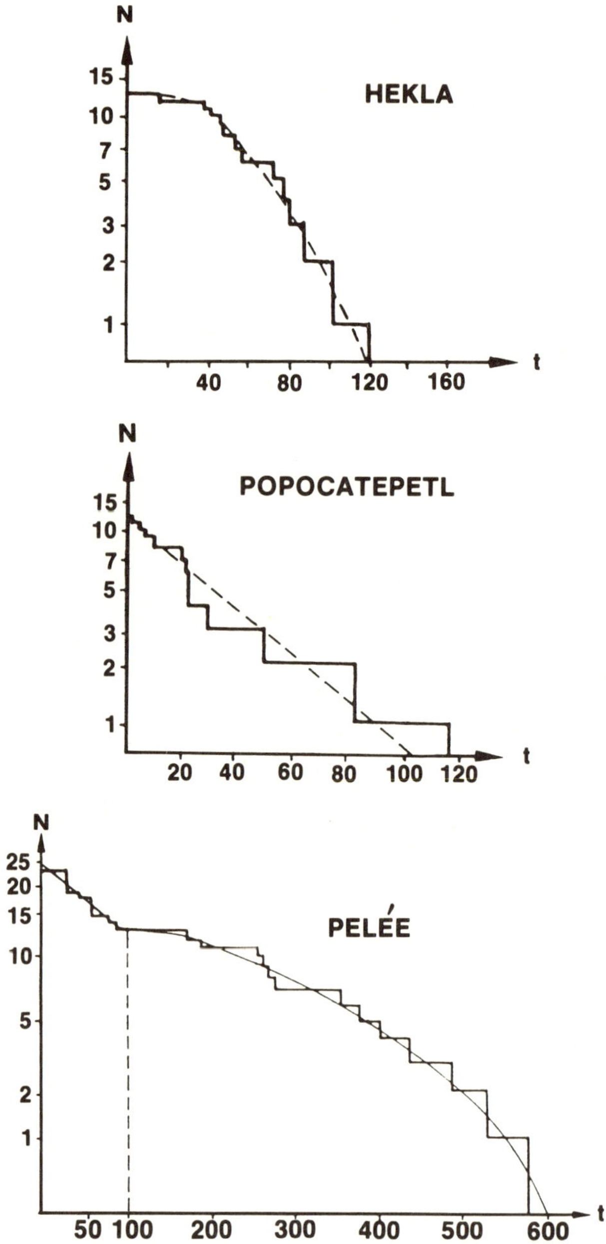

Figure 8.3. Wickman diagrams for three large volcanoes with long records of eruptive activity (Wickman, 1966; Westercamp and Traineau, 1983). N is the number of repose periods, and t is the length of the period of repose in years. The probability of an eruption after a given period of quiescence is proportional to the negative slope of the curve. In theory, a similar diagram could be constructed for any volcanic system for which there are sufficient data. Copyright 1983 by Elsevier Science Publishers.

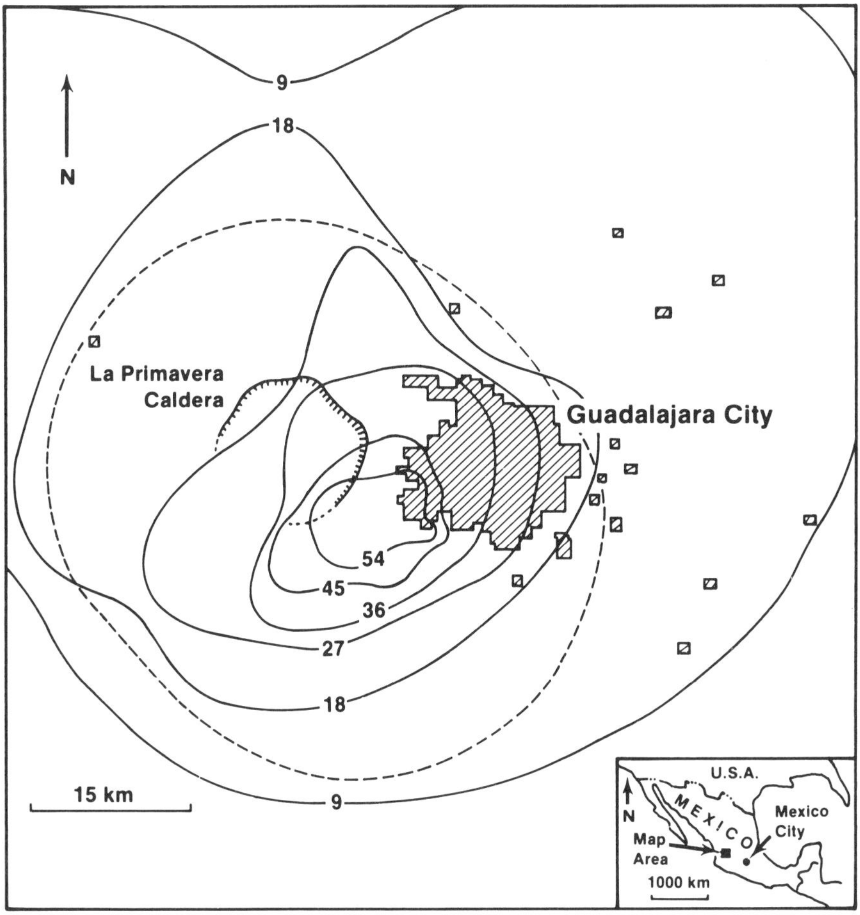

Figure 8.4. Volcanic risk map for the area around Guadalajara, Mexico (Booth, 1979). Isopleths show the percentage probability that the areas they enclose will be buried by over 1 m of ash during the next major eruption. The dashed line encloses the area of risk from lithic missiles at least 0.1 m in diameter. Copyright 1979 by the Geological Society of London.

the rate and distribution of new magmas rising into the crust in the western United States.

Possible Sources of Error in Probabilistic Analyses

Geologic factors may cause the record of events to be incomplete or biased toward certain types of activity, but other sources of error can enter a probabilistic calculation simply through mistaken analysis of the record. In some cases, mathematical equations have been applied to an assumed model without testing

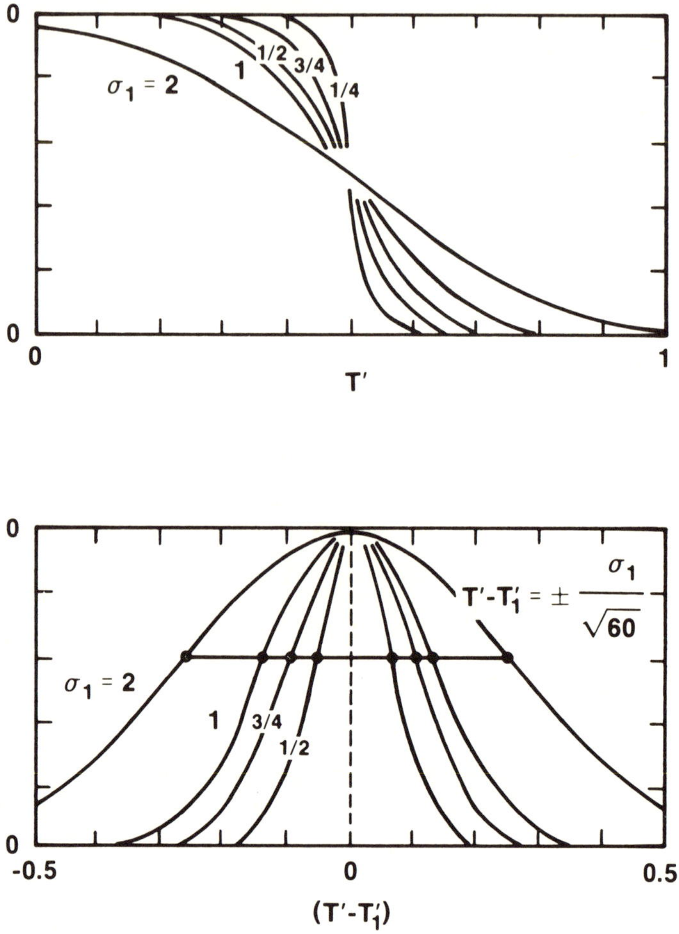

Figure 8.5. A plot of the fraction of crystallization (above) against dimensionless temperature between the liquidus and solidus can be converted to the probability, P_T, that a magma with a given crystallization curve will reach the surface before it becomes too viscous to flow (Marsh, 1981). In the lower diagram, curves of P_T are shown for a variety of such magmas. The form of the curves is defined by a function, sigma, which is essentially the slope of the crystallization curve in the upper diagram. Copyright 1981 by Springer-Verlag.

whether their use is appropriate. A perturbation analysis must be performed to determine what causes the probabilities to change and by how much.

Perhaps the most commonly cited source of error comes from assuming a simple Poisson distribution of events. In his original study, Wickman (1966) found that the probability density functions (pdf's) for the repose periods of many volcanoes fit a Poisson model, but about 20% do not. Although he showed how the model can be tested, this is rarely done. Perhaps the most rigorous recent analysis of this kind is that used by Ho (1990), who presented several ways that the Poisson model can be amended in order to treat the data of volcanoes having more complex records.

POSSIBLE INTERACTIONS BETWEEN VOLCANIC PROCESSES AND WASTE REPOSITORY

A number of workers have discussed the potentially damaging effects of volcanism. Kates (1978) dealt with methods of evaluating all types of natural hazards, but did not stress rare volcanic events. Bailey et al. (1983) outlined regions of past and possible future volcanism in the United States and measures to alleviate various types of volcanic hazards. Martin and Davis (1982) edited a symposium volume dealing with methods of anticipating and alleviating volcanic hazards. They included individual reports and accounts of round-table discussions of a great variety of hazards. Tilling and Bailey (1984) reviewed volcanic hazards in regions of active volcanism in the United States and discussed the problem of dealing with potential hazards in regions with little or no recent activity.

In addition, a few workers have attempted to predict the damage that might occur if volcanism takes place through or in the immediate vicinity of a high-level-waste repository. Crowe (1980) evaluated the disruptive effects of igneous activity on a repository and indicated ways in which these effects might be reduced by judicious siting and by adapting the geometry of the facility to the structural trends of the region. Crowe (1981) also outlined possible ways in which an igneous intrusion with or without a surface eruption could result in a release to the environment. Crowe, Amos, et al. (1982) showed how the possibility of disrupting a repository can be estimated from the relative sizes of typical feeder dikes and the area of the repository. Logan et al. (1982) calculated potential releases of radioactive waste, exposures to human beings, and health effects that might follow the intersection of a repository in southern Nevada by feeder dikes for a basaltic cinder cone.

SUMMARY

Probability estimates for igneous events capable of affecting a repository could have a wide range of reliability depending on the character of the site and the degree to which its tectonic and magmatic evolution can be projected into the future. Calculations can be precise for simple, well-documented volcanic centers with long records of historical activity, but much less precise for structurally complex regions with older and less frequent eruptions.

Analysis of these geologic relations and the degree to which a possible event would result in dispersal to the environment will entail so much expert judgment that, in the end, the considered opinion of informed specialists may be more realistic than numerical calculations based on incomplete data and uncertain projections of the tectonic and magmatic regime. This is particularly true of tectonically stable regions, such as much of the eastern United States, where no Cenozoic igneous events have been detected. An attempt to calculate probabilities of a random eruption or intrusion, such as those that have occurred at widely spaced intervals in the past, would have less value than a considered expert opinion based on a thorough analysis of the geologic setting and comparisons with similar geologic conditions elsewhere.

9

Geochemistry

Malcolm D. Siegel
Heinrich D. Holland
and Carolyn Feakes

Geochemical aspects of performance assessment include descriptions of the chemical states of the various subsystems of a nuclear-waste repository and descriptions of processes leading to changes in the values of state variables. This chapter discusses the use of compartment or box models to describe geochemical systems in probabilistic risk assessments. These models divide geologic systems into reservoirs linked by processes of mass transfer. Box models can be used to describe phenomena on both macroscopic and microscopic scales and are compatible with the probabilistic methods described in Chapter 1 of this volume. The use of probabilistic methods to estimate properties of the reservoirs is described. The box models currently used in probabilistic risk assessments require simplified representations of geochemical processes. The potential that the failure to include important interactions in the probabilistic risk assessments may produce misleading results is also discussed.

Probabilistic aspects of processes important for the disposal of nuclear waste can be divided into two broad types of problems. One involves the probability of occurrence of discrete events; here catastrophic events will dominate the calculated risk and, to a first approximation, probabilistic aspects of geochemical processes probably can be ignored. The second type of problem involves probabilistic and statistical aspects of continuous random variables (J. C. Davis,

1973). Most geochemical parameters have a large number of possible values. The range of possible values may be narrow, but within that range, specifying or predicting the exact value or chemical state of the system may be difficult. The focus in this chapter will be to summarize the methods that can be used to provide the probability density functions (pdf's) associated with the parameter uncertainty for use in probabilistic risk assessment calculations.

Techniques used to describe the pdf's of geochemical variables can be included in the categories delineated in Chapter 1. Mechanistic models based on theoretical frameworks can be used alone in axiomatic probabilistic methods or in conjunction with Monte Carlo-type modeling approaches. When an understanding of important physicochemical mechanisms is not available, empirical observations often are processed using frequentist approaches. Geochemical studies can be carried out by deterministic or stochastic methods (see Chapter 4, this volume) depending on the spatial and temporal scales of interest. For example, the behavior of chemicals in the atmosphere and ocean on scales of miles and weeks are best treated stochastically; long-term behavior of chemicals in large systems, however, can be treated deterministically if input and output functions describing mass transfer are known adequately.

BOX MODELS FOR GEOCHEMICAL SYSTEMS

Box or compartment models have been used with considerable success to describe the present state of oceans and ocean-atmosphere systems (Holland, 1978). These models also have been used to "retrodict" the operations of these systems—to define, for example, the composition of Earth's atmosphere and circulation of the oceans during the height of the last ice age. In risk assessment, these models may be used to predict the future state of geologic systems.

Predictions of geochemical systems require a knowledge of the source function of the disturbance and the response functions of the system. Reliability of geochemical forecasts depends on the precision with which the source function is known and on understanding the system into which the disturbance is propagated. Several possibilities for arrangements of reservoirs and material transfers between them exist (Figure 9.1). The time over which predictions can be made confidently depends on such factors as transfer rate between reservoirs and mixing and residence times within reservoirs. Atmospheric and surface-water geochemical systems are discussed below first; then the same principles are applied to more complex geochemical systems associated with nuclear-waste repositories.

Atmosphere and Surface Waters

In some instances, predictions can be made with a great deal of certainty for a large part of Earth and for long periods of time. Helium is released into the atmosphere from the lithosphere, but escapes slowly from Earth's atmosphere into space. The residence time of He in the atmosphere (10^6 years) is short compared with the age of Earth; therefore the He content of the present-day atmosphere depends on the balance between the rate of supply of He from the lithosphere to the atmosphere during the past several million years and the ease with which He has been able to leak from the atmosphere into space. The system describing its behavior (Figure 9.1b) is easy to model. Major changes in the He content of the atmosphere can take place on a time scale of 10^6 years. Predictions of the He content of the atmosphere on a million-year time scale are somewhat uncertain; but near-constancy of the He content of the atmosphere on a time scale of 10^3 to 10^4 years can be predicted.

Predicting environmental effects of fossil-fuel burning is much less certain and illustrates some of the difficulties inherent in making most geochemical predictions. Products of fossil-fuel burning can be divided into two groups: those that have a residence time in the atmosphere considerably shorter and those that have a residence time considerably longer than the mixing time of the atmosphere. Among the former, SO_2 and nitrogen oxides are the most important; CO_2 is the most important product in the second group. The behavior of SO_2 and NO_X is distinct from that of CO_2 in several ways. Because SO_2 and NO_2 are oxidized and washed out of the atmosphere in a time that is short compared with the mixing time of the atmosphere, their deposition after injection at a particular point tends to be local or regional, rather than global. The removal of these gases from the atmosphere depends on local meteorologic conditions during the few days after injection of a given batch of SO_2 and NO_X. Because most power plants operate for many years, behavior of the contaminant plume can be observed under a variety of meteorologic and climatic conditions, and the intensity of H_2SO_4 and HNO_3 deposition can be predicted with considerable certainty on a weekly, monthly, or yearly basis (e.g., Wilson et al., 1980). This makes it possible to predict acid rain deposition with a fair degree of precision as a function of the intensity of the source and the meteorology of the region.

Assessing the effects of CO_2 injection by fossil-fuel burning differs fundamentally from the problems outlined above. About half the CO_2 now injected into the atmosphere is accumulating; the CO_2 content of the atmosphere has been rising steadily since the first careful measurements were made in the late 1950s (e.g., Sundquist, 1985). The residence time of CO_2 in the atmosphere is clearly greater than the mixing time of the atmosphere, making the CO_2 content reasonably constant over the entire globe; the mean CO_2 content of the atmosphere reflects the rate of global input by fossil-fuel burning and the rate of global removal of CO_2 from the atmosphere by various mechanisms. Predicting atmospheric CO_2 therefore depends only slightly on the source term at a given time and place and is essentially independent of local atmospheric conditions. The course of pCO_2 can only be predicted with assurance if the worldwide rate of

CO_2 injection is known for the next few hundred years and if important removal mechanisms are understood quantitatively and worldwide.

Although these examples deal with the atmosphere as the dispersive medium, similar principles apply to geochemical predictions of the fate of pollutants in rivers, lakes, and aquifers. Rivers, lakes, and aquifers are self-cleaning: water flows through them, and a polluted section will be cleansed by the flow of fresh water. If a source of pollution is removed from a river, its consequences will disappear on a time scale determined by the flow of clean water and of sediments within the river (Figures 9.1c, 9.1d, and 9.1e).

Predicting the fate of pollutants in surface waters is complicated by the complexity of interactions among pollutants, dissolved constituents of the water, sediments, and biota. In lakes, metals such as copper, lead, and zinc become adsorbed on the surface of clay minerals (e.g., Foerstner et al., 1979, Chapter E). This speeds their removal from the water column, but concentrates them within sediments and can speed their entry into the biosphere. Living organisms can concentrate metals in the upper parts of the food chain. Even with a full complement of chemical and biological data, precise geochemical predictions may be impossible, because the rates of material transfer processes can vary widely over short periods. For example, in many river systems, most sediment erosion takes place during a small fraction of their history during periods of flooding. If the time interval between such floods is longer than the period of time during which a spill of toxic materials affects a river system, the damage due to such spills may be strongly affected by their timing with respect to major floods. In this sense, probability of discrete events enters geochemical predictions of the pollution of rivers and lakes.

Applications of Box Models to Transport of Radionuclides by Ground Water

Transport by ground water is considered to be the most likely mechanism for moving radionuclides from a repository to the accessible environment. Systems of reservoirs and material-transfer processes (Figure 9.1) can be used to describe geochemical processes on several temporal and spatial scales in ways that can be used in probabilistic risk assessments of waste repositories. For example, the long-term, large-scale behavior of the repository and accessible environment might consider Reservoir 1 (Figure 9.1b), the source, to correspond to the engineered facility; Reservoir 2 to correspond to the dispersive media (i.e., controlled zone); and Reservoir 3 to represent the accessible environment. The rate of material transfer between Reservoirs 1 and 2 represents the radionuclide source term. The rate of radionuclide transfer from Reservoir 2 to Reservoir 3 depends on the ground-water flow rate and the intensity of the interactions between the radionuclides and the dispersive media. Additional reservoirs

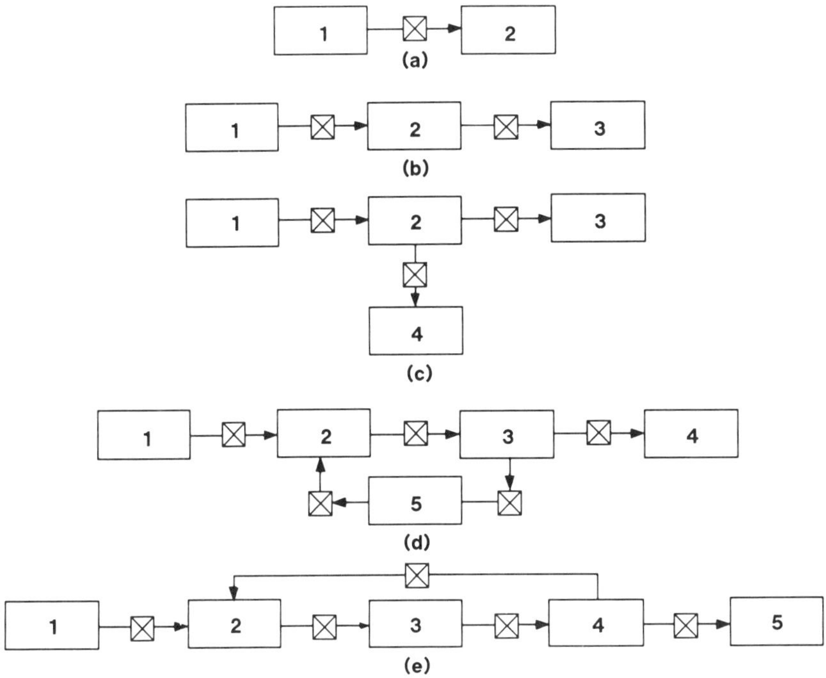

Figure 9.1. Some simple arrangements of reservoirs and material transfer among them (Holland, 1978). Copyright 1978 by John Wiley & Sons.

(Figures 9.1c through 9.1e) could describe transport of radionuclides from the repository (Reservoir 1), through the disturbed and undisturbed zones (Reservoir 2), to aquifers and rivers in the accessible environment, and finally into the food chain.

Geochemical and related physical processes occurring along the path of solute transport in the thermally undisturbed zone of the repository could be described using similar models (Figures 9.1c and 9.1d). Reservoir 2 could represent water along the flow path; radionuclides could be transferred from the fluid to permanent sinks (Reservoir 4, Figure 9.1c), or they could take part in various material transfers that occur serially or in parallel with their release to the accessible environment. On the microscopic scale, radionuclides in the ground water could be considered as one reservoir and radionuclides bound to the surrounding rocks as a second reservoir. Chemical processes such as ion exchange or precipitation and dissolution would represent mass transfer mechanisms. These two examples are related: the mass-transfer process in the large-scale model is broken down in turn into its own system of reservoirs and material-transfer processes when modeled on the microscopic scale. The number of reservoirs depends on the nature of the transmissive medium. Fractured-porous media can be represented by three reservoirs: a fracture fluid reservoir, a pore fluid reservoir, and a rock matrix reservoir. Consideration of the interaction between the fracture fluids and the fracture-lining minerals would require inclusion of an additional reservoir. The deterministic transport models described in Chapter 4 of this volume can thereby be viewed within this structure of geochemical box models.

The complexity of a model chosen to describe a system depends on the types and quality of data available to the modeler and the intended use of the analysis. Data describing the chemical speciation and reaction mechanisms of radionuclides are lacking for most environmental systems of concern. Therefore, in transport models, complex mass transfer mechanisms are lumped together into a single constant describing the distribution of the solute between the aqueous and solid phases (cf. Chapter 4, this volume). When sorption is the dominant chemical interaction, a distribution constant (K_d) is used to describe the intensity of the solute-rock interaction as

$$K_d = \frac{\text{moles radionuclide/g sorbent rock}}{\text{moles radionuclide/ml water}}$$

Retardation factors describe the capacity of the rock matrix or fracture lining to adsorb the radionuclides. They depend on the solution/solid ratio of the aquifer and take the general form

$$R = 1 + \frac{\text{moles radionuclide/gm or cm}^2\text{ sorbent rock in unit volume}}{\text{moles radionuclide/ml water in unit volume}}$$

The mathematical form of the retardation factor depends on the nature of the transmissive media: porous matrix, fractured impermeable rock, or fractured porous matrix (Neretnieks and Rasmuson, 1983; Erickson et al., 1986). The retardation factor is also used to calculate the apparent diffusion coefficient (D_a) and advective velocity (v_r) of the solute from the diffusion coefficient (D_e) and advective velocity (v_f), respectively, of a nonsorbing (conservative) tracer:

$$R = \frac{v_f}{v_r} = \frac{D_e}{D_a}$$

The use of retardation factors in solute transport models has been widely criticized in recent years. The nature of these criticisms and the implications for the inclusion of geochemical phenomena in a probabilistic risk assessment model is discussed in a later section.

Probabilistic Applications

Geochemical box models have been used in probabilistic analyses of the transport of radionuclides in both surface and ground waters. The techniques used to model transport of radionuclides in aquifers include codes such as SWIFT (Reeves and Cranwell, 1981) and NWFT/DVM (Campbell et al., 1981). These and other similar codes, which are well suited for use in probabilistic analyses, are described in Chapter 4 of this volume and will not be reviewed here. Probabilistic analyses of transport through large systems containing surface waters, aquifers, and other subsystems are discussed below.

The Environmental Transport Model (Campbell et al., 1978; Helton and Kaestner, 1981; Helton and Iman, 1982) was designed to be used in Monte Carlo simulations in conjunction with Latin hypercube sampling (Iman et al., 1980). Statistical techniques such as stepwise regression are used to identify variables that strongly influence radionuclide concentrations in complex geochemical and ecological systems.

The Environmental Transport Model represents radionuclide movement by a box model. Radionuclides in an area under consideration are divided into a number of compartments, and radionuclide movement between compartments is represented by a system of linear differential equations. Radionuclides that are in physical regions with different characteristics (e.g., hydrological, climatic, chemical) are placed in different compartments. The distribution that results from various possible radionuclide movements between compartments is determined. Such movements could result from physical transport of a radionuclide from one geographic zone to another or from radioactive decay. Each zone is divided into nonintersecting subzones: (1) ground water, (2) soil, (3) surface water, and (4) sediment (Figure 9.2). Each subzone is assumed to have two phases: (1) liquid (water) and (2) solid (a material that sorbs radionuclides). Representing radionuclide decay chains requires multiple compartments for different radionuclides and is considerably more complicated. Solving the system of differential equations that represents movement of radionuclides between compartments yields the radionuclide distribution in the area under consideration.

Use of box models like the Environmental Transport Model requires description of the compartments and rates of radionuclide movement between compartments. In solute transport codes such as SWIFT, rates of radionuclide movement by both advection and diffusion are controlled by geochemical interactions as described by the retardation factor. The most important geochemical variables for probabilistic analyses using current radionuclide transport models are radioelement solubilities and sorption distribution coefficients (K_d) (Siegel et al., 1983, 1989; Cranwell et al., 1987). In Monte

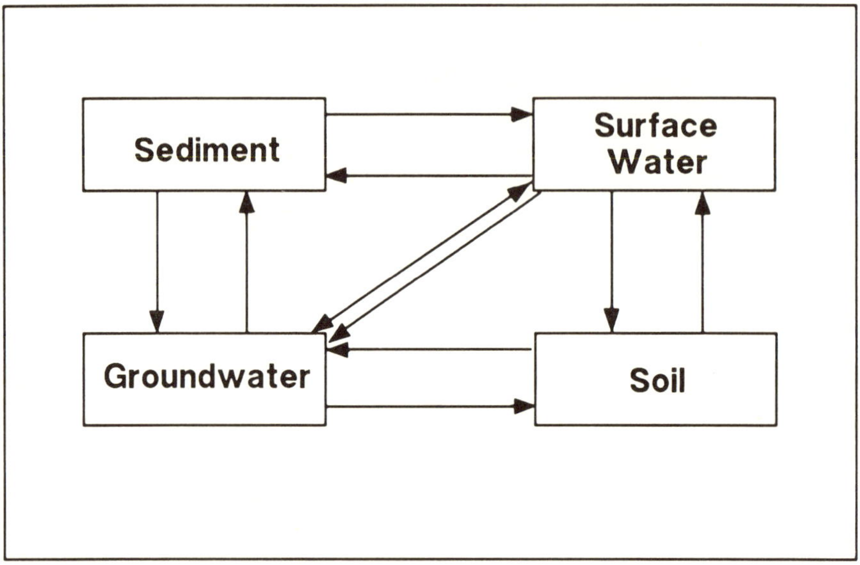

Figure 9.2. Division of a zone into subzones (Helton and Kaestner, 1981). Arrows represent potential directions of movement for water and solid material between the subzones of a zone. Radionuclide movement should follow the same pattern, because these flows dominate radionuclide transport.

Carlo simulations, these characteristics are specified by probability density functions.

In contaminated ground waters, the concentrations of many important radionuclides in nuclear wastes quickly reach maxima, whose values are determined by the solubility of either a metastable solid or the least soluble compound of the radionuclides in the affected ground waters. The maxima and minima for pdf's of solubilities of radionuclides could correspond to the metastable and stable phase solubilities. These values may differ by several orders of magnitude (Schramke et al., 1987); log uniform distribution shapes have been assigned to the solubility pdf's in some risk assessment calculations (Siegel et al., 1983; Cranwell et al., 1987) accordingly.

Because of the diversity of solutions, minerals, and radionuclides that will be present at potential repository sites, a large body of empirical radionuclide sorption data has been generated. These data are at best valid for specific experimental conditions and cannot be extrapolated to other conditions. Methods used to specify pdf's for K_ds for use with sampling techniques such as Latin hypercube sampling have been described by Siegel et al. (1983, 1989) and Tien et al. (1985). Data bases of K_d values that can be used to estimate pdf's for various geologic media were summarized by Barney (1981a, 1981b), Tien et al. (1985),

Serne and Muller (1987), Siegel and Leigh (1989), Bayley et al. (1990), and DOE (1988).

Attempts have been made to find functional relations between experimental variables and the measured sorption coefficients (K_ds). Several of these studies were summarized by Serne and Muller (1987) and are useful in constructing pdf's for K_ds. Mucciardi and Orr (1977) and Mucciardi (1978) used both linear (polynomial regression of first-order independent variables) and nonlinear (multinomial quadratic functions of paired independent variables, termed the Adaptive Learning Network, or ALN) techniques to examine effects of several variables on sorption coefficients. The dependent variables considered included cation exchange capacity (CEC) and surface area (SA) of the solid substrate and solution variables, Na, Ca, Cl, HCO_3, time, pH, and Eh. Mucciardi and Orr (1977) showed that the ALN technique provided consistently better prediction (better correlation and smaller absolute error) of K_ds than the linear regression technique. Techniques such as these allow the modeler to construct a narrow probability density function for K_ds.

Field and laboratory observations of rock and water compositions are used to estimate properties of reservoirs, but invariably data are collected from a relatively small number of erratically spaced locations. The next section discusses the use of statistical techniques such as factor analysis and kriging to estimate important rock and water properties.

ESTIMATING PROPERTIES OF THE DISPERSIVE MEDIUM

Dispersal of radionuclides from repositories depends strongly on the physical and chemical properties of rocks surrounding repositories. These properties will control the efficacy of a number of geochemical processes that can retard the migration of radionuclides to the accessible environment. For example, the presence of clays and zeolites along flow paths may increase the likelihood that radionuclides will be retarded by sorption and ion exchange. The existence of reducing zones may lead to the removal of radionuclides from ground waters by electrochemical reduction. The construction of pdf's of geochemical parameters for use in probabilistic risk assessment requires estimates of the spatial variability of these properties.

A number of multivariate statistical techniques have been used to describe the compositions of rocks and waters in areas comparable in size to the controlled zones of repositories and to estimate probability density functions for variables used in Monte Carlo simulations. Geostatistical techniques could be used to estimate some of the important properties of rocks within a proposed repository. Several multivariate techniques, e.g., multiple analysis of variance, factor analysis, and discriminant analysis, have been used to determine whether ground waters in different parts of basins are connected by flow paths.

Geostatistics

A number of geostatistical methods have been developed by industry for estimating ore grades and ore reserves in mineral deposits (e.g., Brooker, 1975a, 1975b; Journel, 1973). These techniques can be applied directly to the problem of defining the physical and chemical properties of rock units within and surrounding repositories. Commonly used techniques include inverse distance weighting, the method of polygons, trend surface analysis, and kriging.

Inverse distance weighting and similar methods, such as inverse distance squared weighting, allocate a weight to a sample point according to the inverse of the sample distance from the point at which a particular property is to be estimated. Estimates based on inverse distance weighting are often more accurate than unweighted estimates and can be as accurate as kriging when a parameter of interest is continuous. When the distribution of a parameter is erratic, the inverse distance method quickly loses reliability, and kriging is the preferred method of analysis (Brooker, 1980). Inverse distance weighting methods alone do not estimate the error associated with the estimate or take into account sample location, and they assume spatial isotropy of property variation.

In the method of polygons, the area of interest is divided into polygons. Boundaries of the polygons are perpendicular bisectors of lines between all points in the sample set. Each polygon is assigned a value of the parameter of interest equal to that of the sample upon which it is centered. This method frequently overestimates or underestimates the value of an area (Royle, 1980) and provides no estimate of variance.

Trend surface analysis is a third, commonly used technique (Krumbein, 1959; Watson, 1972). This method distinguishes systematic regional changes from variations due to local effects by calculating a moving average. The method assumes no spatial relationships in the residual local variation.

A fourth and probably most useful technique for modeling geologic systems involves statistical approaches such as kriging that accurately assess spatial variations in geologic units and allow confidence limits to be placed on estimates of spatial variability (Matheron, 1963; David, 1977; Whitten, 1977; Journel, 1980). Geostatistical theory is based on the concept that properties of rocks at points close together will be more similar than those of rocks that are a greater distance apart. The most commonly used techniques involve four steps (Journel, 1989): (1) collect data within an area deemed homogeneous enough to warrant statistical averaging within it; (2) calculate a measure of the spatial variability at units of distance h between sample locations (i.e., the experimental semi-variogram); (3) smooth and complete the partial experimental image thus obtained into a model semivariogram; and (4) use the model with regression techniques such as kriging to estimate the variation of the geologic parameter within the area of interest. Semivariograms are graphs showing the change in the variance of a measured variable at successively increasing sample spacings (Whitten, 1977). Once an accurate semivariogram has been developed, kriging can be used to produce a linear, unbiased estimate of a property at an unknown point or within an area or volume. A key result of kriging is that the method keeps the variance estimate to a minimum. Strictly speaking, however, the kriging variance depends on the sample locations and not on the sample values.

The applicability of geostatistical techniques to estimating variability of physical and chemical parameters within rock volumes that are being considered for repositories tends to be limited by the availability of data for the construction of semivariograms. The number of data points required to characterize a semivariogram accurately depends to some extent on the complexity of the geologic setting. One measure of this complexity, the "nugget effect," is expressed as the ratio of the local (random) property variance to the variance of the property over larger spatial distances. A deposit that has an underlying linear semivariogram without an appreciable nugget effect can be characterized adequately by relatively few data. Conversely, if the current data suggest that a mixture of spherical models is required to describe the experimental semivariogram, then more data on a finer spatial scale may be needed to resolve the complex structure of the parameter of interest.

Clark (1980) and DeMarsily (1986) have discussed the effects of different-sized data sets on semivariograms. A rule of thumb is that 25 to 50 sample pairs per semivariogram interval is an appropriate target range for number of data used in preparing semivariograms. In conventional geostatistical calculations (usually involving ore-reserve estimates), the data set often consists of measurements down 50 or more regularly spaced boreholes. Borehole data for proposed repository sites are usually much sparser than this and tend to be distributed erratically. When the number of boreholes is considerably greater than 3 but considerably less than 50, geostatistical techniques may be useful and should be tried.

Unfortunately, published work on developing reliable semivariograms from such limited data sets is sparse (Clark, 1979). In one of the few available studies, Brooker (1977) addressed the consequences of using kriging techniques with few data points. He monitored the reliability and robustness of experimental semivariograms constructed between the early stages of a drilling program (33 sample pairs) and its final stages (595 sample pairs). When only 33 pairs were available, the semivariogram could be fitted to a spherical model with sill and range values close to the semivariogram developed with 595 pairs. However, the semivariogram based on only 33 sample pairs indicated that there was no intrinsic random variance (nugget effect) in the data set. This in turn produced a misleadingly small kriging variance and suggested that more than the warranted confidence had been placed on the kriging estimate. An increase in the number of sample pairs to 92 improved the semivariogram significantly. Thereafter, the model was improved only slightly by adding an additional 400 sample pairs. Brooker (1980) suggested that when few data are available, important information concerning the magnitude of the nugget effect can be gained by drilling several closely spaced holes on a cross pattern. The variogram cross is a useful way to get a scatter of intersample spacings; however, the additional drilling costs may not always be justified by the improved resolution of the semivariogram.

Application of Geostatistical Techniques to Repositories

The application of geostatistical methods to characterize the variability of repositories must be considered on a case-by-case basis. None of the proposed repository sites is particularly well characterized. However, as demonstrated by Brooker (1977), limited data sets may contain useful information, particularly if geologic reasoning is used in conjunction with statistical modeling. Semivariograms could provide insights into the geologic variability of sites being considered for disposal of nuclear wastes. Mineralogical and hydrological sampling programs proposed for the high-level repository at Yucca Mountain, Nevada, will be guided by geostatistical techniques (U.S. DOE, 1988).

When few data are available, a semivariogram could be subject to several different interpretations. Areal (or lateral) variograms should be constructed between boreholes to assess the lateral variability of geologic units. The parameters of interest might include the thickness, porosity, and sorptive mineralogy of the units. "Down-hole" semivariograms can be used to assess vertical variability within single boreholes and can be compared with vertical semivariograms from adjacent and distant boreholes. The presence of anisotropy in formation parameters can be detected and measured by constructing separate semivariograms for sample pairs with different orientations (east-west, north-south, etc.) (Huijbregts and Matheron, 1977; David, 1977).

If samples chosen for computing a semivariogram are averages of values for a length of core, the overall variability expressed in the semivariogram will be less than if individual, unaveraged analyses are used. Effects of averaging can be removed by auxiliary functions, which determine the semivariogram sill for point samples (Clark, 1979). When the data set of sample points does not follow a normal or lognormal distribution, estimating a robust semivariogram becomes difficult. Cressie and Hawkins (1980) have evaluated several geomathematical techniques to solve such problems and have demonstrated several methods that produce robust variograms.

In some cases, "soft" data may be used to supplement measured property values ("hard" data). Such soft information may include semiquantitative data characterized by an inequality constraint (e.g., porosity greater than 5%) or a form of a prior probability distribution (e.g., a geologic inference that a particular rock type prevails in a stratigraphic interval where no core was recovered). Journel (1986, 1989) described how this type of information can be used in the soft kriging approach. Techniques also exist for using nonparametric data (i.e., rank order) to estimate local and global spatial distributions (Journel, 1983).

Various kriging methods can be used once an experimental semivariogram has been constructed and modeled. Journel (1989) and Isaaks and Srivastava (1989) described many of the available methods and provided references. Techniques such as multivariate co-kriging and random kriging address specific geologic situations that are frequently encountered. Co-kriging allows two variables to be used to increase the accuracy of estimates and to reduce estimation variance. Random kriging has been developed for use where sample locations are nonuniformly distributed (Marechal and Serra, 1970). The accuracy of these kriging techniques compared with that of other estimation procedures can be checked by using "jackknife" or "boomerang" tests (Chung, 1984). In these tests a known value is dropped from the sample set, and its value is estimated by several procedures. The difference between actual and estimated values is a measure of the accuracy of the procedures that were used in arriving at the estimated values (Sandefur and Grant, 1980).

For some data sets, the choice of kriging method may not be as important as the assumptions about the homogeneity of the variance and covariance of geologic parameters in the region of interest. The kriging system can be written with or without a nonbias condition, depending on whether the sample probability density functions are assumed to be independent of sample location (i.e., stationary). Boufassa and Armstong (1989) compared six geostatistical estimators (linear kriging, lognormal kriging, and disjunctive kriging, each with and without a nonbias condition) using data from a polymetallic deposit. They found that the differences between the estimators with and without the nonbias conditions were greater than between the different kriging methods. This underscores the importance of collecting enough data during repository-site characterization to support the choice of the appropriate stationarity model.

The geostatistical techniques outlined above may not be useful for estimating all important properties of proposed repositories. The frequency distribution of open fractures of sufficient width to permit free flow of solutions from the repository would be difficult to estimate confidently using currently available geostatistical methods. A small number of these fractures might be sufficient to affect the safety of a repository.

Multivariate Statistical Techniques

Multivariate statistical techniques such as factor analysis, multivariate analysis of variance, and discriminant analysis are used to find simple structures within large complex data sets characteristic of regional hydrochemical studies. Excellent discussions of factor analysis have been given by J. C. Davis (1973) and Klovan (1975). When the objective is to find correlations among a large number of variables, R-mode factor analysis is used; when similarities among samples are examined, Q-mode factor analysis is appropriate. Factor analysis can be used in studies of water-quality data to identify individual water masses within a closed basin and to recognize correlations among concentrations of solutes and hydrogeologic properties. According to Drever (1982), R-mode factor analysis is the most widely used statistical technique in geochemistry, particularly where no relationships among variables can be preassumed. R-mode analysis reduces a large number of correlated variables to a smaller number of factors that will have high communalities (i.e., will account for the variance of original data set fairly completely) and whose composition (or factor loadings) are interpretable in terms of identifiable sources or processes. Examples of sources may be dissolution of evaporites (which may produce factors with high loadings for Na, Cl, and SO_4), dissolution of dolomite aquifers (factors with high loadings of Mg and Ca), and dissolution of igneous (granitic) rocks (high loadings for SiO_2 and K).

Factor analysis can be a first step in modeling the history and future evolution of a hydrochemical system. Results of factor analysis can suggest the identities of mechanisms that have controlled ground-water chemistry in the past and might control it in the future. If only a few interpretable factors account for most of the variance of the water-quality and hydrogeologic data, estimates of water composition at unsampled locations within the basin might be made accurately, and it may be possible to predict future water quality in the area. If no simple structure can be found in the data, water chemistry is probably unpredictable at new locations at this time and at all locations in the future.

Factor analysis has been used to explain the origin of natural waters. R- and Q-mode factor analysis gave insight into the nature of the chemical and physical controls that were exerted on formation waters in the West Canada Sedimentary Basin following an initial period of compaction (Hitchon et al., 1971). Reeder et al. (1972) investigated factors controlling water quality in surface waters of the MacKenzie River Basin, Canada, with a combination of Q- and R-mode factor analysis. Drever (1982) gave examples of the uses of factor analysis in several kinds of hydrochemical systems.

Applications to Repositories

Unpublished studies (Lehman and Quinn, 1982) for the high-level waste repository site originally proposed for Hanford, Washington, used R-mode factor analysis of water-chemistry data to delineate local- and regional-scale flow patterns. They interpreted the factors in terms of possible mechanisms that control water chemistry: sedimentary origin, reactions with basalt, and contamination from irrigation waters.

Siegel, Robinson, and Myers (1990) used R-mode factor analysis to describe solute relationships in the vicinity of the Waste Isolation Pilot Plant in southeastern New Mexico. They identified three major factors that could be related to probable chemical reactions controlling the ground-water chemistry. The first factor was dominated by Na, K, Mg, Br, and Cl, and the second factor included Ca, bicarbonate alkalinity, and Sr. These factors were interpreted to represent dissolution of halite and associated carbonates and sulfates. The third factor contained two groups of correlated elements: Mg, bicarbonate alkalinity, and SiO_2 form one group; pH, B, and Li form another group that is inversely correlated to the first group. This pattern of element association was tentatively interpreted to reflect ion exchange, silicate hydrolysis, and carbonate diagenesis.

Several other multivariate statistical and graphical techniques have been used with chemical analyses of basalt ground waters to test the hypothesis (U.S. DOE, 1984) that vertical flow of ground waters between various basalt flows in the Pasco Basin is minor compared with lateral flow. Williams and Associates (1983) used the techniques of stepwise multiple analysis of variance, canonical analysis, and discriminant analysis to show that compositions of waters from vertically adjacent basalt units at Hanford were statistically distinct.

COMPLEXITY OF GEOCHEMICAL REACTIONS AND THE VALIDITY OF PROBABILISTIC ANALYSES

The previous sections have summarized modeling approaches amenable to Monte Carlo simulations (box models) and parameter estimation techniques (e.g., kriging) required to construct probability density functions for input to probabilistic risk assessments. The capabilities of modern computers place limitations on the complexity of models and number of iterative computations that can be used in Monte Carlo simulations. The approximations that are made in the simplified models used in probabilistic analyses may limit the validity of the calculations under certain conditions. If any significant interactions among geochemical and hydrologic parameters are not included in the model, the probabilistic risk assessments and sensitivity analyses may be misleading. In currently available methods, the most serious limitations are the use of radionuclide solubilities or sorption distribution constants whose values are not coupled to changes in solution chemistry during transport.

Even the most robust available geochemical models do not include complete descriptions of geochemical phenomena (Muller and Duda, 1982; Siegel et al., 1989; U.S. DOE, 1988), and available data bases are not complete or validated (Phillips et al., 1988; Delany and Lundeen, 1989). The failure of available probabilistic models to include important processes must be evaluated in light of uncertainties in current understanding of geochemical processes. Components of this uncertainty include (1) imprecision in available thermodynamic data due to experimental error, (2) weaknesses in aqueous models used to interpret solubility data, and (3) use of invalid assumptions in transport models for radionuclides. Examples can be found in descriptions of radionuclide release from waste into ground waters and in descriptions of interactions between radionuclides and minerals along potential flow paths.

Radionuclide Release: Source Term and Solubility

Many analyses of the interaction of aqueous solutions with canisters and with various waste forms indicate the coupling between radionuclide solubilities, source term, and solution chemistry (e.g., Barney et al., 1984; Alexander et al., 1985; Neretnieks and Rasmuson, 1983; Chambre and Pigford, 1985; Zavoshy et al., 1985). The rate of transfer of radionuclides from solid wastes into solutions depends on composition of the solutions, nature of the waste form, temperature, and rate of solution flow. At slow flow rates, the rate of radionuclide release may depend on limits set by the solubility of compounds of the several radionuclides; at rapid flow rates the rate-limiting step may be the rate of dissolution of the waste form (e.g., Kerrisk, 1984; Barkatt et al., 1985).

Some examples of the behavior of uranium may illustrate these propositions. Grandstaff (1976) studied the rate of oxidation and dissolution of the mineral uraninite in aqueous solutions and demonstrated the effects of composition of this mineral, O_2 and CO_2 content of the ambient atmosphere, and temperature and composition of the aqueous solutions on the rate of uraninite solution. The rate of oxidation and dissolution was reasonably rapid; in most experiments, saturation of the solutions with respect to a solid phase containing U^{6+} occurred in a matter of days or months. Similar periods seem to be required to saturate synthetic ground waters with respect to U^{6+} compounds in the presence of spent UO_2 fuel. Studies by Forsyth et al. (1986) suggest that the concentration of uranium in simple, dilute ground waters might be predicted with some certainty; however, this is probably not true for concentrated ground waters such as those that might be encountered in waste repositories in salt or for ground waters that are saturated with one or more of the many complex mineral phases that contain UO_3 (e.g., Alwan and Williams, 1980; O'Brien and Williams, 1983; Magalhaes et al., 1985).

The concentration of uranium is extremely low in solutions that are sufficiently reducing so that uranium is in the tetravalent oxidation state. Thermodynamic data for tetravalent uranium species are correspondingly uncertain. For example, Ryan and Rai (1983) have shown that the data of Gayer and Leider (1957) for the solubility of UO_2 are too high by at least four orders of magnitude, and Tremaine et al. (1981) have pointed out the extreme sensitivity of the surface of UO_2 and aqueous U(IV) to oxidation. The behavior of uranium therefore depends critically on the oxidation state of ground waters at sites where they react with nuclear waste. If ground waters are reducing at these sites, the concentration of uranium after equilibration with UO_2 or other U(IV)-containing phases will normally be in the parts-per-billion range, whereas oxidizing ground waters can pick up more than one part per million uranium.

The behavior of uranium during the interaction of ground water with spent fuel illustrates many of the problems of predicting the behavior of other long-lived radionuclides under similar circumstances. With the exception of thorium, actinide elements all exhibit more than one valence state in the normal range of near-surface and shallow subsurface environments. Like uranium, they form strong complexes with oxygen-containing ligands (oxides, hydroxides, phosphates, carbonates, sulfates) and fluorides; their affinity for sulfur and heavy halides is weak. In many environmental aqueous systems, hydroxide and carbonate are probably the most important ligands; complexes with fluorine may also be significant (e.g., Allard et al.; 1984, Cleveland et al., 1983a, 1983b). The strength of many complexes of these ligands with UO_2^{2+} is reasonably well known; however, uncertainties in the stability constants of many potentially important complexes of the other actinides are large. In addition, the kinetics of complex formation will also greatly affect solubility (U.S. DOE, 1986b; Schramke et al., 1987).

Carbonate complexes of Pu may be important in the transport of this element in ground waters, because actinides that normally exhibit low solubilities in near-neutral solutions can be complexed by carbonate ligands and form solutions of significantly higher actinide concentrations. The solubility of Pu(IV) hydrous oxide is a function of carbonate concentration and the value of the association quotient of $PuCO_3^{2+}$. Estimates of the value of the association constant range from 10^{52} M^{-5} (Moskvin and Gel'man, 1958; Moskvin, 1971) to less than 10^{13} M^{-1} (Silva, 1985). Additional experiments are needed to settle the controversy regarding the value of this and other association quotients that are potentially important in the transport of actinide elements from repositories to surface environments.

Thermochemical equilibrium models have been used to predict the solid mineral phases that control the concentration of various radionuclides and major ions in waters from basalt, tuff, and salt (Kelmers et al., 1984; Kelmers, Kessler, et al., 1985; Kelmers, Seeley, et al., 1985; Deutsch et al., 1982a, 1982b; U.S. DOE, 1988; Schramke et al., 1987). Results of various equilibrium codes (WATEQ2, PHREEQE, MINTEQ, and EQ3/EQ6) have shown general agreement in predicting stable solid phases that may control the solubility of various radionuclides. Major differences in results have been attributed to the use of different thermodynamic data for solid phases and aqueous species. The types of uncertainties in the data that lead to lack of consensus in choice of thermodynamic constants have been illustrated above.

Compilations of data by Langmuir (1978), Lemire and Tremaine (1980), and Phillips et al. (1988) for solution-mineral equilibria involving uranium and other elements in the waste are extremely useful but still incomplete. These data bases have not been validated against new experimental data and may contain data drawn from inconsistent aqueous speciation models (Tripathi, 1983). Rossotti (1974) and Garvin et al. (1976, 1981) have outlined the requirements for compilation of a self-consistent data base but these have not been realized yet for radioelements important in waste management. Phillips et al. (1988) suggested that uncertainties in thermodynamic data could be propagated through solubility and speciation calculations using the Latin hypercube sampling technique and efficient speciation codes such as HYDRAQL (Papelis et al., 1988). However, the task of assigning a correlation structure to the data base required for such calculations may be prohibitively large. A first attempt could be made in which the uncertainties in the equilibrium constants are assumed to be independent. However, this is not a valid assumption: many equilibrium constants are conditional on the particular experimental conditions and values of other constants used to solve the set of simultaneous equations describing the chemical system under investigation.

Radionuclide Retardation

Kerrisk (1984) and Siegel et al. (1989) have shown that a number of elements in nuclear wastes are apt to be released into ground water at a sufficiently rapid rate that their presence in ground water will exceed EPA release limits, unless they are retarded or removed en route to the surface. The efficiency of most removal mechanisms is difficult to predict. Sorption and cation exchange on minerals are observed for many radioelements. Uranyl ion, UO_2^{2+}, can be adsorbed on clays and on the surfaces of other minerals (e.g., Borovec, 1981; Ho and Doern, 1985;

Siegel, Leckie, et al., 1990) as well as by humic acids (Borovec et al., 1979; Idiz et al., 1986). The fraction of a particular ion that is removed during passage from a repository to the surface, however, can only be estimated if a large number of pertinent data are available. The quantity of ion exchangers along the route must be known, as well as their accessibility, their loading prior to reaction with contaminated ground water, and changes in this loading during the flow of contaminated ground water. The composition of the contaminated ground water must also be known, as well as temporal and spatial changes in ground-water composition.

At the proposed nuclear-waste repositories at Yucca Mountain and the Waste Isolation Pilot Plant, considerable uncertainty is associated with the composition of waters in the strata above the sites (Tien et al., 1985; U.S. DOE, 1988; Siegel, Robinson, and Myers, 1990; Siegel, Leckie, et al., 1990). The composition of ground water is important to the calculated speciation and sorption of uranium, plutonium, and americium. The dependence of the K_d on the concentrations of the groundwater can be illustrated by considering the sorption of uranyl (UO_2^{2+}) onto a surface site (SOH) of a mineral:

$$SOH + UO_2^{2+} = SOUO_2^{+} + H^{+}$$

with a binding constant K_{ads} defined for the reaction. The concentration of UO_2^{2+} available to complex with the surface site will be affected by complexation reactions with other ligands such as carbonate. The K_d in a system containing the uranyl ion and its hydroxo and carbonato complexes can be calculated as

$$K_d = \frac{K_{ads}(SOH)\,C}{(H^{+})\,[1 + \sum_{ijk} (UO_2^{2+}]^{i+1}\,(OH)^{j}(CO_3^{2-})\,]^{k}}$$

where C = units conversion constant. The effects of relatively small changes in the composition of the ground waters may results in order-of-magnitude changes in the K_d. Figure 9.3 shows that the K_d of uranium in systems containing several competing ligands can be sensitive to the concentration of other cations.

Dissolved organic carbon may form strong complexing ligands that increase the aqueous concentration limits of most actinides (Olofsson and Allard, 1983). Complexes such as EDTA and fulvic and humic acids complex with most actinides under different valence states and may decrease the extent of sorption onto rocks (e.g., Figure 9.3). The K_d for a radionuclide may not be constant over postulated flow paths and the different migration rates of competing and complexing ligands may cause the K_d at any point along the flow path to change during the history of the waste repository (Cederberg, 1985). Such complexity

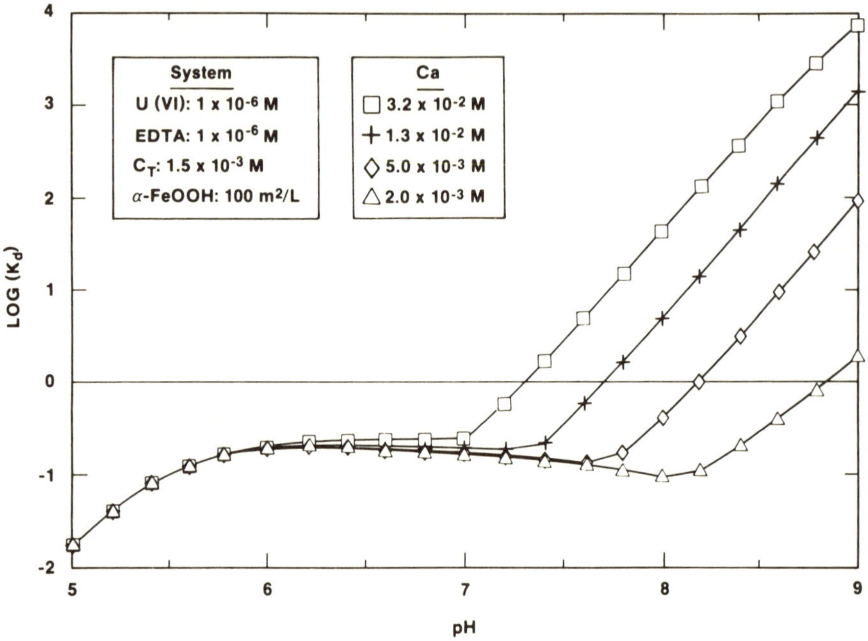

Figure 9.3. Calculated K_d for sorption of U(VI) onto a goethite substrate as a function of pH at fixed total carbon concentration, in the presence of EDTA (after Leckie, 1989). K_ds are shown for several levels of calcium concentration. Surface area of the substrate (α-FeOOH) was 100 m^2/L; total carbon (C_T) was 1.5 x 10^{-3} M; total uranium concentration was 10^{-6} M.

has not been addressed in sorption experiments or in probabilistic calculations using simple geochemical box models. Coupled reaction/transport codes such as TRANQL (Cederberg, 1985) and HYDROGEOCHEM (Yeh and Tripathi, 1989) can represent changing K_ds. However, attempts to use such codes in Monte Carlo simulations are limited by the required computing time (Siegel and Leigh, 1989).

Three mechanisms for removing radionuclides from ground waters have been discussed: direct precipitation, adsorption, and cation exchange. Co-precipitation is a fourth mechanism that may be important in some settings. Fracturing of rocks within and near repositories is apt to expose fresh rock to alteration by ground water. New minerals that may form along such fractures may incorporate radionuclides or act as additional sites for adsorption and ion exchange. The relatively high temperatures of ground water near high-level-waste repositories not long after closure may speed their reaction with fresh rocks (Benson and Teague, 1979, 1980; Teague, 1980; U.S. DOE, 1988). Modeling the development of postclosure fractures and the reaction of contaminated ground waters with fresh rocks exposed along these fractures is difficult. The kinetics of rock-water interactions are complicated functions of temperature, ground-water composition, and physical and chemical properties of the rock. The temperature and composition of ground waters near repositories will change along their flow path, and the mineralogy and thickness of the coating of secondary minerals at any given point along their flow path will change with time.

SUMMARY

Geochemical predictions must be based on an understanding of the steady-state chemical and physical characteristics of the system as well as estimates of the frequency of certain major perturbations. Radionuclides can be dispersed from a waste repository into surface environments by several mechanisms, of which transport in aqueous solutions is the most likely and important. Removal of radionuclides from ground waters during passage through the geologic barriers to the surface may be important for the safety of repositories.

The central principles underlying predictions of the state of geochemical systems with box models and the limitations of such predictions have been presented in this chapter. Such models can be used to describe solute transport in ground water and between surface waters and aquifers. They are compatible with the structure of probabilistic methods described elsewhere in this volume. Mass-transfer processes in large-scale models can be broken down into subsystems of reservoirs and material-transfer processes which are modeled on the microscopic scale.

The precision of geochemical forecasts is apt to be limited both by a lack of data and understanding and by intrinsic uncertainties due to short-term variabilities and fluctuations in the affected system. Field and laboratory observations of rock and water compositions are used to estimate properties of reservoirs, but invariably data are collected from a relatively small number of erratically spaced locations. Statistical techniques such as factor analysis and kriging may be used to estimate important rock and water properties. The applicability of geostatistical techniques tends to be limited by the availability of data for the construction of semivariograms.

Use of box models requires description of the rates of radionuclide movement between compartments and partitioning of radionuclides between subsystems within compartments. The most important geochemical variables for probabilistic analyses using current radionuclide box models are radioelement solubilities and sorption distribution coefficients (K_d). In Monte Carlo simulations, these characteristics are specified by pdf's.

Probabilistic techniques are apt to be useful in predicting at least some parameters that are important for assessing the elimination or retardation of radionuclides. There is usually, however, not enough detailed information on the mineralogy, chemistry, texture, and structure of rocks between repositories and the surface to permit precise forecasts to be made of the interaction of radionuclides in contaminated ground waters with the rock units along their flow path. The problem is exacerbated by a current lack of good thermodynamic and kinetic data for the chemistry of some radionuclides. In currently available methods, the most serious limitations are the use of radionuclide solubilities or sorption distribution constants whose values are not coupled to changes in solution chemistry during transport. If all significant interactions among geochemical and hydrologic parameters are not included in the model, the probabilistic risk assessments and sensitivity analyses may be misleading. It is important to evaluate the sensitivity of probabilistic risk calculations to the failure of real systems to conform to the assumptions of the transport models case by case.

10

Resource Exploration

John W. Harbaugh

Resource exploration or exploitation activities here means any action, such as drilling boreholes or sinking shafts, in the search for mineral commodities. *Mineral commodity* commonly means any material commodity extracted from Earth's crust, including oil and natural gas, coal, metals, and nonmetallic materials of all sorts, including salt, clay, stone, and so forth. Geothermal energy, either in the form of steam or hot water or as "dry heat," could be included as a mineral or "extractable" commodity, although this chapter generally ignores geothermal energy. Ground water in the usual sense (i.e., potable water at ambient temperatures at relatively shallow depths) is commonly excluded as a mineral commodity. However, ground water in the form of brines, or even of waters of moderate salinity, is included as a mineral commodity if at depths generally below those at which potable ground water is extracted, and if it is potentially valuable for its dissolved solids. Exploitation by drilling disposal wells (e.g., wells for disposal of brines produced in oil wells) might or might not be included in exploration or exploitation activity. Exploration and exploitation are part of a continuum, with no well-defined distinction between them.

Limits on what sets a mineral commodity apart from a "noncommodity" are desirable. Concentration at some specified multiple of ambient or background concentration has been suggested as necessary to the definition. Such a definition may be acceptable for metallic ore deposits (e.g., gold and iron), but it is not generally applicable to nonmetallic deposits, whose value may depend on aesthetic qualities (e.g., precious and semiprecious stones) or chemical quality and volume (e.g., limestone). At least one mineral commodity, dry heat, is universally present in the crust.

THE BASIC TREE OF INTERDEPENDENT PROBABILITIES

Probability trees are useful in considering whether a radioactive-waste repository is likely to be breached. The probability that a radioactive-waste repository will be breached during exploration should be considered in context with other chance events with which it is interdependent. Probability trees or other logic diagrams are well suited to show such relationships. A probability tree diagrams a sequence of chance events, generally in succession from left to right. Complementary probabilities, which must sum to 1.0, pertain to branches stemming from each fork. Branches fork repeatedly at "chance forks," finally terminating at branch tips representing a terminal event. The probability of reaching a specific tip or terminal event can be calculated by progressively working from the trunk of the tree to the tip, or vice versa, multiplying probabilities attached to the specific branch followed at each of the forks. The multiplicative probabilities attached to all routes that lead to tips must sum to 1.0, provided that all possible branches are included. The probability of reaching a particular terminal event, such as breaching a repository, can be calculated by summing all the multiplicative probabilities attached to branch tips that represent that terminal event.

Trees may vary in complexity. A simple tree may represent gross probabilistic relationships, but the same basic tree can be expanded, ad infinitum, to represent probabilistic relationships at any level of detail. A simple tree showing breaching of a repository by all possible alternatives (Figure 10.1) might consist of six terminal events, three involving breaching of the repository by exploration. Of these three, two are deliberate and one is inadvertent. The probability of breaching is calculated by multiplying the probabilities attached to each branching sequence that leads to breaching as a terminal event and summing those that terminate in breaching.

The probability of breaching can be calculated only if probabilities are assigned to each branch. Thus, probabilities must be assigned to all branches that involve deliberate or careless breaching of a repository [Figure 10.1, branches (e) and (g)]. Although assigning such probabilities may be protested as futile because human activity is involved, some value must be assigned to evaluate the tree. A small value, even zero, may be used. Setting values for "deliberate-breach" branches is a matter of subjective judgment, because objective data do not exist by which these may be estimated.

Exploration activity is critical in two separate situations (Figure 10.1). In

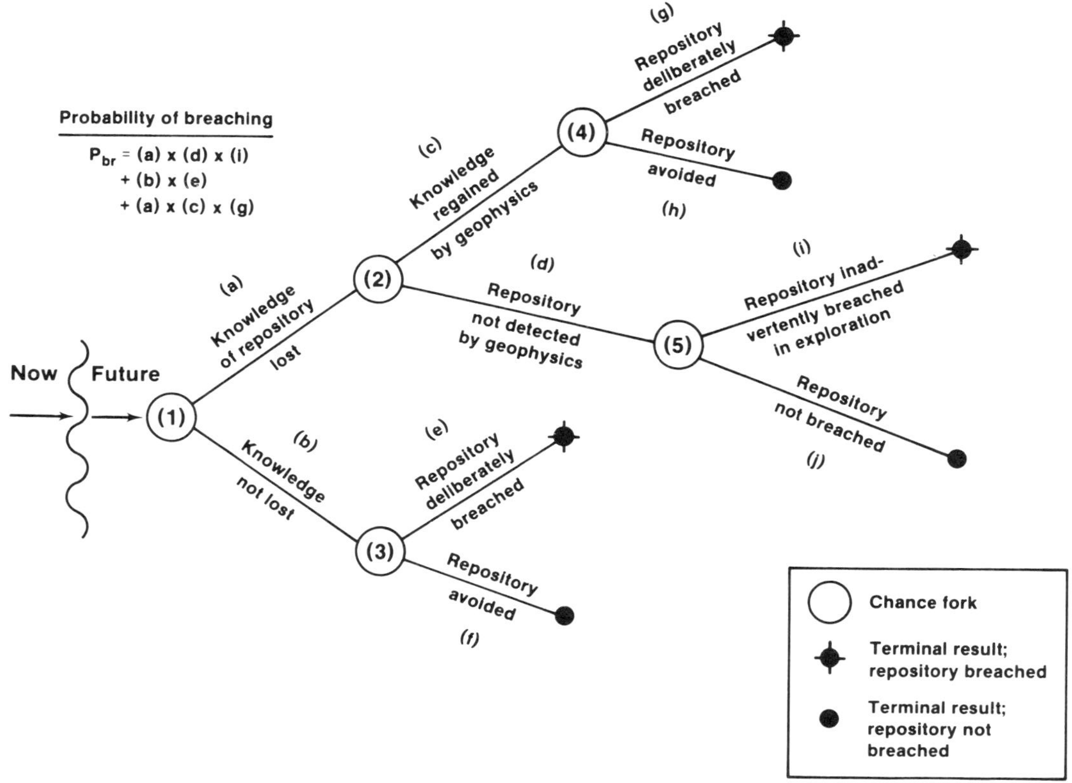

Figure 10.1. Basic probability tree involving breaching vs. not breaching a repository by human activity. Numbers identify individual chance forks and permit successive extensions of the tree (Figures 10.6 and 10.8) to be linked together. See text for discussion.

branch (c), knowledge of a repository that had been previously lost [represented by branch (a)], is regained using geophysical techniques. Objective procedures for estimating the probability to be assigned to branch (c) seem impossible to devise. Like the probability of branch (a), the probability for branch (c) will be largely subjective, involving assumptions about the efficacy of geophysical and geochemical procedures in the distant future. In branch (i), a repository that is unknown is inadvertently breached in exploration.

One ameliorating factor may be the availability of estimates of the likelihood that repository sites will have physical, chemical, or biological manifestations at or near the surface. For example, increases in heat flow or radioactivity may indicate the presence of a repository.

Branch (d) involves failure to detect the repository site by geophysics or geochemistry in advance of drilling. The subbranches involve either not breaching (j) or inadvertent breaching by exploration (i). Branch (d) and its subbranches can be detached and examined, excluding the rest of the tree. The probability attached to branch (d) is the complement of branch (c). The two sum to 1.0, and one necessarily determines the other. Because branch (c) seems to be indeterminate by objective methods, branch (d) is also indeterminate, and the probability attached to (d) apparently must be assigned as a matter of subjective judgment.

The probability tree, which can be expanded to as great a level of detail as desired, is representative of the kind of analysis that is essential. The tree represents the interplay between human activity and geologic endowment and requires that probabilities be supplied for each branch, if overall probabilities attached to breaching are to be estimated. One possible expansion (Figure 10.2) possesses more detail than the previous tree (Figure 10.1), but it is still much simpler than more-detailed representations that could be constructed. Except for the last chance fork on the right (Figure 10.2), the chance forks lead either to terminal events or to another chance fork.

THE REPOSITORY'S PROBABILITY ENVELOPE

The three-dimensional geometry of a repository and possible locations of commodities in rocks that surround the repository bear strongly on the probability of the repository being intersected by future exploratory drilling. However, not only must the shape of the repository itself be considered, but also that of a *probability envelope* surrounding the repository. The probability envelope can be defined as the continuum of probabilities, in three-dimensional space, that radioactive waste will have migrated into the surrounding rocks. The envelope can be represented by contours drawn on planes that intersect the envelopes

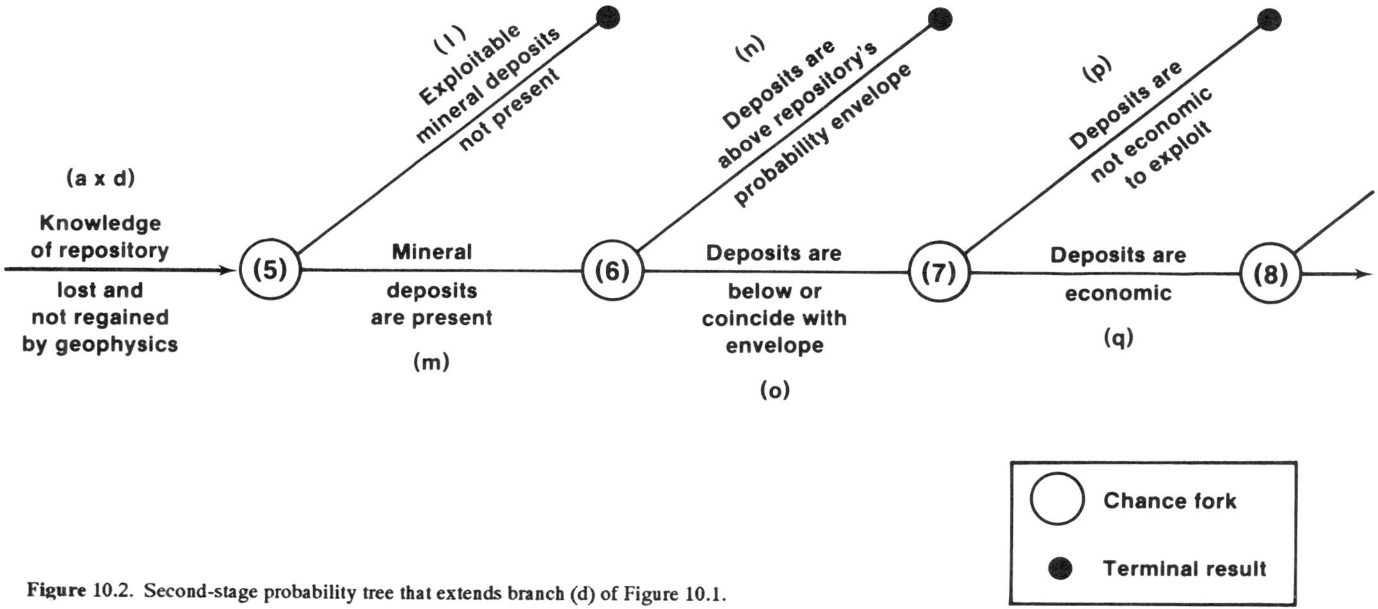

Figure 10.2. Second-stage probability tree that extends branch (d) of Figure 10.1.

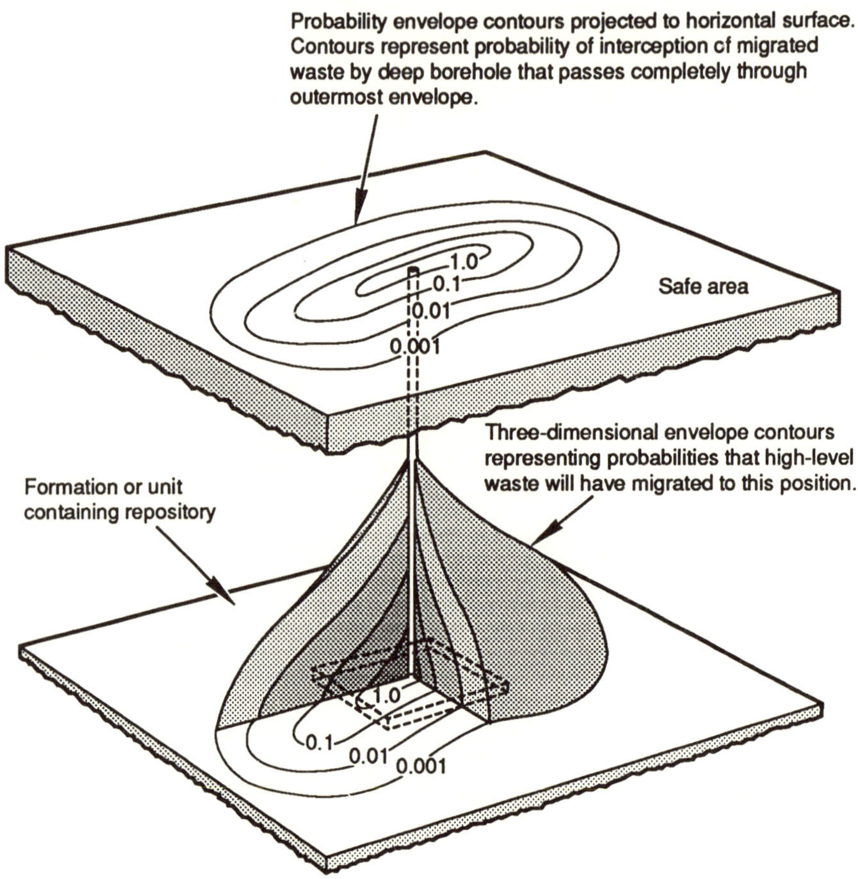

Figure 10.3. Hypothetical probability envelopes surrounding repository. Maximum probability values that have been projected to a horizontal surface define safe vs. unsafe areas for vertical boreholes that completely penetrate the outermost envelope.

(Figure 10.3). Probabilities will decrease away from the repository. Probably the decrease would vary greatly with respect to direction in three-dimensional space and would be sensitive to local geologic features, including bedding surfaces, faults, igneous intrusions, and joints. The probability envelope should expand with time. If the repository's probability envelope decreases exponentially, for example, zero probability will never be reached. Although a truly exponential decrease would be unrealistic, the example shows that some probability cutoff, beyond which the rock volume surrounding the envelope can be considered to be "safe," must be assigned.

The shape and specific values of the probability envelope will be assigned by repository-site engineers, with input from experts in rock physics, hydrogeology, and geochemistry. Although simple in concept, the probability envelope will be complex to calculate; it will derive point probability values from composite probability estimates for various chemical concentrations, ground-water flow velocities and directions, stratigraphic variation, rock properties, and other geologic and engineering factors affecting radionuclide migration. Defining the probability envelope is essential for estimating probabilities of its inadvertent unbiased intersection by boreholes.

The shape of the probability envelope affects the probability of its intersection by boreholes. If a vertical borehole is drilled from a surface location above the envelope, and if the borehole is sufficiently deep that it intersects the largest probability value at that geographic location, the probability of the borehole intersecting the envelope can be mapped by projecting the maximum probabilities within the envelope to a horizontal surface (Figure 10.3). The horizontal surface could be the land surface, and contours on it can thereby define the area that is safe versus the area that is unsafe.

Borehole depth affects the probability of intersecting the envelope. The envelope may be thought of as consisting of a large number of layers, like a large, irregular onion that could be successively peeled, revealing progressively increasing probabilities until the core (the repository) is finally reached. Boreholes drilled to some specified depth that is less than the minimum depth of the cutoff layer of the envelope (i.e., the outermost layer of the onion), but above the repository itself, will have probabilities of intersection that are less than those drawn on the horizontal plane (Figure 10.3). These probabilities can be mapped by passing a plane representing the specified intermediate depth through the envelope (Figure 10.4). These probabilities can be projected to the land surface, defining safe and unsafe areas.

The probability of a borehole intersecting the envelope can be calculated readily. A borehole that is shallower than the minimum depth of the probability envelope's cutoff value is a special case, with zero probability of intersection.

The concept of a "unit area" must be introduced. Suppose that a unit area is defined as an area outlined by a square. The size is arbitrary, but it would be convenient if the square encompassed the projection of the probability envelope, with some area to spare (Figure 10.5). If a vertical borehole is drilled in an unbiased manner within the unit area, either at random or as part of a grid, the probability of intersecting the probability envelope is given by the proportion of the unit's area occupied by each of the different levels of the probability envelope.

About two-thirds of the unit area in the hypothetical example (Figure 10.5) is "safe," that is, below the arbitrarily chosen cutoff level of 0.1%. For the remaining third, the proportion of the area occupied by each subenvelope must be calculated and the area multiplied by the respective value of the subenvelope. If

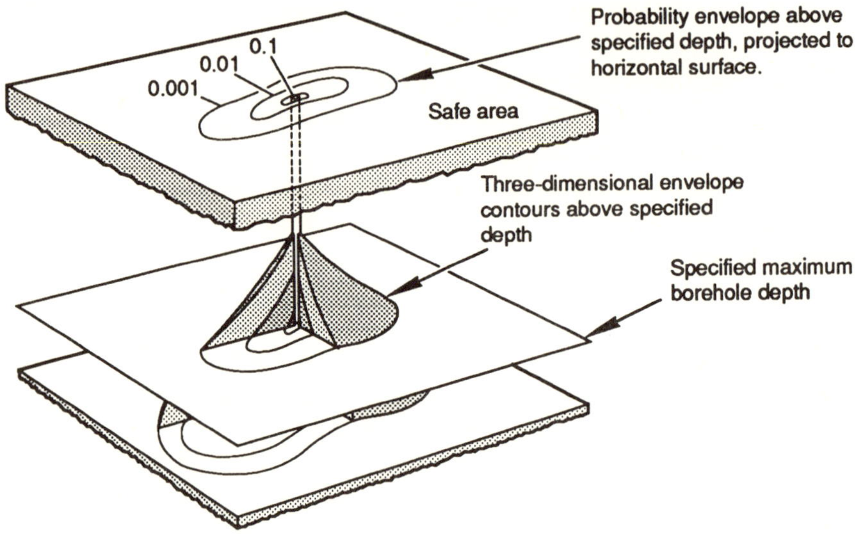

Figure 10.4. Effect of setting depth limit for borehole such that borehole penetrates only part of envelope.

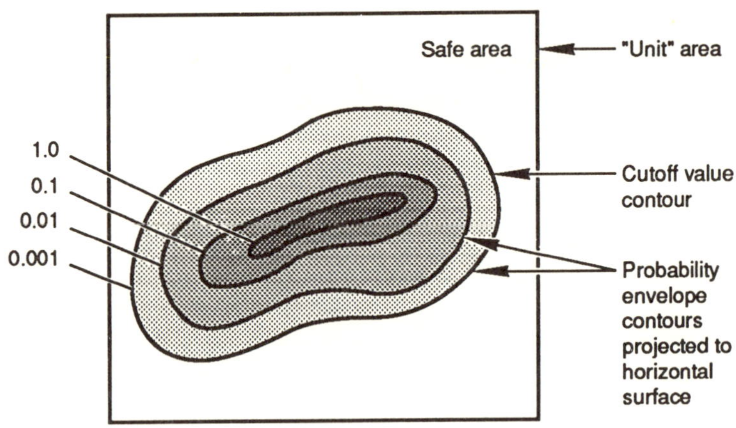

Probability of intersection within unit area

$$P = \text{Number of holes} \times \frac{\Sigma \text{ Area of subenvelope} \times \text{Probability envelope value}}{\text{Unit area}}$$

values of the probability envelope decrease exponentially with distance away from the repository, a slight problem in carrying out the calculations is encountered. We cannot readily compute a mean value for such a subenvelope and multiply it by its area, and the presumed irregularity of the envelope would rule out classical integration procedures. A Monte Carlo procedure to obtain the probability could be used, particularly if the shape of the probability envelope is intricate. The projection of the probability envelope could be digitized by superimposing a grid over the projection and encoding the value of the envelope of each point. Then a succession of x and y coordinate values (e.g., north-south and east-west) drawn at random from a uniform distribution and the encoded probability values combine to yield the probability of intersection on a unit-area basis. Alternatively, if the envelope were represented by contouring procedures that use grids, it would be simple to integrate them numerically.

FUTURE ECONOMICS AND EXPLOITABILITY VERSUS DEPTH

Whether to consider future economics in deciding if a mineral resource is potentially exploitable is a dilemma. The classical definition of an ore deposit (or for that matter, of any mineral commodity) is that it must be economically exploitable. This definition often is used loosely. A deposit of gold may shift back and forth between an "ore" and a "non-ore" with variations in the price of gold. Assessing the likelihood of potentially exploitable mineral commodities being present near repository sites requires a definition of economic exploitability. Undiscovered deposits near repository sites must be regarded as having some probability of being potentially economic for exploitation in the future, if future exploratory drilling is to be undertaken in search of them.

One approach is to consider future demand for a specific mineral commodity on a relative basis, conveniently on a scale of 0.0 to 1.0, so that it may be combined multiplicatively as if it were a probability estimate (Figure 10.6). If "relative demand" for a commodity is zero, then regardless of its presence, no incentive to search for it will exist. If relative demand is 1.0, multiplication will not change the overall probability calculations. Relative demand for specific commodities doubtless will vary with time, but because relative demand for the next century cannot be objectively predicted, forecasts for a 10,000-year period are little better than guesses.

An additional quasi-economic factor is the effect of depth on economic exploitability. Exploitation costs generally increase exponentially with depth, as do exploration costs. Ideally, depth should also be considered as an economic factor, defining a relative "exploitability/depth index" on a scale of 0.0 to 1.0. An exploitability/depth index estimate can also be used as a multiplier. A value of

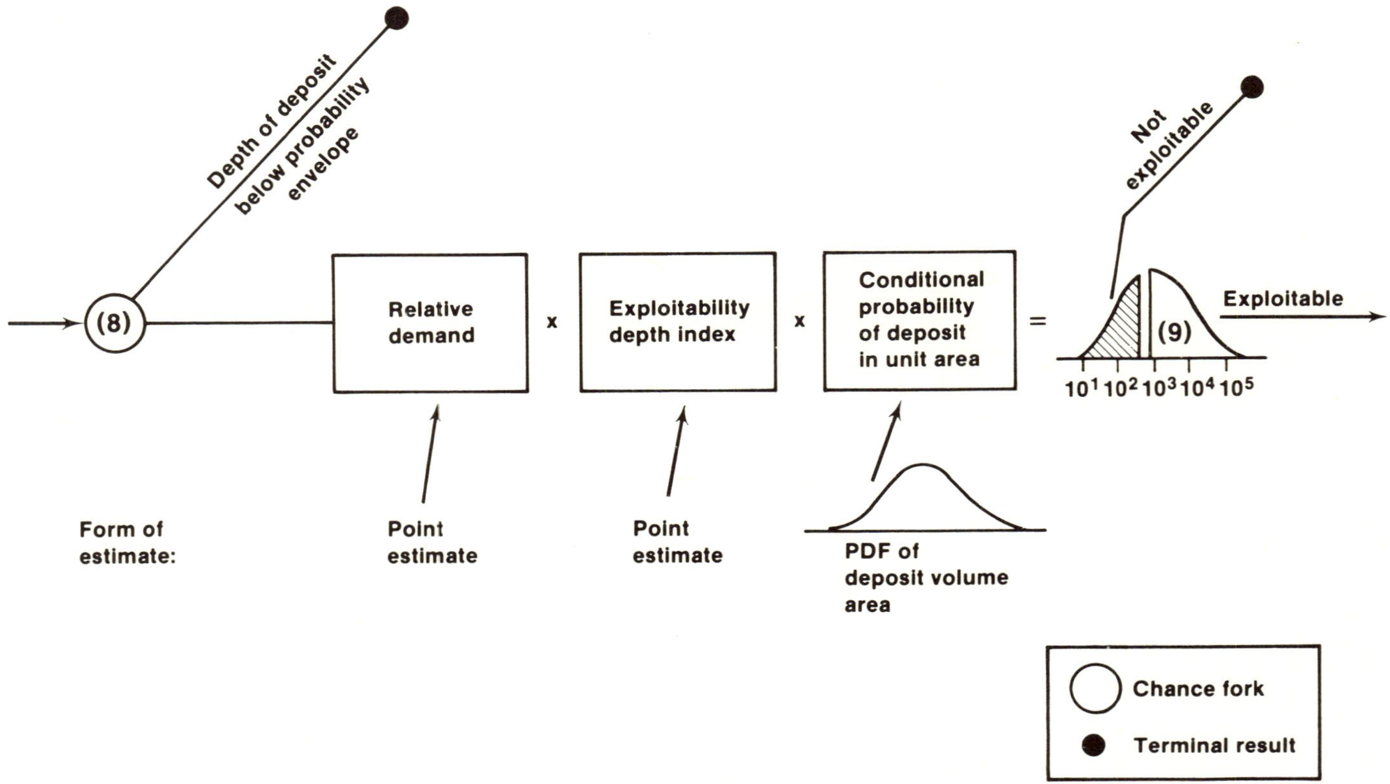

Figure 10.6. Multiplicative roles of relative demand, exploitability/depth index, and conditional probability of commodity's occurrence.

1.0 has no effect, but lesser values cause the exploitability estimate to decrease in proportion, culminating in zero at some depth below which the commodity is not economically exploitable.

An exploitability/depth index combines engineering capabilities and economics of deep drilling or deep mining. Current costs of mining or of oil and gas production with depth can be estimated, but extrapolating estimates that apply today over the next 10,000 years is difficult. Conservatively setting the exploitability/depth index values to 1.0 raises the specter of the probable presence of various nonmetallic deposits at any repository site.

ESTIMATING CONDITIONAL PROBABILITIES FOR SPECIFIC MINERAL COMMODITIES

Three factors govern the probability that an exploitable mineral commodity is present in a given unit area (Figure 10.6): the commodity's relative demand, its exploitability/depth index, and the conditional probability that the specific commodity (or specific aggregation of commodities) occurs within the unit area. The probability is "conditional" in that it depends on present knowledge or inferences concerning the geology of the area.

Suppose that several principal groups of rocks are known or inferred to be present beneath a repository within the unit area containing the repository site. These groups might include a Precambrian basement complex of schists and gneisses with granitic intrusions; Paleozoic marine shales, limestones, and sandstones; Mesozoic rocks; and Cenozoic volcanic flows and tuffs. On the basis of this knowledge or inference, pdf's that pertain to size of deposits could be assigned, conditional upon knowledge of the geology (Figure 10.7). If these pdf's are confined to individual groups of rocks, they will reflect the influence of lithology on the assessment. For example, the pdf for oil and gas should be more favorable for Paleozoic marine rocks than pdf's for the other two groups. The pdf for metallic ores may be greatest in the Precambrian basement.

Pdf's can be expressed on a unit-rock-volume basis. The rock volume can be estimated by multiplying the unit area times the average thickness of the group. An arbitrary depth limit for the Precambrian basement complex must be assumed.

If individual pdf's are obtainable, a theoretical procedure for obtaining a composite pdf combines the conditional probability for the commodity, its relative demand, and its exploitability/depth index (Figure 10.6). The resulting cumulative pdf could be separated into exploitable and nonexploitable components, provided that the boundary separating the two categories could be defined.

A major question is whether conditional pdf's can be obtained more or less

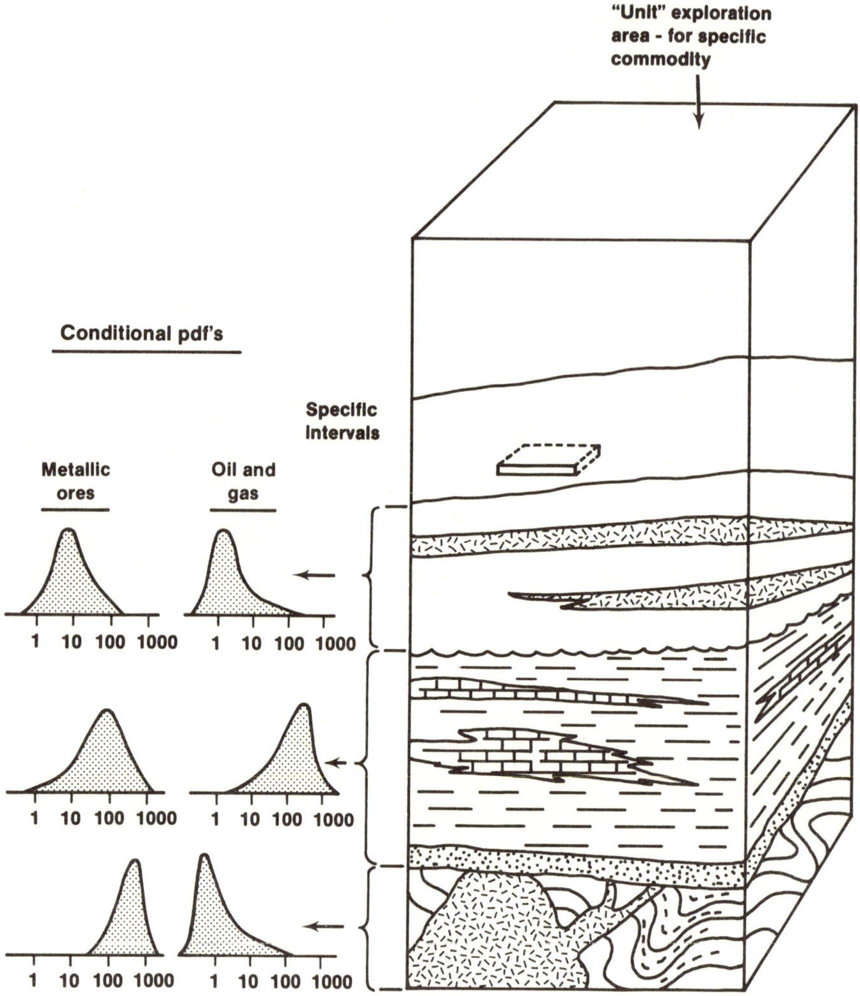

Figure 10.7. Hypothetical diagram to illustrate probability distribution functions, for aggregates of commodities, that are conditional on specific geologic groups of rocks and their volumes, within unit area.

objectively or only as expert guesses. Desirable general specifications for estimating pdf's follow:

1. Geologic analogs are known, preferably from within the region containing a repository, or perhaps in several regions outside the region that contains a repository.

2. Geologic analogs should contain rocks of the same general lithology, structure, and age.

3. The analogs should have been deposited under similar conditions and within equivalent tectonic settings.

4. Postdepositional (or postorigin if not sedimentary rocks) history of analog rocks should be similar, including depth of burial.

5. The analog region should be more or less maturely explored and exploited, permitting frequency distributions of commodity volumes to be established.

6. Frequency distributions of commodity volumes should be expressible on both a per-unit-area and a per-unit-rock-volume basis.

Conditions 1 through 6 commonly are difficult to satisfy, and severe compromises may be essential in an attempt to use conditional pdf's. Furthermore, segregating mineral commodities to obtain pdf's for individual commodities is extremely difficult. Next, however, consider procedures for relating pdf's, on a unit-volume or unit-area basis, to probability distributions pertaining to the number of exploratory holes needed to reasonably prove the presence or absence of an exploitable commodity or aggregate of exploitable commodities. If feasible, these procedures would yield estimates of the probability that a repository envelope will be intersected in drilling.

ESTIMATING PROBABILITIES FOR ALL COMMODITIES COMBINED

Probabilities for all exploitable commodities combined must be estimated. This is a difficult task. If the probabilities attached to different groups of commodities are assumed to be independent, then a discrete, binomial distribution can be used, provided that treating each exploitable commodity category as being either present or absent is acceptable. Suppose that three categories are established: fuels, metals, and nonmetals. A modified binomial distribution can be written

$$(P_f + A_f)(P_m + A_m)(P_n + A_n)$$

where P is presence; A is absence; and f, m, and n are fuels, metals, and nonmetals, respectively. Expansion yields

$$P_fP_mP_n + P_fP_mA_n + P_fA_mP_n + P_fA_mA_n +$$

$$A_fP_mP_n + A_fP_mA_n + A_fA_mP_n + A_fA_mA_n$$

Each term in the expansion yields the probability attached to that specific combination. The probabilities of all terms sum to 1.0. The categories must be statistically independent, however, which is a difficult requirement to satisfy. Literature that bears on the statistical independence of given categories of mineral resource commodities is virtually nonexistent.

Alternatively, continuous pdf's that pertain to the sizes or values of deposits could be used, provided that various commodity units could be transformed into a single unit, common to all. The only unit capable of serving in this fashion is a monetary unit, e.g., present or future U.S. dollars. If the pdf's are transformed so that they are expressed in monetary units, they can be combined by multiplication, employing simple Monte Carlo procedures. A single pdf would be obtained for all commodity categories. The presumption of independence between the various distributions that are subsequently combined is essential, just as with the binomial distribution.

PROBABILITY OF INTERSECTION OF PROBABILITY ENVELOPE IN EXPLORATORY DRILLING

Calculating the probability that one or more exploratory holes will intersect the probability envelope is simple in principle, but more complex in practice. If holes are located without bias and no minimum distance between them is specified, the probability of intersection by a hole drilled at random within the unit area containing the repository is the sum of the successive products obtained by multiplying the area of each subenvelope (expressed as a proportion of the unit area) times the probability attached to the subenvelope (Figure 10.8). The issue then becomes how many exploratory holes will be drilled within the unit area. The situation seems complex, however:

1. Will records be kept by future explorationists, so that the presence of a previously drilled exploratory hole may either encourage or discourage subsequent holes being drilled in the immediate vicinity?

2. If the result of an exploratory hole is discouraging (for example, a dry hole drilled in search of oil and gas), does it "block out" a specific subarea, inhibiting subsequent drilling within the subarea to the depth drilled?

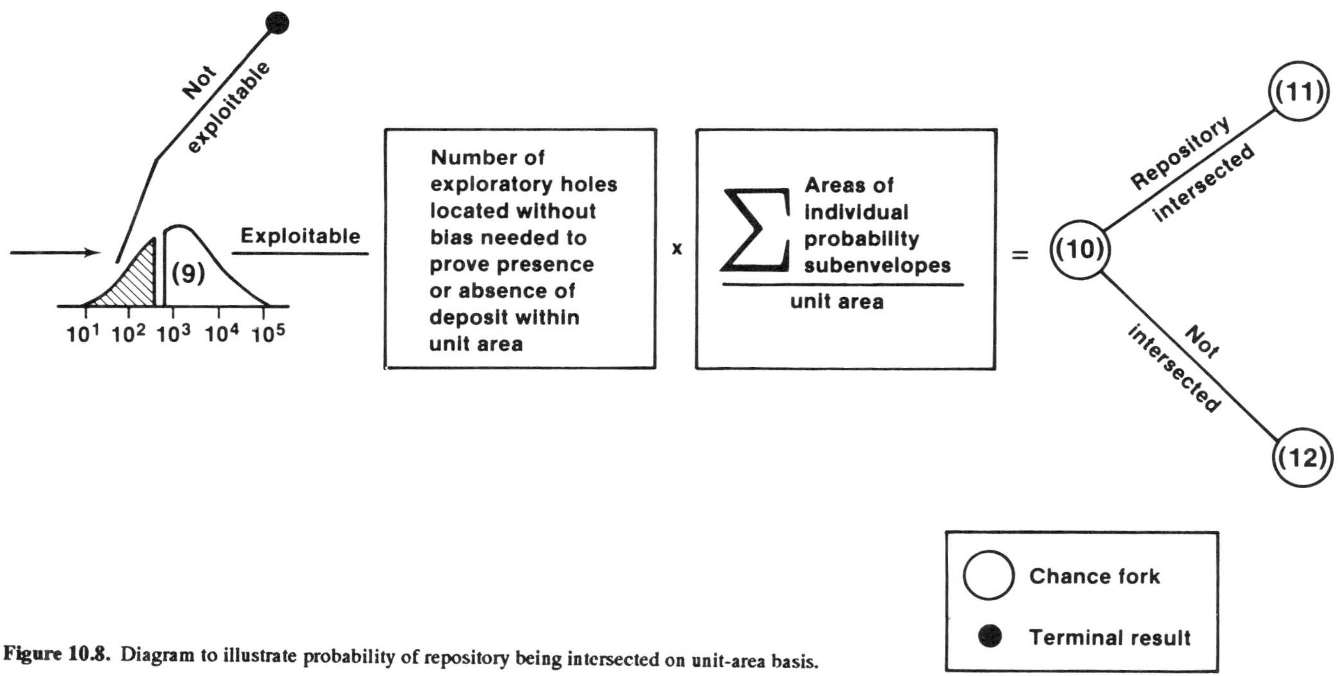

Figure 10.8. Diagram to illustrate probability of repository being intersected on unit-area basis.

3. If a subarea is more or less "blocked" from further drilling, how big is the area, and what is its shape (a circle, square, or hexagon)?

4. Will future explorationists be guided by geology and geophysics? Presumably they will be. If so, will the presence of geologic "leads" encountered in drilling (in contrast to direct discovery of an exploitable commodity) influence decisions to drill or locations of subsequent exploratory holes?

These questions seem unanswerable, except by speculation. Given a specific set of detailed assumptions, a procedure for estimating probabilities of intersection could be designed. Explicit guidelines are needed if an event tree of sufficient complexity is to be considered.

SOURCES FOR FREQUENCY DISTRIBUTION OF SPECIFIC COMMODITIES OR CATEGORIES OF COMMODITIES EXPRESSABLE ON A UNIT-AREA BASIS

Evaluating a repository site in terms of its location's general potential for future mineral exploration requires estimates of pdf's for specific groups of mineral commodities that may be present in the area. These pdf's should be expressed on a unit-area basis if probabilities of intersection of a repository by a borehole are to be estimated. Ideally, pdf's should be on a unit-rock-volume basis, which in turn can be reexpressed on a unit-area basis. Sources for pdf estimates for individual commodities are past production data of some individual commodities; production statistics, in turn, can be manipulated to yield projections of the resource base with respect to specific commodities.

Oil and Gas

Estimated pdf's are needed for oil and gas. Production statistics for oil and gas provide a starting point for estimating pdf's for specific repository sites. Most states in the United States compile production statistics for individual oil and gas fields. For example, the Kansas Geological Survey publishes an annual report on oil and gas produced in Kansas, with statistics on the amount produced during the year and cumulative totals for each field. From this information, frequency distributions of field size based on cumulative production are readily tabulated. Generally, cumulative production figures are more or less lognormally distributed. If plotted in log-probability form, an ideal lognormal distribution is a straight line, which is convenient (but hazardous) for making projections for large fields with small probabilities.

The U.S. Department of Energy has prepared unpublished estimates of ultimately recoverable oil and gas on a field-by-field basis. These estimates extrapolate decline rates to estimate ultimate recovery per field. This is a "quick-and-dirty" method at best, but it yields numerical estimates, however imprecise. Other groups have estimated ultimate recovery using decline rates plus other engineering considerations (e.g., Barlow and Haun, Inc., 1978).

If the desired estimate is of ultimately recoverable oil and gas in a region in terms of the region's resource base, production data from a relatively mature area can be used to extrapolate decline rates of existing fields to obtain their ultimately recoverable hydrocarbons, and also extrapolate progressive changes in field-size distribution to the oil and gas obtainable from as-yet undiscovered fields (Harbaugh and Ducastaing, 1981). Such a procedure is useful for estimating the resource base in regions that have been well explored. Bloomfield et al. (1979) presented statistics on the relationship between oil-field sizes and sequence of discovery.

If the area for which a resource base is estimated is circumscribed, the regional estimate can be divided by the area of the region to put the estimate on a unit-area basis. The estimate can be made on a unit-volume basis by incorporating the thickness of the interval that has been explored. One problem, however, is that few areas have been explored uniformly with respect to depth. Intensity of exploration generally decreases downward, because more shallow holes are drilled than deep holes. No great technical problems occur in obtaining unit-area and unit-volume estimates of recoverable hydrocarbons in much of the United States, providing that appropriate analogs for specific proposed repository sites can be selected.

Metals

Estimates of ore deposits near a repository site are also needed. The procedures are the same in principle as for oil and gas, namely, to find a region or regions that are more or less analogous geologically and have been substantially explored. Here enormous difficulties are encountered. Regions that have been explored for ore deposits with the same intensity as certain regions that produce oil and gas are almost impossible to find. For example, in central Kansas, counties such as Russell and Barton Counties have been thoroughly explored for oil and gas; most of the Paleozoic sedimentary rocks that overlie the Precambrian basement have been thoroughly sampled by drilling. Although some small oil and gas fields probably remain to be found in these counties, the contribution of these remaining undiscovered fields to the total petroleum resource base can be readily estimated. No region that has been explored to a comparable extent for metals is known. Areas may have been relatively thoroughly explored for ore deposits close to the surface (i.e., within a few hundred feet); some mining districts, such as the Butte district or the Southeast Missouri lead district, have been relatively intensely explored at greater depths locally.

Setting aside the problem of exploration intensity, do regions exist that could provide analogs for ore deposits in groups of rocks inferred to be present beneath repository sites? Selecting appropriate analogs for nonmetallic deposits (excluding fuels) is even more challenging.

THE UNIT-REGIONAL-VALUE APPROACH

An alternative to the above procedures is simply to combine all mineral commodities according to their monetary value (expressed in constant dollars) and to measure monetary value for the crust as a whole, treating it on a unit-area basis. This is the "unit-regional-value" (URV) approach championed by John C. Griffiths of Pennsylvania State University. The URV procedure has the advantage of reducing or avoiding difficult assumptions regarding geologic analogs and difficulties of dealing with different mineral commodities.

The URV approach is a measure of human activity, but it also provides some measure of the endowment of Earth's crust with exploitable mineral commodities. The records available for URV computations go back only a century and a half or so, at best. Mankind has used mineral resources for many millenia; the value of commodities produced over the centuries is incalculable, although much of the mineral production has taken place in the past two centuries. In spite of its shortcomings, a URV procedure may be suitable for assessing the resource potential of proposed repository areas.

Griffiths (1978) provided an excellent generalized overview of mineral resource assessment using the unit-regional-value concept. He summarized the URV for each state in the United States, for aggregate production from 1905 through 1972. The URV for the 50 individual states forms a lognormal distribution, inasmuch as URVs, expressed as dollars of mineral production per square kilometer, provide a reasonable approximation to a normal distribution when their logarithms are plotted using the 1967 dollar as the standard. The URVs of the 50 states varied enormously, ranging from a maximum of \$810,179/km^2 for Pennsylvania to a minimum of \$2,739/km^2 for Alaska (the figure for Alaska is now greater by virtue of oil production from Prudhoe Bay).

These figures are not directly conditional on geology, and they span only a short segment of human history, about two-thirds of a century. URV figures reveal that there is great diversity in the extent to which mineral resources have been exploited to date. An important question is whether these differences from state to state also reflect differences in mineral endowment. For example,

Alaska's small URV reflects a relatively small exploration effort per unit of area, and therefore Alaska's potential on a URV basis may be much larger than past production figures might suggest.

A major point of the URV approach is that all categories of mineral resources are lumped together. Contrast the first four states in the United States in terms of their URVs. Pennsylvania and West Virginia are first and second, respectively; coal and petroleum have been their predominant commodities. Louisiana is third (principally petroleum), and New Jersey is fourth (principally sand and gravel). South Africa's overall URV (Menzie, 1977) lies close to that of the United States as a whole. In the United States, petroleum is the largest single commodity, with fuels overall accounting for two-thirds of the URV. In South Africa, however, gold, diamonds, and coal are first, second, and third in importance, respectively.

A number of other countries have been surveyed to determine their URVs, including Canada (Labovitz, 1978), New Zealand (Watson, 1977), Australia (Engelder, 1979), United Kingdom and Ireland (Walsh, 1979), Israel (Gill and Griffiths, 1984), and Venezuela (Arteaga, 1978). Singer (1971) has examined counties in California. A marked relationship exists between geologic diversity in a region and the number of mineral commodities produced. Griffiths et al. (1980) plotted geologic diversity (expressed as the number of time-petrographic units present) versus the number of mineral commodities. A pronounced linear relationship is evident, although considerable scatter is present. The plot included states in the United States, states in South Africa, Rhodesia, and New Zealand. Given a measure of a region's geologic diversity, a conditional estimate of the number of mineral commodities expected to be present may be obtained.

HOW MUCH EXPLORATORY DRILLING SHOULD BE EXPECTED?

The penultimate question is, how much exploratory drilling should be expected per unit area at proposed repository sites in the forthcoming 10,000 or 100,000 years, in light of the estimated resource potential for each site? No answer can be provided without assumptions about relative demand and exploitability/depth index. If commodities are aggregated in the form of unit-area-value or unit-volume-value, the problem remains of determining what specific area a vertical borehole is capable of testing to establish whether the volume that underlies a specific area is or is not exploitable.

At this point, a formidable, if not impossible, task arises. At present, neither borehole geophysics technology, nor geology per se, can establish a minimum "area-of-influence" that has been explored. If it could be determined, then the number of boreholes required to test exhaustively a unit area or volume could be calculated. If assumptions are made about sizes and shapes of deposits in plan view and whether a deposit will be detected without fail when it is intersected by an exploratory hole, then an "area of influence" for boreholes can be calculated. Singer and Drew (1976) described the method for such calculations, using simple geometrical assumptions.

Deposits of some commodities can be approximated by simple geometrical shapes in plan view. For example, oil and gas fields are characterized by definite, measureable areal expanse, as are large ore deposits such as porphyry copper deposits. Seams of coal and beds of evaporites also have appreciable areal expanse. Some other deposits, however, have small areal expanse, such as steeply dipping gold-bearing quartz veins, and defy approximation by simple geometric forms. Areal dimensions of different deposits might be measured and the areal expanses of the commodities used in preparing URV estimates might be described statistically, but the task would be formidable and perhaps impossible.

SUMMING UP THE LITERATURE: ARE PROCEDURES AND DATA AVAILABLE FOR THE TASK?

In focusing on the ultimate question of whether available procedures and data permit the probability of breaching of a repository site by exploration activities to be calculated, the relevance of the available literature must be determined. Ideally, a repository site should have little potential for producing mineral commodities in the future. A site's resource potential must be forecasted in terms of information presently available.

Data Bases Pertaining to Mineral Commodities

Most available data bases pertain either to production of mineral commodities or are relevant to exploration for them. Few data bear directly on sizes of deposits of various mineral commodities.

Various data can be used to estimate undiscovered resources. The Accelerated National Oil and Gas Resource Evaluation (ANOGRE) system of the U.S. Geological Survey (Meyer, 1977) estimates volumes of undrilled sedimentary rock that could contain hydrocarbons and multiplies by a factor derived from other areas representing the ratio of known recoverable hydrocarbons to volume of rocks that have been drilled. Subjective judgment must be used to estimate rock volume and suitability of rocks as potential hosts for oil and gas. Given an estimate of the volume of sedimentary rocks and their characteristics, a point estimate of hydrocarbon volume could be generated for rocks near a repository site. Presumably persons in oil and gas resources assessment within the U.S.

Geological Survey could supply conversion factors that relate rock volume to hydrocarbon volumes.

Many papers discuss attributes of data files that bear on mineral resource assessment. Meyer (1977) thoroughly reviewed petroleum resource assessment in the United States. Eckstand (1977) described computer files used by the Geological Survey of Canada for metals-resource appraisal. Bonham-Carter and Chung (1983) described a comprehensive procedure for integrating geologic data in uranium resource assessment, but the procedures could be used for other commodities. Chung (1983) described the computerized data-handling and analysis system used by the Geological Survey of Canada. Labovitz et al. (1977) presented a computer program to standardize mineral commodity data.

Many of the available geologic, geochemical, and geophysical data have only indirect relevance to this problem. Seismic surveys may be purchased from geophysical contractors; petroleum production data for individual wells can be purchased from Dwights; scout-ticket data for individual oil and gas wells containing stratigraphic "tops," drill-stem test data, and so forth can be purchased from Petroleum Information ("PI"); and a large quantity of geochemical exploration data (ground-water and stream sediment) resulting from the National Uranium Resource Evaluation (NURE) Program can be obtained from the DOE. Geological surveys of most states, as well as the U.S. Geological Survey, maintain geologic records, files, and subsurface well data of various forms that may be obtained at minimal or no cost. Data from these sources probably will not be of great value in estimating occurrence probabilities of mineral commodities at specific repository sites, because these data have little direct bearing on deposit size, grade, or geologic occurrence.

John C. Griffiths (personal communication) has assembled a large data base for estimating unit regional values. His data base includes digitized (and transformed) geologic map data that may be used to provide conditioned estimates of URV.

Procedures for Estimating Probabilities of Intersecting Deposits (or Radioactive-Waste Repositories) by Boreholes

There is an extensive literature of procedures for estimating probabilities associated with patterns of boreholes and for estimating measures of resource exhaustion related to density of drilling. Grid drilling procedures (which assume unfailing detection if a borehole intersects a target) use simple geometrical probability concepts. Singer and Wickman (1969) have produced sets of probability tables that pertain to elliptical targets "located" with square, rectangular, and hexagonal point nets. Singer (1975) discussed the relative efficiency of square and triangular grids. Shurygin (1976a, 1976b) and J. H. Davis (1976) have discussed relative efficiencies of various rectangular grids. Drew (1979) thoroughly discussed pattern drilling and procedures for Bayesian revision at successive stages in a multistage drilling program. Omre (1983) discussed procedures for combining subjective prior information with Bayesian revision dependent on results in grid drilling.

If realistic assumptions can be made about the area that an individual borehole "exhausts" with respect to a mineral resource commodity, the extent of exhaustion, with contours of probabilities, can be mapped. Drew et al. (1979) prepared petroleum exhaustion maps for the Denver Basin, using various assumptions as to target (i.e., oil field) sizes and shapes. Collins (1985) has dealt with relationships between drilling density and petroleum discovery rates through time in Kansas. Schuenemeyer, Bawiec, and Drew (1980) and Schuenemeyer, Drew, and Bawiec (1980) dealt with statistical procedures to forecast regional petroleum discovery rates.

Available procedures for calculating borehole intersection probabilities use simple shapes, such as circles and ellipses. If the shape and size (in plan view) of deposits that might occur in a repository site area could be specified, procedures to calculate the intersection probabilities could be developed. But there is little incentive to extend present procedures without first defining shapes and sizes of deposits that might be present. For example, Singer's (1976) program for calculating intersection probabilities could be extended to use various target shapes. Deffeyes et al. (1982) dealt with some of the problems concerned with other shapes and their packing.

Procedures for Assessing Undiscovered Resources

Many articles deal with resource assessment procedures. White and Gehman (1979) provided an excellent overview of methods in use for petroleum resource assessment. The general categories of methods include geologic analogy, Delphi, areal yield, volumetric yield, geochemical materials balance, numbers of oil fields and sizes in a region, Monte Carlo summation of plays and prospects, and extrapolation of discovery rates. Meyer (1978) described volumetric methods used by the U.S. Geological Survey.

A number of papers deal with statistical estimation procedures for oil-field sizes, for example, maximum-likelihood estimation procedures for the size distribution of North Sea oil fields of Smith and Ward (1981) or the discovery-process models of Schuenemeyer, Drew, and Bawiec (1980).

Historical data (e.g., Cargill et al., 1980) can be used to predict changes in sizes of metallic deposits according to discovery sequence, log-log grade/tonnage plots providing a relatively simple means for projection.

Procedures for estimating conditional probabilities on an areal basis have been developed. Probabilities are conditional upon one or more geologic variables that can be mapped continuously. Harbaugh et al. (1977) described an application based on the properties of subsurface structure-contour maps to generate conditional probabilities of oil occurrence. Davis and Harbaugh (1983) used a regression procedure to generate pdf's of oil and gas field sizes conditional upon area of structural closure mapped seismically in part of the Louisiana and Texas outer continental shelf. Berlanga and Harbaugh (1981) devised procedures for mapping oil-exploration outcome probabilities conditional upon seismic structure maps, with an example application in the Tabasco Basin, Mexico. Agterberg (1974) described a procedure for mapping occurrence probabilities for metallic deposits, conditional upon proportions of various rock types mapped in part of the Canadian shield. Chung and Agterberg (1980) dealt with regression procedures for estimating mineral commodity occurrences from geologic map data.

Many statistical procedures for estimating undiscovered oil and gas resources deal with field-size statistics (e.g., Lee and Wang, 1983, 1985; Kaufman and Wang, 1980; Meisner and Demirmen, 1980; Drew et al., 1980; Root and Schuenemeyer, 1980; Harbaugh and Ducastaing, 1981). Although procedures vary widely, they all require an historical base of previously discovered fields. These procedures are not conditional upon geology, except in the sense that they can be focused on specific regions, including geologic basins.

One of the most thorough regional, statistical studies of mineral-resource potential is that of Lambie et al. (1983). This study pertained to the whole of the California Desert Region, which spans southeast California and extreme east-central California, and made use of virtually all regional geologic, geophysical, and geochemical data that were publicly available. The data included digitized geologic map units (taken by superimposing a transparent grid over maps of the California Division of Mines and Geology 1:250,000 scale geologic quadrangle map series), faults as mapped, regional gravity measurements, lineaments observed on Landsat and Skylab imagery, NURE geochemical prospecting data, NURE airborne gamma ray surveys, and mineral commodity data, consisting of both production statistics and information on occurrence when only individual locations have been reported.

The mineral-occurrence data came from various sources, including (1) DOE's locations of uranium claims (maintained at DOE's office in Reno, in open-file status with the general designation *Preliminary Reconnaissance Reports for Uranium*); (2) mineral resources reports prepared by the California Division of Mines and Geology, e.g., that by Saul et al. (1971) for Riverside County; (3) locations of mines and mineral occurrences prepared by the Santa Fe Southern Pacific Corporation (1987); (4) locations of mines and occurrences in the *Bureau of Mines Mineral Industry Location System* (MILS) (U.S. Bureau of Mines,

1977); and (5) compilation by the U.S. Bureau of Land Management (1979) of mines and prospects. In all, these data bases provided much of the mineral-occurrence information relevant to the region. Other data were also used (Lambie et al., 1983).

The data were used to produce contour maps of the California Desert area showing the probability of occurrence of four specific groups of metallic deposits: (1) gold only, (2) iron only, (3) copper, lead, silver, and zinc combined, and (4) all of these metals combined. Mapping was carried out using geographic cells 2 km by 2 km in extent, each cell receiving values that express the probability that it has been correctly classified within each of these four groups. These probability values were contoured. As explained below, the probability estimates pertain only to occurrence of deposits (e.g., presence or absence) and do not pertain to sizes or frequencies of deposits. The probability of correct classification was calculated using discriminant function analysis (DFA). The computer program used in the study was described by Nie (1975). (Abry [1975] also used DFA to produce contour maps of probabilities of occurrences of undiscovered oil fields in specific size categories in the Tatum Basin of New Mexico, which is part of the much larger Permian Basin.)

Results of the California Desert study, although expressed as numerical probabilities, are qualitative. A specific geographic cell can be classed as prospective for the occurrence of various metals as defined for each of four groups, but no measure of the magnitude of deposits that may be present is given. The probability of correct classification in each of the four groups is conditional upon available geologic information. Thus, there is strong similarity to the general objectives of Griffiths (1978) in attempting to establish relationships by which regional geology may be used to enhance prediction of resource occurrences. Griffiths (1978) and his students (e.g., Labovitz and Griffiths, 1982) have worked principally with information extracted from geologic maps, whereas the California Desert Region study of Lambie et al. (1983) used other geologic data in addition to map data.

SUMMARY

Analysis using probability trees makes clear that probabilities of breaching a nuclear-waste repository by exploratory drilling are interdependent with probabilities pertaining to future human behavior and history. The probabilities of breaching by drilling cannot be calculated in isolation from these influences. Any forecast of future activity must, however, consider the resource potential of a repository site based on available information.

Objective estimates of mineral-resource potential on a unit-area or unit-rock-volume basis cannot be produced, if a variety of mineral commodities are to be considered. These estimates cannot be prepared for the near future, much less for the next 10,000 years. The only widely available resource data that allow a reasonable estimate on a rock-volume basis pertain to oil and gas fields in some regions that have been maturely explored. Available data pertaining to mineral production of commodities, including metals, stone, brines, and sand and gravel, are difficult to translate to a total resource inventory on a unit-area basis, and their translation to a unit-rock-volume basis invokes assumptions about depth versus exploitability.

The future economics of mineral-resource exploration cannot be objectively treated but cannot be ignored. If depth is ignored as an economic factor, the possible presence of mineral deposits at substantial depths, which may be exploitable in the future even though they are not today, presents difficulties. Thus, probability estimates of future economic factors must be largely subjective.

Areal shapes and sizes of some deposits are only poorly known, and existing information does not permit a statistical inventory to be made. Areal shape and size statistics are essential if intersection probabilities by boreholes are to be calculated accurately. Information on areal shapes and sizes of oil fields is more or less available, but detailed statistical studies of shapes of oil fields have not been made. Procedures for estimating intersection probabilities for point nets (e.g., grid drilling) are available only for targets defined by circles or ellipses, although procedures for other shapes could be developed.

If a "probability envelope" can be established for a repository site, calculating probabilities of intersection by vertical boreholes of prescribed depths is readily handled by Monte Carlo procedures. The probability of intersection by slanted boreholes could also be calculated.

No method is known by which the area of exhaustion (or more properly, rock-volume of exhaustion) of a borehole can be determined for most commodities. If assumptions are made concerning areal shapes and sizes of mineral resource targets, and if detection is clearly established if the target is intersected, procedures could be devised that would yield the number of boreholes that would be required to exhaust the resource potential of a unit area, assuming boreholes extend to some limiting depth.

Objectively estimating the probability of intersection of a repository's probability envelope by future exploratory drilling is currently impossible. This conclusion has been derived from an abstract analysis of the problem and from the available literature. The issue of future exploration is largely unaddressed in the literature, except that resource potential studies are germane to future exploration.

If it is feasible to drill an extensive series of exploratory holes beyond the periphery of a repository probability envelope to assess the geology and resource potential near a proposed site to some specified depth, then the mineral resource potential could be estimated subjectively on the basis of a wealth of objective geologic information. If such drilling is considered seriously, it should be preceded by such regional resource studies as URV analysis.

11

Conclusions

C. John Mann
and Regina L. Hunter

The probabilistic standard for disposal of high-level and transuranic nuclear waste in the United States has been a beneficial stimulus to geologists and other earth scientists in several ways. First, the standard has forced us to describe natural events and processes in probabilistic terms, which differs dramatically from our more traditional deterministic approach to the science. Except in a few fields, like mining and hydrology, geologists have been tardy to embrace the probabilistic treatment of most geologic phenomena and parameters and in so doing have lagged behind other sciences. Nevertheless, numerous probabilistic techniques have been developed for application to geologic questions, and in many instances, the necessary data are available to answer these questions. In other instances, the probabilistic approach has indicated where new data should be obtained. This literature review brings together many of the probabilistic techniques that have been applied to geologic questions. We hope that this examination will stimulate fresh thinking about applications of probabilistic techniques in geology.

Second, we have been forced by the nuclear-waste disposal standard to examine more closely and in greater detail many components, variables, and interactions existing in and between geologic phenomena and, similarly, to reexamine traditional geologic concepts, approaches, and methods of analysis. In many cases this reexamination has provided us with greater insight into nature and how it operates. Third, extensive and critical use of probabilities in geologic applications has demonstrated clearly the shortcomings of using index parameters such as means, medians, modes, or ranges in lieu of actual probability density functions. Use of index parameters alone, rather than the more appropriate density function, discards information and produces an inherently larger uncertainty in the final result. Finally, a probabilistic approach to geology provides us with a new perspective on our existing philosophies, biases, and bases of inferences that can only be beneficial in advancing our science. This new perspective inevitably will provide us with a new focal point from which to address old problems in our science and seek new and better answers.

EVENTS AND PROCESSES

Probabilistic analyses of geologic repositories have demonstrated the need to treat geologic events and processes differently, although in actuality the distinction between them may not be clear and is more likely to depend on the purpose of the analyst and the interval examined than on any inherent characteristic of the event or process itself. An event is a result of some process, but the same igneous eruption that is treated as an event by a stratigrapher may be treated as a process by a volcanologist. Nature provides a full spectrum of phenomena between simple, discrete events that result from processes of extreme brevity through complex, continuous events that result from processes acting over long durations. The former phenomena suddenly introduce easily recognized geologic changes; the latter appear to be entirely process, because the event is indistinct and the geologic situation changes constantly but slowly.

It may be difficult to decide whether a phenomenon is an event or a process, and the decision may depend on the particular application and on the detail required in an analysis. Nonetheless, a distinction is desirable and necessary because of the manner in which probabilities normally are established for each. Events that appear as a discrete, recognizable entity can be treated best as a random variable, and probabilities can be attached to them through empirical density functions established from similar events in nature. That is, actual occurrences are the basis for the probability; we do not necessarily have to understand the phenomena involved. Although an event is considered a random variable mathematically, this should not be construed to mean that the event is random geologically. Excellent examples are known that clearly demonstrate that many events are not random, such as volcanism and faulting. However, other geologic events, such as meteoritic impacts and storm floods, appear to be random geologically.

Geologic processes, conversely, are largely deterministic responses to the physical characteristics of a geologic system; unfortunately some characteristics may not be well known and their attributes probably vary spatially and temporally. These processes could be treated deterministically (in theory) if the parameters were adequately understood and if values for all variables were known accurately for all points in space and time. Most processes are too complex and current understanding of them is too limited to permit simple deterministic analysis. Two alternatives to deterministic analysis exist. Geologic processes may be treated stochastically, i.e., randomness is considered to be an inherent feature of the important parameters, or the analysis may view the process as deterministic and only the data as uncertain. Both of these alternative treatments have been considered in this study. Stochastic analysis or modeling is not inferior to deterministic analysis; correctly formulated probabilistic treatment is equivalent or comparable to deterministic treatment, as has been demonstrated clearly in hydrology. Stochastic modeling can be superior to deterministic modeling of some natural systems in two important ways. First, it can handle natural variations inherent in all geologic processes and material in a simple and accurate manner. Second, it is better able to predict accurately when all data are not available and all variables are not well known and understood.

Processes differ fundamentally from events in another important way. Processes may occur continuously, regardless of what events occur, and thus are inherent in all geologic systems. Rates vary, however, depending on specific geologic environments and conditions. When a process is progressing slowly, events associated with that process may be occurring at frequencies that place them outside the concern of a particular analysis. Events may or may not occur during a specified period; even a large probability of occurrence does not guarantee occurrence.

Uncertainties associated with processes and the final probabilistic prediction may be larger than those for events, because values for variables in processes may be difficult to determine and because inherent spatial and temporal variations may be poorly understood. Correctly designed sampling plans for determining accurate values for variables in mathematical modeling of geologic processes are important.

Although simple in concept, separating geologic events from processes may be difficult in practice. For those cases which are not clear, appropriate treatment may present a significant problem. Drilling for an economic deposit of a natural resource appears to be an event; but how can it be treated as an event when a probability for future occurrences cannot be based on past occurrences? Perhaps treatment as a process would be more appropriate. Conversely, climatic change clearly is a process because it occurs continuously. Nonetheless, on the time scale of concern for geologic repositories and its effect upon geologic processes, it geologically is an "event" that triggers changes in the geologic system. Can it be treated satisfactorily as an event? These are types of questions that arise when economic and temporal constraints exist in probabilistic treatment of geologic events and processes.

Probability enters all scientific activity, often in subtle ways. In addition to the natural variability of geologic systems, phenomena, processes, and events, subjectivity is inherent in all human activity through the decisions or selections that must be made in all we do. Examples in scientific activity include parameter values that are selected for use in analyses; number of samples that are collected to determine a data value; the data base used; method of analysis employed; accuracy of measurements; care taken to eliminate gross errors, bias, and systematic errors; and even the philosophy and experience of the individual. All these forms of subjectivity, which can never be eliminated but ideally can be minimized, express themselves in an analysis as an inherent probabilistic component. Subjectivity is present regardless of the type of analysis, accuracy of the computer, or validity of the operational concept; each of these, in turn, will be supplying additional variations that also enter the final results as a probabilistic component. These variations are even present in deterministic analyses, where they routinely have been ignored. Nevertheless, deterministic analysis generally and ubiquitously is considered in science to be superior to stochastic or probabilistic analysis, because it is thought to result in a unique and "exact" value.

METHODS OF DETERMINING PROBABILITIES

Although ideally all probabilities determined for geologic events and processes should be based on objective, empirical data, this is impossible in practice. Our knowledge of actual events and processes is so limited and data are so sparse that probabilities for some variables must be determined by subjective methods. Consequently, geologists must use a full range of methods that extend from objective probabilities based on extensive observations and empirical density functions through Bayesian methods to purely subjective probabilities established by expert judgment and Delphi methods. In general, uncertainty associated with each probability increases greatly from the objective to subjective basis.

Determining probabilities for geologic events and processes resolves into three major stages. First, individual probabilities must be established in the most accurate and realistic manner possible for each elementary component, variable, or event in the system. In addition, an accurate or realistic uncertainty must be associated with that elementary probability. When empirical distribution functions are the basis for the initial probability assignment, the first stage is relatively simple and straightforward for both probability and associated uncertainty. However, as fewer data exist for an objective assignment of probability and greater subjectivity enters into a determination, the best course to follow becomes less clear and more difficult to identify. Second, all individual components of an event or process must be combined (1) in a geologically correct fashion, that is each component must carry a weighting factor comparable to its importance and contribution in producing the final result, and (2) in a mathematically proper manner that establishes as accurately as possible the probability of the final result and its associated uncertainty. Third, individual events and processes must be combined in a manner that will mimic as accurately as possible the complex systems and interactions occurring in nature.

Methods for determining probabilities can be divided into axiomatic approaches, frequentist approaches, modeling approaches, and expert judgment. Which approach is most suitable in a particular application depends on the complexity of the system producing the event or process whose probability is to be determined, how well the system is understood, and the number and nature of data available. The use of any of these approaches implicitly assumes that an event or process under consideration is random. Geologic events and processes for the most part are not truly random phenomena, however, because they rarely occur repeatedly under identical conditions. Most scientists would agree that many geologic events and processes actually are deterministic phenomena and could be treated as such if they were understood adequately and enough information were available to calculate these deterministic relationships. However, because of complexity in these phenomena and the lack of understanding of true relationships, many of them must be treated as probabilistic phenomena. Expert opinion in assigning probabilities is more subjective and uncertain than the use of axiomatic, frequentist, or modeling approaches. In the absence of adequate data for objective probability determinations, probabilities estimated by experts may be used for performance assessment.

The preferred order and basis by which probabilities of geologic events and processes should be estimated might be

Objective estimates obtained from
- Edf's derived from local data,
- Edf's derived from regional data,
- Edf's derived from global data,
- Probability density functions or cdf's derived from theoretical basis,
- Limited observational data (too incomplete to establish edf or cdf;

Objective estimates (without distribution functions) established by
- Empirical Bayesian methods,
- Bayesian methods,
- Expert judgment;

Subjective estimates established by
- Delphi methods,
- Relative-probability estimation.

The exact order of preference might be given differently by different authors. All of these methods are used in estimating probabilities of events and processes in the fields considered here: thermomechanical behavior, mining and tunneling, hydrology, climatology, seismicity and tectonics, seismic hazard, volcanology, geochemistry, and resource exploration.

PROBABILITY FUNCTIONS

Four probabilistic functions of natural events and processes are of interest: empirical distribution function (edf), cumulative distribution function (cdf), probability density function (pdf), and complementary cumulative distribution function (ccdf). Cdf's would provide exact probability determinations if they were available. Unfortunately, cdf's rarely are known for naturally occurring variables; in real applications, they must be estimated. The estimator for a cdf that is most free of assumptions is an edf derived from geologic data gathered locally. However, local data rarely will be adequate for this purpose and, hence, dependence will normally fall upon generic edf's derived from regional or, less satisfactorily, global data for the same geologic event or process. Generic data are less desirable than local data because geologic conditions rarely, if ever, will be identical for different geologic sites or times.

Lacking an edf or a cdf derived from empirical data, the next best estimate of a pdf or cdf normally will be derived from theoretical considerations based on knowledge of the geologic event or process. Historically, some distributions predicted theoretically have been reasonably close to subsequently observed data distributions; nonetheless, their accuracy directly depends upon current understanding of a particular geologic event or process. In some instances, knowledge was poor at the time predictions were made, and theoretically derived cdf's have been significantly in error as a result.

Often geologic data will be so incomplete that an accurate edf or cdf cannot be determined from either available data or our existing knowledge of the event or process. Even so, some estimates of objective probabilities of future occurrences still may be obtained on the basis of expert judgment and extremely limited data. These probability estimates will be objective determinations made without benefit of an edf or cdf. An example commonly encountered in repository performance assessments is frequency of movement along a recent fault; total data obtained from field studies may consist of only two or three values. Uncertainty in these objective probability estimates typically will be great.

Even for geologic events and processes that cannot be assigned probabilities through normal objective means, additional techniques embodying great uncertainty remain. Subjective probabilities may be established using the expert judgment of several specialists from appropriate fields of geologic knowledge. Great uncertainties invariably are associated with subjective probabilities established in this manner because they represent, for example, a poor data base or one that cannot be quantified easily and is not capable of experimental verification.

Another technique is assignment of relative probabilities to events and processes on the basis of general knowledge and individual or group understanding of them. These are subjective, order-of-magnitude estimates. Although relative probabilities will have the greatest uncertainties associated with them and they are the least desirable of all probability assessments, they may be sufficiently informative to be helpful in assessing performance risks. The basis for both probability and uncertainty estimates and the methods by which they were obtained must be stated clearly and explicitly so that others are able to freely evaluate the validity of each probability and uncertainty estimate. Even such ill-defined probabilities as those generated subjectively may be adequate for some parts of the performance assessment, for example, where the estimated probability is very small or very large.

UNCERTAINTY, SENSITIVITY, AND EXTREME VALUES

Probabilistic approaches to nuclear-waste disposal have demonstrated also the importance of uncertainty analysis, sensitivity analysis, and extreme values. The uncertainty associated with the data used in generating each probability estimate must be specified. These uncertainties about the values of the probability estimates (as opposed to uncertainty about, say, occurrence, which is represented by the probability estimates) must be propagated through the analysis of risk, so that a realistic measure of uncertainty in total risk is obtained. Combining uncertainties may be accomplished by various means, including Monte Carlo, Maximus, and Bayesian techniques. The final uncertainty is just as important to decision-makers as the final probability, because their judgment regarding environmental safety must be based on both aspects. Each case must be considered independently, because a larger probability of releasing radionuclides to the environment with an associated smaller uncertainty may be more desirable than a smaller probability of release with an associated larger uncertainty.

Uncertainty analysis reveals major sources of uncertainty in the geologic system and indicates where the greatest diminishment of uncertainty is possible. From this information, the most efficient use of effort to decrease uncertainty can be determined. A large uncertainty for an important variable automatically signals that additional research and refinement of probabilistic analysis is desirable or necessary.

Sensitivity analysis similarly identifies which parameters are contributing most importantly to a final result or behavior of a system. It can be used to most efficiently improve the final probability accuracy by indicating where additional data may be desirable, where more detailed modeling may be necessary, where an improved method of analysis may be beneficial, and where other various improvements, or contrarily where less detail, should be made to accomplish the primary goals of analysis. Sensitivity analysis is extremely valuable as new, complex natural systems are investigated and modeled, because it provides rapid insight into interactions between variables and identifies relative importance of various components of a system.

Rare events, extreme values, or those values contained in the tails of probability density functions are important in all risk assessments, because they commonly will present the greatest hazard and most extreme cases that may be encountered. They also are least common and least likely to be recognized or recorded in empirical data. Furthermore, they are most sensitive to errors in fitting a prediction curve to empirical data that will be necessary if they are to be used in risk assessment analysis. No completely satisfactory method exists for addressing this problem. Considerable attention should be directed by earth scientists to extreme values, to better forms of analysis of extreme values, and to identification of pdf's associated with various and specific types of geologic events and processes.

Probabilities assigned to geologic events and processes, as well as risk assessment generally, demonstrate again the crucial importance of explicit statement of all assumptions made in an analysis and data bases used in determining probabilities and uncertainties for geologic events and processes. This classic fundamental of all science—a clear statement of what has been done, what was used and not used, what assumptions were made or are inherent in the analysis, and what constraints existed—unfortunately is largely ignored or merely implied today in the earth sciences. Risk assessments of all types, a primary use for probabilities in geology, require constant updating as new data, new concepts, new methods of analysis, and improved understanding of nature become available. In addition, explicit statement of limitations in the analysis, data used, accuracy, and precision will permit meaningful, productive discussion of conflicting viewpoints and opinions and will result in more accurate safety assessments and determination of validity in opposing approaches to common geologic problems.

GENERAL ASPECTS OF DETERMINING PROBABILITIES IN SPECIFIC FIELDS

Thermomechanical effects on geologic systems have not been considered probabilistically in the past. However, deleterious effects in the HLW repository are uncertain and must be modeled probabilistically; deterministic analyses that take into account the uncertainty in the data are most commonly used. Thermomechanical theory, which is well advanced and understood, has dealt only with intact rock that contains no discontinuities, such as bedding, joints, fracture, and faults. Much less is known about how discontinuities modify thermomechanical predictions. Thermal properties of various lithologies are well established by abundant data and are easily measured for specific sites. Accurate and realistic predictions for rock masses will depend largely on successful modeling of the discontinuities revealed through field exploration data and their expected modification to thermomechanical behavior.

A good basic understanding exists in tunneling and mining engineering of structural performance and associated uncertainty in geotechnical projects. Numerous deterministic and probabilistic methods have been used for predicting performance. Interaction mechanisms, such as temperature changes, permeability changes, and other factors that may alter stresses with time, will require more extensive study and modeling for long performance (thousands of years). Most serious limitations appear to reside in making comprehensive risk assessments and in making predictions exceeding short-term experience and data by the two to three orders of magnitude necessary for geologic repositories. Uncertainties will be larger because experience is limited. Generally, initial-value accuracy is a direct function of the thoroughness of data collection prior to and during construction.

Quantitative methods and procedures are well developed for probabilistic predictions in ground-water hydrology; they are well established, accurate, and dependable. Water flow through saturated porous media is well understood and may be treated either deterministically or stochastically with equal dependability. Flow through fractured media is less well understood but relatively well modeled stochastically. Hydrological parameter values are well known and easily obtained for specific sites through adequate data collection. Greatest uncertainty and lack of knowledge are found in spatial variations of these parameters; data collection never can be sufficiently extensive to learn all parameter values without error.

Climatic models designed to predict changes beyond the immediate future (weather) do not exist and no comprehensive effort has been directed toward this goal. Although some elements of such a model now exist individually, their integration into a general climatic prediction model has not been accomplished. In addition, problems remain unsolved concerning boundary conditions, dust loading and CO_2 concentrations, variations in solar energy output, details of slowly varying components, and relative importance of forced and free variations in the climatic system. Spatial resolution is poor, system details are poorly understood, hydrospheric circulation and thermal characteristics are known inadequately, and better prediction of ice volumes will be required. Data sets are needed for past records of boundary conditions and slowly varying components, for spatial variability in the present system, and for validation purposes to test model simulations.

Reasonable and accurate predictions of probabilities for tectonic behavior at specific sites are possible today only where local areas have been tectonically active more recently than 100,000 years ago and where a good stratigraphic and seismic-data record has documented faulting movements. Where the data are sparse or where little or no activity has occurred, prediction will be difficult and undependable; uncertainty will be great. No way currently is recognized for predicting new fault movements where none has already occurred. To the extent that prediction is possible, quantification has been based on probabilistic models and standard statistical analysis. Few quantitative data are available for prediction of magnitudes. Generally, inadequate data availability, no consensus about quality and significance of available data, uncertainty as to current understanding of tectonic processes, and an inability to assign meaningful and generally accepted probabilities limit the usefulness of probability assignments to tectonic activity in aseismic areas.

Probabilistic analysis, assessment, and prediction for future seismic hazard is well established and routine practice today. However, which of several developed prediction methods to use in any given instance is not always clear and the confidence associated with each is not well established. Like tectonic prediction, seismic-hazard predictions depend strongly on the available data and frequency of past occurrences. Probabilistic assignments in seismically active areas generally are good for more frequent events but diminish rapidly for less frequent events and aseismic areas; data on extreme events are severely limited.

Probabilities of volcanic activity are derived from knowledge about past occurrences. Probabilistic predictions for areas of recent activity will be good to excellent generally, whereas in stable portions of Earth's crust that have undergone no recent volcanic activity, predictions may be extremely poor and unrealistic. The stratigraphic record provides the basic data required for probabilistic determination and it may range from nonexistent to excellent. An inherent assumption is that future volcanic activity will continue at the same intensity and frequency as past activity; this assumption has been questioned recently and may not be valid. Large uncertainties normally will be attached to probabilities for volcanic events and processes. Prediction for initiating new volcanic activity is virtually nonexistent today; however, encouraging and potentially useful methods are being explored that may make this possible in the near future.

Probabilistic analysis of geochemical behavior generally is well developed and methods are well established. Many data exist and new data are determined easily for specific sites. The greatest uncertainty in prediction arises form natural variability in the geologic media and environments in which chemical laws and processes will be operating. Because of these inherent variations and extreme sensitivity of certain chemical reactions and interactions to slight environmental changes, accurate prediction with small uncertainties may not always be possible. In nuclear-waste applications, well-established K_d and retardation factors measured in the laboratory may not be exactly suitable for field applications, whereas field determinations for these factors require a number of assumptions; in both cases, uncertainties of probabilities based on these factors will be increased.

Although reasonably objective probabilities of existing resources and potential resources are possible today, no objective basis is known for predicting human activity, economic conditions beyond the immediate few years, or technological advances that are essential for objective prediction of future exploration for natural resources. Thus subjective probability assignments by expert opinion or Delphi methods will dominate any probabilistic analysis; this condition undoubtedly will prevail indefinitely into the future. Data bases for objective assignments of probabilities for resources are generally poor, except for petroleum deposits, because they do not contain adequate data on size of deposits, grades, distribution, and variability. Thus, an exploratory assessment of natural resources might be necessary to gain adequate local data for any specific site.

References

Abou-Sayed, A. S., C. E. Brechtel, and R. J. Clifton, 1978, "In-Situ Stress Determination by Hydrofracturing: A Fracture Mechanics Approach," *Jour. Geophys. Research*, v. 83, pp. 2851-2862.

Abry, Claude G., 1975, "Geostatistical Model for Predicting Oil: Tatum Basin, New Mexico," *Bull. Amer. Assoc. Petroleum Geologists*, v. 59, pp. 2111-2122.

Acharya, H. K., 1984, "Changes in Volcanic and Seismic Activity Prior to Great Earthquakes," *Earthquake Notes*, v. 55, 11 pp.

Adamchuk, Y. V., Y. P. Fedorovskiy, P. P. Firstov, K. P. Kim, A. V. Storcheus, and P. P. Titov, 1984, "A Method for Continuous Measurement of the Bulk Activity of Emanations for Predicting Earthquakes and Eruptions of Volcanoes Using Plastic and Combined Scintillational Detectors," *Inst. Atomic Energ. Prepr.*, 3894/14, 31 pp.

Adams, John, 1980, "Active Tilting of the United States Midcontinent; Geodetic and Geomorphic Evidence," *Geology*, v. 8, n. 9, pp. 442-446.

Adams, John, 1981, "Earthquake-Dammed Lakes in New Zealand," *Geology*, v. 9, n. 5, pp. 215-219.

Adem, J., A. Berger, Ph. Gaspar, P. Pestiaux, and J. P. van Ypersele, 1984, "Preliminary Results on the Simulation of the Climate During the Last Deglaciation with a Thermodynamic Model," *in* A. Berger, J. Imbrie, J. Hays, G. Kukla, and B. Saltzman, *eds., Milankovitch and Climate—Understanding the Response to Astronomical Forcing*, Part 2, D. Reidel Co., Dordrecht, Holland, pp. 527-537.

Agbabian Associates, 1981, "Application of a 3-Dimensional Finite Element Analysis for the Design of Large Underground Intersection," Final Report R-7814-5159, Agbabian Associates, El Segundo, CA, 398 pp.

Aggarwal, Yash P., Lynn R. Sykes, John Armbruster, and Marc L. Sbar, 1973, "Premonitory Changes in Seismic Velocities and Prediction of Earthquakes," *Nature*, v. 241, n. 5385, pp. 101-104.

Agnew, Duncan Carr, and William L. Ellsworth, 1991, "Earthquake Prediction and Long-Term Hazard Assessment," *U.S. National Report to International Union of Geodesy and Geophysics, Reviews of Geophysics, Supplement*, American Geophysical Union, pp. 877-889.

Agterberg, F. P., 1970, "Autocorrelation Functions in Geology," *in* D. F. Merriam, *ed., Geostatistics*, Plenum, New York, pp. 113-141.

Agterberg, F. P., 1973, *Geomathematics*, Elsevier, New York, 569 pp.

Agterberg, F. P., 1974, "Automatic Contouring of Geological Maps to Detect Target Areas for Mineral Exploration," *Internat. Assoc. Math. Geol. Jour.*, v. 6, pp. 373-395.

Ahnert, F., 1970, "Functional Relationships Between Denudation, Relief, and Uplift in Large Mid-Latitude Drainage Basins," *Amer. Jour. Sci.*, v. 268, n. 3, pp. 243-263.

Aitchison, J., and I. R. Dunsmore, 1975, *Statistical Prediction Analysis*, Cambridge Univ. Press, Cambridge, 273 pp.

Aki, K., 1966, "Generation and Propagation of G-waves from the Niigata Earthquake of June 16, 1964," *Bull. Earthquake Research Inst.*, Tokyo University, v. 44, pp. 23-88.

Aki, K., 1967, "Scaling Law of Seismic Spectrum," *Jour. Geophys. Research*, v. 72, n. 4, pp. 1217-1231.

Aki, K., 1968, "Seismic Displacements Near a Fault," *Jour. Geophys. Research*, v. 73, n. 16, pp. 5359-5376.

Aki, K., 1981, "A Probabilistic Synthesis of Precursory Phenomena," *in* D. W. Simpson and P. G. Richards, *eds., Earthquake Prediction, An International Review*, M. Ewing Series, v. 4, Amer. Geophys. Union, Washington, DC, 680 pp.

Aki, K., 1982, "Strong Motion Prediction Using Mathematical Modeling Techniques," *Bull. Seismol. Soc. Amer.*, v. 72, pp. S1-S13.

Aki, K., 1985a, "An Overview and Issues: Review of Simulation Procedures," *in* R. E. Scholl and J. L. King, *eds., Strong Ground Motion Simulation and Earthquake Engineering Applications*, Publication No. 85-02, Earthquake Engineering Research Institute, Berkeley, CA, pp. 11-1-11-9.

Aki, K., 1985b, "Technical Evaluation: Source Effects," *in* R. E. Scholl and J. L. King, *eds., Strong Ground Motion Simulation and Earthquake Engineering Applications*, Publication No. 85-02, Earthquake Engineering Research Institute, Berkeley, CA, pp. 33-1-33-7.

Aki, K., 1987, "Magnitude-Frequency Relation for Small Earthquakes; A Clue to the Origin of f_{max} of Large Earthquakes," *Jour. Geophys. Research B*, v. 92, n. 2, pp. 1349-1355.

Aki, K., and P. G. Richards, 1980, *Quantitative Seismology*, W. H. Freeman and Co., San Francisco.

Aldrich, M. J., Jr., and A. W. Laughlin, 1982, *Orientation of Least-Principal Horizontal Stress: Arizona, New Mexico, and the Trans-Pecos Area of West Texas, Stress Data and References*, LA-9158-MAP, Los Alamos National Laboratory, Los Alamos, NM.

Alexander, D. H., M. J. Apted, A. M. Liebetrau, A. E. van Luik, R. E. Williford, and P. G. Doctor, 1985, "Conceptual Model for Deriving the Repository Source Term," *in* C. M. Jantzen, J. A. Stone, and R. C. Ewing, *eds., Scientific Basis for Nuclear Waste Management VIII*, Proc. Materials Research Soc. Symp., v. 44, pp. 439-450.

Allais, M., 1957, "Method of Appraising Economic Prospects over Large Territories," *Management Sci.*, v. 3, pp. 285-347.

Allard, B., V. Olofsson, and B. Torstenfelt, 1984, "Environmental Actinide Chemistry," *Inorganica Chimica Acta*, v. 94, pp. 205-221.

Allen, Clarence R., 1975, "Geological Criteria for Evaluating Seismicity," *Geol. Soc. Amer. Bull.*, v. 86, n. 6, pp. 1041-1057.

Alonzo, E. E., 1976, "Risk Analysis of Slopes and Its Application to Slopes in Canadian Sensitive Clays," *Geotechnique*, v. 26, pp. 453-472.

Alwan, A. K., and P. A. Williams, 1980, "The Aqueous Chemistry of Uranium Minerals, Part 2. Minerals of the Liebigite Group," *Min. Mag.*, v. 43, pp. 665-667.

Ambraseys, N. N., 1971, "Value of Historical Records of Earthquakes," *Nature*, v. 232, n. 5310, pp. 375-379.

Anagnos, T., 1985, *A Stochastic Earthquake Recurrence Model with Temporal and Spatial Dependence*, Ph.D. Thesis, Department of Civil Engineering, Stanford University, Stanford, CA.

Anagnos, T., and A. S. Kiremidjian, 1984, "Stochastic Time-Predictable Model for Earthquake Occurrences," *Bull. Seismol. Soc. Amer.*, v. 74, n. 6, pp. 2593-2611.

Anderson, J. G., 1976, "Motions Near a Shallow Rupturing Fault: Evaluation of the Effect Due to the Free Surface," *R. Astron. Soc., Geophys. Jour.*, v. 46, n. 3, pp. 575-593.

Anderson, J. G., 1979, "Estimating the Seismicity from Geological Structure for Seismic Risk Studies," *Bull. Seismol. Soc. Amer.*, v. 69, n. 1, pp. 135-158.

Anderson, M. P., 1979, "Using Models to Simulate the Movement of Contaminants Through Groundwater Flow Systems," *CRC Crit. Rev. Environ. Control*, v. 9, n. 2, pp. 97-155.

Anderson, O. L., and P. C. Grew, 1977, "Stress Corrosion Theory of Crack Propagation with Applications to Geophysics," *Rev. Geophys. Space Phys.*, v. 15, n. 1, pp. 77-104.

Andersson, J., 1985, "Predicting Mass Transport in Fractured Rock with the Aid of Field Data," *Internat. Symp. Stochastic Approach to Subsurface Flow*, Montevillargenne, France, pp. 537-545.

Andersson, J., A. M. Shapiro, and J. Bear, 1984, "A Stochastic Model of a Fractured Rock Conditioned by Measured Information," *Water Resources Research*, v. 20, n. 1, pp. 79-88.

Andrews, D. J., 1978, "Coupling of Energy Between Tectonic Processes and Earthquakes," *Jour. Geophys. Research*, v. 83, B5, pp. 2259-2264.

Archambault, C., G. Scarpinati, J. Stoschek, and J. C. Tanguy, 1982, "Analyse et surveillance thermiques de volcans actifs," *Bull. PIRPSEV*, No. 66, 27 pp.

Archuleta, R. J., 1985, "Simulating the Near Source Ground Motion of the 1979 Imperial Valley Earthquake," *in* R. E. Scholl and J. L. King, *eds., Strong Ground Motion Simulation and Earthquake Engineering Applications*, Publication No. 85-02, Earthquake Engineering Research Institute, Berkeley, CA.

Archuleta, R. J., and G. A. Frazier, 1978, "Three-Dimensional Numerical Simulation of Dynamic Faulting in a Half-Space," *Bull. Seismol. Soc. Amer.*, v. 68, n. 3, pp. 541-572.

Aric, K., R. Gutdeutsch, H. Friedmann, F. Hernegger, C. Y. King, A. Chalupka, E. Wild, and G. Dirniger, *eds.*, 1984, "Radon Measurements for Earthquake Prediction Research," *Progress Report, 1983*, Vienna Univ., Austria, 23 pp.

Arnett, R. C., R. G. Baca, J. A. Caggiano, S. M. Price, R. E. Gephart, and S. E. Logan, 1980, *Preliminary Hydrologic Release Scenarios for a Candidate Repository Site in the Columbia River Basalts*, RHO-BWI-ST-12, Rockwell Hanford Operations, Richland, WA.

Arteaga, Daisy M., 1978, "Unit Regional Value of the Mineral Resources in Venezuela," Master of Engineering Thesis, Pennsylvania State Univ., 91 pp.

Arthur D. Little, Inc., 1980, *Technical Support of Standards for High-Level Radioactive Waste Management, Volume D, Release Mechanisms*, EPA 520/4-79-007D, U.S. Environmental Protection Agency.

Asaoka, A., 1978, "Observational Procedure of Settlement Prediction," *Soils and Foundations*, v. 18, n. 4, pp. 87-100.

Ashley, D. B., D. Veneziano, H. H. Einstein, and M. H. Chan, 1981, "Geological Prediction and Updating in Tunnelling—A Probabilistic Approach," *Proc. 22nd U.S. Symp. Rock Mech.*, pp. 391-396.

Askins, R. C., 1980, *Advanced Statistical Techniques for Seismic Hazard Attenuation Law Estimation*, M.S. Thesis, Department of Civil Engineering, Massachusetts Institute of Technology, Cambridge, MA.

Atkinson, G. M., 1984, "Attenuation of Strong Ground Motion in Canada from a Random Vibrations Approach," *Bull. Seismol. Soc. Amer.*, v. 74, n. 6, pp. 2629-2653.

Atkinson, J. H., and D. M. Potts, 1977, "Stability of a Shallow Circular Tunnel in Cohesionless Soil," *Geotechnique*, v. 27, n. 2, pp. 203-215.

Bada, J. L., 1985, "Amino Acid Racemization Dating of Fossil Bones," *Ann. Rev. Earth and Planetary Sci.*, v. 13, pp. 241-268.

Bada, J. L., and R. Protsch, 1973, "Racemization Reaction of Aspartic Acid and Its Use in Dating Fossil Bones," *Proc. Nat. Acad. Sci. USA*, v. 70, pp. 1331-1334.

Baecher, G. B., 1972, "Site Exploration: A Probabilistic Approach," Ph.D. Thesis, Massachusetts Institute of Technology, Cambridge, MA.

Baecher, G. B., 1979, "Search Strategies in Geotechnical Engineering," *in* O. G. Ingles, *ed., Third Internat. Conf. Applications of Stat. and Probability in Soil and Structural Engineering*, Univ. of New South Wales, pp. 23-38.

Baecher, G. B., 1983a, "Professional Judgment and Prior Probabilities in Engineering Risk Assessment," *Internat. Conf. Applications of Stat. and Probability to Soils and Structural Engineering*, IV, Florence.

Baecher, G. B., 1983b, "Statistical Analysis of Rock Mass Fracturing," *Internat. Assoc. Math. Geol. Jour.*, v. 15, n. 2, pp. 329-348.

Baecher, G. B., 1985, "Statistical Methods in Site Characterization," *in* S. K. Saxena, *ed., Updating Subsurface Sampling of Soils and Rocks and Their In-Situ Testing*, Engineering Foundation, New York.

Baecher, G. B., and H. H. Einstein, 1981, "Size Effect in Rock Testing," *Geophys. Research Let.*, v. 18, n. 7, pp. 671-674.

Baecher, G. B., and T. S. Ingra, 1981, "Stochastic FEM in Settlement Predictions," *Jour. Geotechnical Engineering Div., Proc. Amer. Soc. Civil Engineers*, v. 107, pp. 449-464.

Baecher, G. B., and N. A. Lanney, 1978, "Trace Length Biases in Joint Surveys," *19th U.S. Symp. Rock Mech.*, Stateline, NV, pp. 56-65.

Baecher, G. B., N. A. Lanney, and H. H. Einstein, 1977, "Statistical Description of Rock Properties and Sampling," *in* F. D. Wang and G. B. Clark, *eds., Proc. 18th U.S. Symp. Rock. Mech.*, Amer. Inst. Mining Engineering, Keystone, CO, pp. 5C1.1-5C1.8.

Bailey, R. A., P. R. Beauchemin, F. P. Kapinos, and D. W. Klick, 1983, *The Volcano Hazards Program; Objectives and Long Range Plans*, Open-file Report No. 83-0400, U.S. Geological Survey, 35 pp.

Bakr, A. A., L. W. Gelhar, A. L. Gutjahr, and J. R. MacMillan, 1978, "Stochastic Analysis of Spatial Variability in Subsurface Flows, 1: Comparison of One- and Three-Dimensional Flows," *Water Resources Research*, v. 14, n. 2, pp. 263-271.

Balachandra, M. B., 1978, *The Three Dimensional Structural Analysis of Double-Entry and Single-Entry Coal Mines. Vol. I: Three Dimensional Finite Element Analysis of Crosscut and Entry Intersection of a Double-Entry Coal Mine*, Agbabian Associates, Report No. R-7638-2-4162, El Segundo, CA, 137 pp.

Barazangi, M., and J. Dorman, 1969, "World Seismicity Maps Compiled from ESSA," Coast and Geodetic Survey Epicenter Data, 1961-1967, *Bull. Seismol. Soc. Amer.*, v. 59, n. 1, pp. 369-380.

Barbreau, A., M. Bonnet, P. Goblet, J. Margat, G. DeMarsily, and P. Peaudecerf, 1982, "Long Term Prediction of the Fate of Nuclear Waste Deeply Buried in Granite," *in* G. DeMarsily and D. F. Merriam, *eds., Predictive Geology*, Pergamon Press, Elmsford, NY, pp. 153-170.

Barkatt, A., P. B. Macedo, B. C. Gibson, and C. J. Montrose, 1985, "Modelling of Waste Form Performance and System Release," *in* C. M. Jantzen, J. A. Stone, and R. E. Ewing, *eds., Scientific Basis for Nuclear Waste Management VIII*, Proc. Materials Research Soc. Symp., vol. 44, pp. 3-13.

Barlow and Haun, Inc., 1978, *Appendices to Oil and Gas Production Reserves and Resources Report*, (unnumbered), Minerals Division, Wyoming Department of Economic Planning and Development, 108 pp.

Barney, G. S., 1981a, *Radionuclide Reactions with Groundwater and Basalts from Columbia River Basalt Formations*, RHO-SA-217, Rockwell Hanford Operations, Richland, WA, 47 pp.

Barney, G. S., 1981b, *Evaluation of Methods for Measurement of Radionuclide Distribution in Groundwater/Rock Systems*, RHO-BWI-LD-47, Rockwell Hanford Operations, Richland, WA, 34 pp.

Barney, G. S., J. D. Navratil, and W. W. Schulz, *eds.*, 1984, *Geochemical Behavior of Disposed Radioactive Waste*, American Chemical Society Symp. Ser. 246, Washington, DC.

Barnola, J. M., D. Raynaud, Y. S. Korotkevich, and C. Lorius, 1987, "Vostok Ice Core Provides 160,000-Year Record of Atmospheric CO_2," *Nature*, v. 329, pp. 408-414.

Barouch, E., and G. M. Kaufman, 1975, "A Probability Model for Undiscovered Oil and Gas Resource," *IIASA Conf. Energy Resource Estimates*, Laxenburg, Austria.

Barstow, N. L., K. G. Brill, Jr., O. W. Nuttli, and P. W. Pomeroy, 1981, *An Approach to Seismic Zonation for Siting Nuclear Electric Power Generating Facilities in the Eastern U.S., Final Technical Report, 28 Sept. 1978-28 Sept. 1980*, NUREG/CR-1577, U.S. Nuclear Regulatory Commission, Washington, DC, 366 pp.

Barsukov, O. M., 1972, "Variations of Electric Resistivity of Mountain Rocks Connected with Tectonic Causes," *Tectonophysics*, v. 14, n. 3-4, pp. 273-277.

Bartlein, P. J., 1988, "Late-Tertiary and Quaternary Paleoenvironments," *in* B. Huntley and T. Webb, III, *eds., Vegetation History*, Kluwer Academic Publishers, Dordrecht, pp. 113-152.

Barton, C. M., 1977, "A Geotechnical Analysis of Rock Structure and Fabric in the C.S.A. Mine, Cobar, New South Wales," C.S.I.R.0. Australia Div. of Applied Geomechanics, Technical Paper No. 24, 30 pp.

Barton, N., 1973, *A Review of the Shear Strength of Discontinuities in Rock Masses with Particular Reference to Slope Stability*, Internal report, Norwegian Geotechnical Institute.

Barton, N., 1982, *Modeling Rock Joint Behavior from In Situ Block Tests: Implications for Nuclear Waste Repository Design*, ONWI-308, Office of Nuclear Waste Isolation, Columbus, OH.

Barton, N. R., and S. Bandis, 1982, "Effects of Block Size on the Shear Behavior of Jointed Rock," *Proc. 23rd U.S. Symp. Rock Mech.*, Berkeley, CA, pp. 739-760.

Barton, N. R., R. Lien, and J. Lunde, 1974, "Engineering Classification of Rock Masses for the Design of Tunnel Supports," *Rock Mech.*, v. 6, n. 4, pp. 189-236.

Barton, N., K. Bakhtar, S. Woodhead, and D. Bush, 1983, *Joint Characterization and Modeling at NTS G-Tunnel*, TRE 83-40, Terra Tek report to Sandia National Laboratories, Salt Lake City, UT.

Basalt Waste Isolation Project, 1981, *Supporting Document—Applicability of the Near-Surface Test Facility Program to Repository Design*, RSD-BWI-TI-024, Rockwell Hanford Operations, Richland, WA, 95 pp.

Basalt Waste Isolation Project, 1983, *Principal Borehole Report, Borehole RRL-2*, SD-BWI-TI-113, Rockwell Hanford Operations, Richland, WA, 95 pp.

Basalt Waste Isolation Project, 1985, *Task V, Engineering Study No. 9, Underground Repository Layout*, RHO-BWI-ES-023, Rockwell Hanford Operations, Richland, WA, 474 pp.

Bathe, K.-J., 1976, "Static and Dynamic Geometric and Material Nonlinear Analysis Using ADINA," Report 82448-2, Acoustics and Vibration Laboratory, Mechanical Engineering Department, Massachusetts Institute of Technology, Cambridge, MA, 430 pp.

Bathe, K.-J., 1977, "ADINA—A Finite Element Program for Automatic Dynamic Incremental Nonlinear Analysis," Report No. 82448-1, Acoustics and Vibration Laboratory, Mechanical Engineering Department, Massachusetts Institute of Technology, Cambridge, MA (revised May, 1977), 240 pp.

Bathe, Klaus-Juergen, *ed.*, 1985, *Nonlinear Finite Element Analysis and ADINA, Proceedings of the 5th ADINA Conference*, Mechanical Engineering Department, Massachusetts Institute of Technology, Cambridge, MA, 378 pp.

Bathe, K.-J., and A. Chaudhary, 1984, "On Finite Element Analysis of Large Deformation Frictional Contact Problems," *Unification of Finite Element Methods*, North Holland Publ. Co., Amsterdam.

Bathe, K.-J., and A. Chaudhary, 1985, "A Solution Method for Planar and Axisymmetric Contact Problems," *Internat. Jour. Numerical Methods in Engineering*, v. 21, n. 1, pp. 65-88.

Bayley, Susan E., Malcolm D. Siegel, Malcolm Moore, Stuart Faith, 1990, *Sandia Sorption Data Management System Version 2 (SSDMS II) User's Manual*, SAND89-0371, Sandia National Laboratories, Albuquerque, NM.

Bear, J., 1972, *Dynamics of Fluids in Porous Media*, Elsevier, New York, 764 pp.

Beer, G., and J. L. Meek, 1980, "A 'Boundary Finite Element' for Underground Mining Applications," *in* C. A. Brebbia, *ed., New Developments in Boundary Element Methods*, CML Publications, Southampton, pp. 281-294.

Begin, Z. B., 1983, *An Approach to the Dating of Young Faults Through the Analysis of Gullies Dissecting Fault Scarps*, Publication of Environmental and Hydrogeology Div., Geological Survey of Israel, 13 pp.

Begin, Z. B., D. F. Meyer, and S. A. Schumm, 1981, "Development of Longitudinal Profiles of Alluvial Channels in Response to Base-Level Lowering," *Earth Surface Processes and Landforms*, v. 6, n. 1, pp. 49-68.

Bender, B., 1984, "A Two-State Poisson Model for Seismic Hazard Estimation," *Bull. Seismol. Soc. Amer.*, v. 74, n. 4, pp. 1463-1468.

Benjamin, Jack R., and Associates, 1980, *Additional Probability Analysis of Surface Rupture Offset Beneath Reactor Buildings, General Electric Test Reactor*, JBA-111-013-01, prepared for General Electric Company, San Jose, CA, variously paginated.

Benjamin, J. R., and C. A. Cornell, 1970, *Probability, Statistics and Decisions for Civil Engineering*, McGraw Hill, New York, 864 pp.

Benson, L. V., and L. S. Teague, 1979, *A Study of Rock-Water-Nuclear Waste Interactions in the Pasco Basin, Washington, Part I. Distribution and Composition of Secondary and Primary Mineral Phases in Basalts of the Pasco Basin, Washington*, LBL-9677, Lawrence Berkeley Laboratory, 157 pp.

Benson, L. V., and L. S. Teague, 1980, "Diagenesis of Basalts from the Pasco Basin, Washington—I. Distribution and Composition of Secondary Mineral Phases," (unpublished MS).

Berger, A., 1977, "Power and Limitation of an Energy-Balance Climate Model As Applied to the Astronomical Theory of Paleoclimates," *Palaeogeography, Palaeoclimatology, Palaeoecology*, v. 21, n. 3, pp. 227-235.

Berger, A., 1984, "Accuracy and Frequency Stability of the Earth's Orbital Elements During the Quaternary," *in* A. Berger, J. Imbrie, J. Hays, G. Kukla, and B. Saltzman, *eds., Milankovitch and Climate—Understanding the Response to Astronomical Forcing*, Part 2, D. Reidel Co., Dordrecht, Holland, pp. 3-39.

Berger, A. L., J. Imbrie, J. Hays, G. Kukla, and B. Saltzman, *eds.*, 1984, *Milankovitch and Climate—Understanding the Response to Astronomical Forcing*, D. Reidel Co., Dordrecht, Holland.

Berlanga, Juan M., and John W. Harbaugh, 1981, "Use of Seismic Data for Statistical Estimation of Outcome Probabilities of Complexly Faulted Structures in Tabasco Basin, Mexico," *Bull. Amer. Assoc. Petroleum Geologists*, v. 65, n. 3, pp. 422-437.

Bernreuter, D. L., J. B. Savy, R. W. Mensing, and D. H. Chung, 1984, *Seismic Hazard Characterization of the Eastern U. S.: Methodology and Interim Results for Ten Sites*, UCRL-53527, NUREG/CR-3756, Lawrence Livermore Laboratories, Livermore, CA, 524 pp.

Bernreuter, D. L., J. B. Savy, R. W. Mensing, J. C. Chen, and B. C. Davis, 1985, *Seismic Hazard Characterization of the Eastern United States*, UCID-20421, Vols. 1 and 2, Lawrence Livermore National Laboratory, Livermore, CA.

Berrill, J. B., and R. O. Davis, 1980, "Maximum Entropy and the Magnitude Distribution," *Bull. Seismol. Soc. Amer.*, v. 70, n. 5, pp. 1823-1831.

Bertozzi, G., M. D'Alessandro, F. Girardi, and M. Vanossi, 1977, "Evaluation de la Surete du Stockage de Dechets Radioactifs en Formations Geologiques: une Application Preliminaire de la 'Fault Tree Analysis' a des Formations Salines," *Risk Analysis and Geologic Modelling in Relation to the Disposal of Radioactive Wastes into Geologic Formations*, Proceedings of a Workshop Organized Jointly by the Nuclear Energy Agency, OECD, and the Commision of the European Communities.

Bianchi, L., and D. T. Snow, 1968, "Permeability of Crystalline Rock Interpreted from Measured Orientations and Apertures of Fractures," *Annals Arid Zone*, v. 8, pp. 231-245.

Bieniawski, Z. T., 1973, "Engineering Classification of Jointed Rock Masses," *The Civil Engineer in South Africa*, v. 15, n. 12, pp. 335-344.

Bieniawski, Z. T., 1974, "Geomechanics Classification of Rock Masses and Its Application in Tunneling," *Proc. 3rd Internat. Congress Rock Mech.*, ISRM, Denver, CO, Vol. II A, pp. 27-32.

Bieniawski, Z. T., 1976, "Rock Mass Classifications in Rock Engineering," *Proc. Symp. Exploration for Rock Engineering*, A. A. Balkema, Capetown, South Africa, v. 1, pp. 97-106.

Bieniawski, Z. T., 1979, "The Geomechanics Classification in Rock Engineering Applications," *Proc. 4th Internat. Congress Rock Mech.*, Montreux, Vol. 2, pp. 41-48.

Bingham, C., 1964, "Distributions on the Sphere and on the Projective Plane," Ph.D. Thesis, Yale Univ., New Haven, CN.

Bingham, Felton W., and George E. Barr, 1979, *Scenarios for Long-Term Release of Radionuclides from a Nuclear-Waste Repository in the Los Medanos Region of New Mexico*, SAND78-1730, Sandia Laboratories, Albuquerque, NM, 90 pp.

Bischke, R. E., 1974, "A Model of Convergent Plate Margins Based on Recent Tectonics of Shikoku, Japan," *Jour. Geophys. Research*, v. 79, n. 32, pp. 4845-4857.

Blacic, J. D., and J. M. Christie, 1984, "Plasticity and Hydrolytic Weakening of Quartz Single Crystals," *Jour. Geophys. Research*, v. 89, n. 6, pp. 4223-4239.

Bloom, A. L., W. S. Broecker, J. M. A. Chappell, R. K. Mathews, and K. J. Mesolella, 1974, "Quaternary Sea-Level Fluctuations on a Tectonic Coast: New $^{230}Th/^{234}U$ Dates from the Huon Peninsula, New Guinea," *Quaternary Research*, v. 4, n. 2, pp. 185-205.

Bloomfield, P., K. S. Deffeys, G. S. Watson, Y. Benjamin, and R. A. Stine, 1979, *Volume and Area of Oilfields and Their Impact on Order of Discovery*, Resource Estimation and Validation Project, Dept. of Statistics, Princeton Univ., Prepared for U.S. Department of Energy under Contract No. EI-78-S-01-6540, 53 pp.

Boatwright, J., 1982, "A Dynamic Model of Far-Field Acceleration," *Bull. Seismol. Soc. Amer.*, v. 72, n. 4, pp. 1049-1068.

Bollinger, G. A., 1973, "Seismicity and Crustal Uplift in the Southeastern U.S.," *Amer. Jour. Sci.*, v. 273-A, pp. 396-408.

Bolt, B. A., and N. A. Abrahamson, 1982, "New Attenuation Relations for Peak and Expected Accelerations of Strong Ground Motion," *Bull. Seismol. Soc. Amer.*, 6 Part A, pp. 2307-2322.

Boneva, L. I., D. G. Kendall, and I. Stefanov, 1971, "Spline Transformations," *Jour. Royal Stat. Soc.*, Ser. B, v. 33, pp. 1-70.

Bonham-Carter, G. F., and C. F. Chung, 1983, "Integration of Mineral Resource Data for the Kasmere Lake Area, Northwest Manitoba, with Emphasis on Uranium," *Internat. Assoc. Math. Geol. Jour.*, v. 15, pp. 25-46.

Bonilla, M. G., 1967, *Historic Surface Faulting in Continental United States and Adjacent Parts of Mexico*, Open-file Report, U.S. Geological Survey, 36 pp.

Bonilla, M. G., 1970, "Surface Faulting and Related Effects," in R. L. Wiegel, *ed., Earthquake Engineering*, Prentice Hall, Englewood Cliffs, NJ, pp. 47-74.

Bonilla, M. G., 1979, "Historic Surface Faulting—Map Patterns, Relations to Subsurface Faulting, and Relation to Preexisting Faults," *Proc. Conf. VIII, Analysis of Actual Fault Zones in Bedrock*, Open-file Report 79-1239, U.S. Geological Survey, pp. 36-65.

Bonilla, M. G., 1982, *Evaluation of Potential Surface Faulting and Other Tectonic Deformation*, Open-file Report 82-732, U.S. Geological Survey.

Bonilla, M. G., and J. M. Buchanan, 1970, *Interim Report on World-Wide Historic Surface Faulting*, U.S. Geological Survey Open-file Report, 32 pp.

Bonilla, M. G., R. K. Mark, and J. J. Lienkaemper, 1984, *Statistical Relations between Earthquake Magnitude, Surface Rupture Length and Surface Fault Displacement*, Open-file Report 84-256, U.S. Geological Survey.

Bonilla, M. G., H. A. Villalobos, and R. E. Wallace, 1984, *Exploratory Trench Across the Pleasant Valley Fault, Nevada*, Prof. Paper 1274-B, U.S. Geological Survey, pp. B1-B14.

Boore, D. M., 1983a, "Strong-Motion Seismology," *U.S. National Report to International Union of Geodesy and Geophysics: Contributions in Seismology, Reviews of Geophysics and Space Physics*, v. 21, n. 6, pp. 1308-1318.

Boore, D. M., 1983b, "Stochastic Simulation of High-Frequency Ground Motions Based on Seismological Models of the Radiated Spectra," *Bull. Seismol. Soc. Amer.*, v. 73, pp. 1865-1894.

Boore, D. M., 1986, "Short-Period P- and S-Wave Radiation from Large Earthquakes: Implications for Spectral Scaling Relations," *Bull. Seismol. Soc. Amer.*, v. 76, n. 1, pp. 43-64.

Boore, D. M., and G. M. Atkinson, 1987, "Stochastic Prediction of Ground Motion and Spectral Response Parameters at Hard-Rock Sites in Eastern North America," *Bull. Seismol. Soc. Amer.*, v.77, n. 2, pp. 440-467.

Boore, D. M., and W. B. Joyner, 1982, "The Empirical Prediction of Ground Motion," *Bull. Seismol. Soc. Amer.*, 6 Part B, pp. S269-S286.

Booth, B., 1979, "Assessing Volcanic Risks," *Jour. Geol. Soc. London*, v. 136, pp. 331-340.

Borovec, Z., 1981, "The Adsorption of Uranyl Species by Fine Clays," *Chem. Geol.*, v. 32, n. 1-2, pp. 45-58.

Borovec, Z., B. Kribek, and V. Tolar, 1979, "Sorption of Uranyl by Humic Acids," *Chem. Geol.*, v. 27, pp. 39-46.

Borsetto, M., G. Giuseppetti, S. Martinetti, R. Ribacchi, and T. Silvestri, 1983, "Design and Construction of the Timpagrande Powerhouse, *Rock Mech. and Rock Engineering*, v. 16, n. 2, pp. 85-116.

Boufassa, A., and M. Armstrong, 1989, "Comparison Between Different Kriging Estimators," *Internat. Assoc. Math. Geol. Jour.*, v. 21, n. 3, pp. 331.

Box, George E. P., and Gwilym M. Jenkins, 1976, *Time Series Analysis—Forecasting and Control*, Holden-Day, San Francisco, 575 pp.

Box, G. E. P., and G. C. Tiao, 1973, *Bayesian Inference in Statistical Analysis*, Addison-Wesley, Reading, MA.

Bradley, R. S., 1985, *Quaternary Paleoclimatology: Methods of Paleoclimatic Reconstruction*, Allen and Unwin, Boston, 472 pp.

Bradley, R. S., 1988, "The Explosive Volcanic Eruption Signal in Northern Hemisphere Continental Temperature Records," *Climatic Change*, v. 12, pp. 221-243.

Bradshaw, R. L., and W. C. McClain, *eds.*, 1971, *Project Salt Vault: A Demonstration of the Disposal of High-Activity Solidified Wastes in Underground Salt Mines*, ORNL-4555, Oak Ridge National Laboratory, Oak Ridge, TN.

Brady, B. H. G., 1979, "A Direct Formulation of the Boundary Element Method of Stress Analysis for Complete Plane Strain," *Inter-nat. Jour. Rock Mech., Mining Sci. and Geomech. Abstr.*, v. 16, pp. 235-244.

Brady, B. H. G., and J. W. Bray, 1978, "The Boundary Element Method for Determining Stress and Displacements Around Long Openings in a Triaxial Stress Field," *Internat. Jour. Rock Mech., Mining Sci. and Geomech. Abstr.*, v. 15, n. 1, pp. 21-28.

Bras, R. L., 1975, "Rainfall-Runoff as a Spatial Stochastic Process: Data Collection and Synthesis," Ph.D. Thesis, Massachusetts Institute of Technology, Cambridge, MA.

Bray, J. W., 1967, "A Study of Jointed and Fractured Rock," *Felsmech. Ingenieurgeol.*, v. 5, pp. 117-136 and 197-216.

Brebbia, C. A., and S. Walker, 1978, "Introduction to the Boundary Element Method," *in* C. A. Brebbia, *ed., Recent Advances in Boundary Element Methods*, Pentech Press, London, pp. 1-44.

Bridges, M. C., 1975, "Presentation of Fracture Data for Rock Mechanics," *Proc. 2nd Autr-NZ Conf. Geomech.*, Brisbane, pp. 144-148.

Brillinger, D. R., 1982, "Some Bounds for Seismic Risk," *Bull. Seismol. Soc. Amer.*, v. 72, n. 4, pp. 1403-1410.

Brillinger, D. R., and H. K. Preisler, 1984a, "An Exploratory Analysis of the Joyner-Boore Attenuation Data," *Bull. Seismol. Soc. Amer.*, v. 74, n. 4, pp. 1441-1450.

Brillinger, D. R., and H. K. Preisler, 1984b, "Further Analysis of the Joyner-Boore Attenuation Data," *Bull. Seismol. Soc. Amer.*, v. 75, n. 2, pp. 611-614.

Broccoli, A. J., and S. Manabe, 1987, "The Influence of Continental Ice, Atmospheric CO_2, and Land Albedo on the Climate of the Last Glacial Maximum," *Climate Dynamics*, v. 1, pp. 87-99.

Brooke, J., 1978, "Geomathematical Investigation of Fault Populations at Selected Locations," *19th U.S. Rock Mechanics Symposium*, v. 1, pp. 23-29, Stateline, NV.

Brooker, P. I., 1975a, "Avoiding Unnecessary Drilling," *Austral. Inst. Min. Metall.*, v. 253, pp. 21-23.

Brooker, P. I., 1975b, "Optimal Block Estimation by Kriging," *Austral. Inst. Min. Metall.*, v. 253, pp. 15-19.

Brooker, P. I., 1977, "Robustness of Geostatistical Calculations: A Case Study," *Austral. Inst. Min. Metall.*, v. 264, pp. 61-68.

Brooker, P. I., 1980, "Kriging," *in* A. G. Royle, *ed., Geostatistics*, McGraw-Hill, New York, pp. 41-60.

Brown, A. A., 1960, "Search Theory and Problems of Exploration Boring," *Minerals Ind. Exp. Sta. Bull.*, v. 72, pp. 33-37, Pennsylvania State Univ.

Brown, D. M., 1984, "Stochastic Analysis of Flow and Solute Transport in a Variable-Aperture Rock Fracture," M.S. Thesis, Massachusetts Institute of Technology, Cambridge, MA, 69 pp.

Brown, D. M., and L. W. Gelhar, 1985, "The Hydrology of Fractured Rocks: A Literature Review," Ralph M. Parsons Lab, Draft Report, 76 pp.

Brown, E. T., and G. Hocking, 1976, "The Use of the Three Dimensional Boundary Integral Equation Method for Determining Stresses at Tunnel Intersections," *Proc. Second Australian Tunnelling Conf.*, Australasia Inst. of Min. and Metall., Parkville, Victoria, Australia, pp. 55-64.

Brown, E. T., and J. A. Hudson, *eds.*, 1984, *ISRM Symposium, Proc. Design and Performance of Underground Excavations*, British Geotechnical Soc., London.

Brown, L. D., 1978a, "Prediction of Large-Scale Deformation of the Solid Earth," *in* National Academy of Science, *Geophysical Predictions*, Washington, DC, pp. 19-28.

Brown, L. D., 1978b, "Recent Vertical Crustal Movements Along the East Coast of the United States," *Tectonophysics*, v. 44, pp. 205-231.

Brown, L. D., and J. E. Oliver, 1976, "Vertical Crustal Movements from Leveling Data and Their Relation to Geologic Structure in the Eastern United States," *Rev. Geophys. Space Phys.*, v. 14, pp. 13-35.

Brune, J. N., 1968, "Seismic Moment, Seismicity, and Rate of Slip Along Major Fault Zones," *Jour. Geophys. Research*, v. 75, pp. 777-784.

Brune, J. N., 1970, "Tectonic Stress and the Spectra of Seismic Shear waves from Earthquakes," *Jour. Geophys. Research*, v. 75, pp. 4997-5002.

Brune, J. N., 1971, "Correction," *Jour. Geophys. Research*, v. 76, pp. 5009.

Brunk, H. D., 1978, "Univariate Density Estimation by Orthogonal Series," *Biometrika*, v. 65, pp. 521-528.

Bryson, Reid A., and Brian M. Goodman, 1980, "Volcanic Activity and Climatic Changes," *Science*, v. 207, n. 4435, pp. 1041-1044.

Buckley, J. T., Y. C. Kim, M. S. Nataraja, and D. H. Tiktinsky, 1986, "Evaluating Host Rock Representativeness Through Geostatistics," *Proc. 27th U.S. Symposium on Rock Mechanics*, University of Arizona, Tucson, AZ, pp. 821-826.

Bucknam, R. C., and R. E. Anderson, 1979, "Estimation of Fault-Scarp Ages from a Scarp-Height-Slope-Angle Relationship," *Geology*, v. 7, pp. 11-14.

Bucknam, R. C., S. T. Algermissen, and R. E. Anderson, 1980, "Patterns of Late Quaternary Faulting in Western Utah and an Application in Earthquake Hazard Evaluation," *Proceedings, Conference X, Earthquake Hazards Along the Wasatch and Sierra-Nevada Frontal Fault Zones*, Open-file Report 80-801, U.S. Geological Survey, pp. 299-314.

Budyko, M. I., 1982, *The Earth's Climate: Past and Future*, Academic Press, New York, 307 pp.

Bufe, C. G., P. W. Harsh, and R. O. Burford, 1977, "Steady-State Seismic Slip: A Precise Recurrence Model," *Geophys. Research Let.*, v. 4, pp. 91-94.

Bulau, J. R., B. R. Tittman, M. Abdel-Gawad, and C. Salvado, 1984, "The Role of Aqueous Fluids in the Internal Friction of Rock," *Jour. Geophys. Research*, v. 89, pp. 4207-4212.

Bull, W. B., 1978a, *Geomorphic Tectonic Activity Classes of the South Front of the San Gabriel Mountains, California*, Unpubl. final report, U.S. Geological Survey Contract 14-08-0001-G-394, 59 pp.

Bull, W. B., 1978b, *Tectonic Geomorphology of the Mojave Desert, California*, Unpubl. final report, U.S. Geological Survey Contract 14-08-0001-G-394, 188 pp.

Bull, W. B., C. M. Menges, and L. D. McFadden, 1979, *South Front of the San Gabriel Mountains, Southern California: Tectonic Geomorphology of the Palmdale Uplift Region; Stream Terraces of the San Gabriel Mountains, Southern California*, U.S. Geological Survey Final Contract Report 14-08-001-G-394, 100 pp.

Burns, J. Q., and R. M. Richard, 1964, "Attenuation of Stresses for Buried Cylinders," *Proc. Symp. on Soil-Structure Interaction*, Tucson, AZ, pp. 378-392.

Burridge, R., and L. Knopoff, 1967, "Models and Theoretical Seismicity," *Bull. Seismol. Soc. Amer.*, v. 57, pp. 341-371.

Burton, B. W., 1982, *Geologic Evolution of the Jemez Mountains and Their Potential for Future Volcanic Activity*, LA-8795-GEOL, Los Alamos National Laboratory, Los Alamos, NM, 35 pp.

Busk, H. G., 1929, *Earth Flexures*, Cambridge Univ. Press, Cambridge, MA, 106 pp.

Byerlee, J. D., 1968, "Brittle-Ductile Transition in Rocks," *Jour. Geophys. Research*, v. 73, pp. 4741-4750.

Cacuci, D. G., P. J. Maudlin, and C. U. Parks, 1983, "Adjoint Sensitivity Analysis of Extremum-Type Responses in Reactor Safety," *Nuclear Sci. Engineering*, v. 83, n. 1, pp. 112-135.

Call, R. D., 1984, "Statistical Methods for Designing Pit Slope Geometry," Lecture to the Swedish Mining Research Institute (BeFo), Stockholm.

Call, R. D., 1985, Personal Communication, Call and Nicholas, Inc., Tucson.

Call, R. D., and Y. C. Kim, 1978, "Composite Probability of Instability for Optimizing Pit Slope Design," Preprints for *19th U.S. Symp. Rock Mech.*

Call, R. D., and D. E. Nicholas, 1978, "Prediction of Step Path Failure Geometry for Slope Stability Analysis," Preprints for *19th U.S. Symp. Rock Mech.*

Cameron-Clarke, I. S., and S. Budavari, 1981, "Correlation of Rock Classification Obtained from Borecore and In Situ Observations," *Engineering Geol.*, v. 17, pp. 19-53.

Campbell, D. L., 1978, "Investigation of the Stress-Concentration Mechanism for Intraplate Earthquakes," *Geophys. Research Let.*, v. 5, pp. 477-479.

Campbell, J. E., R. L. Iman, and M. Reeves, 1980, *Risk Methodology for Geologic Disposal of Radioactive Waste: Transport Model Sensitivity Analysis*, NUREG/CR-1377, SAND80-0644, Sandia National Laboratories, Albuquerque, NM.

Campbell, J. E., D. E. Longsine, and R. M. Cranwell, 1981, *Risk Methodology for Geologic Disposal of Radioactive Waste: The NWFT/DVM Computer Code Users Manual*, SAND81-0886, NUREG/CR-2081, Sandia National Laboratories, Albuquerque, NM.

Campbell, J. E., R. T. Dillon, M. S. Tierney, H. T. Davis, P. E. McGrath, F. J. Person, Jr., H. R. Shaw, J. C. Helton, and F. A. Donath, 1978, *Risk Methodology for Geologic Disposal of Radioactive Waste: Interim Report*, NUREG/CR-0458, SAND78-0029, Sandia Laboratories, Albuquerque, NM, 264 pp.

Campbell, K. W., 1981a, "A Ground Motion Model for the Central United States Based on Near Source Acceleration Data," *Earthquakes and Earthquake Engineering: Eastern U.S.*, Ann Arbor Science Publ., Ann Arbor, MI, pp. 213-232.

Campbell, K. W., 1981b, "Near Source Attenuation of Peak Horizontal Acceleration," *Bull. Seismol. Soc. Amer.*, v. 71, pp. 2039-2070.

Campbell, K. W., 1982, "A Preliminary Methodology for the Regional Zonation of Peak Ground Acceleration," *in* M. Sherif, *ed., Proc. Third Internat. Conf. Earthquake Microzonation*, Seattle, WA, pp. 365-376.

Capan, E. C., 1976, "The Difficulty of Assessing Uncertainty," *Jour. Petroleum Technology*, August, pp. 833-850.

Cargill, S. M., D. H. Root, and E. H. Bailey, 1980, "Resource Estimation from Historical Data, Mercury, a Test Case," *Internat. Assoc. Math. Geol. Jour.*, v. 12, pp. 489-522.

Carr, T. R., 1982, "Log-Linear Models, Markov Chains, and Cyclic Sedimentation," *Jour. Sedimentary Petrology*, v. 52, pp. 905-912.

Carr, W. J., 1974, *Summary of Tectonic and Structural Evidence for Stress Orientation at the Nevada Test Site*, Open-file Report 74-0176, U.S. Geological Survey, 53 pp.

Carrera, J., S. P. Neuman, and J. Casanovas, 1985, "On Optimum Parameterization in Aquifer Inverse Problems," *Internat. Symp. Stochastic Approach to Subsurface Flow*, Montevillargenne, France, pp. 235-247.

Carter, J. P., and J. R. Booker, 1983a, "The Effects of Ground Creep on a Lining Installed in a Circular Tunnel," Research Report No. R439, School of Civil and Mining Engineering, Univ. of Sidney, 43 pp.

Carter, J. P., and J. R. Booker, 1983b, "The Behaviour of a Lined Circular Tunnel in Viscoelastic Ground," Research Report No. R451, School of Civil and Mining Engineering, Univ. of Sidney.

Casagrande, A., 1965, "Role of the Calculated Risk in Earthworks and Foundation Engineering," *Jour. Soil Mech. and Foundation Engineering*, Amer. Soc. Civil Engineers, v. 91, no. SM4.

Castle, R. O., J. P. Church, and M. R. Elliott, 1976, "Aseismic Uplift in Southern California," *Science*, v. 192, n. 4236, pp. 251-253.

Cederberg, G., 1985, "TRANQL: A Ground-Water Mass-Transport and Equilibrium Chemistry Model for Multicomponent Systems," Ph.D. dissertation, Stanford University, Stanford, CA.

Chadwick, O. A., S. Hecker, and J. Fonseca, 1984, *A Soils Chronosequence at Terrace Creek: Studies of Late Quaternary Tectonism in Dixie Valley, Nevada*, Open-file Report 84-0090, U.S. Geological Survey, 32 pp.

Chambre, P. L., and T. H. Pigford, 1984, "Prediction of Waste Performance in a Geologic Repository," *in* G. L. McVay, *ed., Scientific Basis for Nuclear Waste Management VII*, Proc. Materials Res. Soc. Symp., v. 26, North-Holland, New York, pp. 985-1008.

Chan, L. Y., and P. E. Goodman, 1983, "Prediction of Support Requirement for Hard Rock Excavations Using Keyblock Theory and Joint Statistics," *Proc. 24th U.S. Symp. Rock Mech.*, pp. 557-576.

Chan, M. H., 1980, "A Geological Prediction and Updating Model in Tunneling," M.S. Thesis, Massachusetts Institute of Technology, Cambridge, MA, 236 pp.

Chan, T., V. Guvanasen, and J. A. Keith Reid, 1985, "Numerical Modelling of Coupled Thermo-Hydro-Mechanical Processes in Nuclear Fuel Waste Disposal," *Internat. Symp. on Coupled Processes Affecting the Performance of a Nuclear Waste Repository*, Lawrence Berkeley Laboratory, Berkeley, CA, pp. 235.

Chandrakant, S., and J. T. Christian, 1977, "Introduction, Numerical Methods and Special Topics," *in* C. S. Desai and J. T. Christian, *eds., Numerical Methods in Geotechnical Engineering*, McGraw-Hill, New York, pp. 1-64.

Chang, C.-Y, M. S. Power, I. M. Idriss, P. G. Sommerville, W. Silva, and P. C. Chen, 1986, *Engineering Characterization of Ground Motion, Task II: Observational Data on Spatial Variations of Earthquake Ground Motion*, NUREG/CR-3805, Vol. 3, U.S. Nuclear Regulatory Commission, 295 pp.

Chappell, J., and H. H. Veeh, 1978, "Late Quaternary Tectonic Movements and Sea-Level Changes at Timor and Atauro Island," *Geol. Soc. Amer. Bull.*, v. 89, pp. 356-367.

Chappellaz, J., J. M. Barnola, D. Raynaud, Y. S. Korotkevich, and C. Lorius, 1990, "Ice-Core Record of Atmospheric Methane Over the Past 160,000 Years," *Nature*, v. 345, pp. 127-131.

Chapple, W. M., and T. E. Tullis, 1977, "Evaluation of the Forces that Drive the Plates," *Jour. Geophys. Research*, v. 82, pp. 1967-1984.

Chen, W., and P. Molnar, 1983, "Focal Depths of Intracontinental and Intraplate Earthquakes and Its Implications for the Thermal and Mechanical Properties of the Lithosphere," *Jour. Geophys. Research*, v. 88, pp. 4183-4214.

Chester, D. K., 1986, "A Statistical Analysis of Flank Eruptions on Etna Volcano, Discussion and Reply," *Jour. Volc. Geoth. Research*, v. 28, pp. 385-395.

Chiburis, E. F., 1981, *Seismicity, Recurrence Rates and Regionalization of the Northeastern U.S. and Adjacent Southeastern Canada*, NUREG/CR-2309, U.S. Nuclear Regulatory Commission, Washington, DC.

Chinnery, M. A., 1979, "An Investigation of Maximum Possible Earthquakes," *in* U.S. Nuclear Regulatory Commission, *Investigations of the Seismological Input to the Safety Design of Nuclear Power Reactors in New England*, NUREG/CR-0563, U.S. Nuclear Regulatory Commission, Washington, DC, 67 pp.

Chirlin, G. R., and G. Dagan, 1980, "Theoretrical Head Variograms for Steady Flow in Statistically Homogeneous Aquifers," *Water Resources Research*, v. 16, pp. 1001-1015.

Chung, C. F., 1983, "SIMSAG: Integrated Computer System for Use in Evaluation of Mineral and Energy Resources," *Internat. Assoc. Math. Geol. Jour.*, v. 15, pp. 47-58.

Chung, C. F., 1984, "Use of the Jackknife Method to Estimate Autocorrelation Functions (or Variograms)," *in* G. Verly and others, *eds., Geostatistics for Natural Resources Characterization*, Nato ASI series C, v. 122, Pt. 1, pp. 55-71.

Chung, C. F., and F. P. Agterberg, 1980, "Regression Models for Estimating Mineral Resources from Geological Map Data," *Internat. Assoc. Math. Geol. Jour.*, v. 12, pp. 473-488.

Cividini, A., G. Maier, and H. Nappi, 1983, "Parameter Estimation of Static Geotechnical Model Using a Bayes' Approach," *Internat. Jour. Rock Mech. and Mining Sci.*, v. 20, pp. 215-226.

Claiborne, H. C., and F. Gera, 1974, *Potential Containment Failure Mechanisms and Their Consequences at a Radioactive Waste Repository in Bedded Salt in New Mexico*, ORNL-TM-4639, Oak Ridge National Laboratory, Oak Ridge, TN.

Clark, I., 1979, *Practical Geostatistics*, Applied Science Publishers Ltd., Great Britain, 129 pp.

Clark, I., 1980, "Part 2: The Semivariogram," *in* A. G. Royle, *ed., Geostatistics*, McGraw-Hill, New York, pp. 29-40.

Clark, M. M., K. K. Harms, J. J. Lienkaemper, P. S. Harwood, K. R. Lajoie, J. C. Matti, J. A. Perkins, M. J. Rymer, A. M. Sarna-Wojciki, R. V. Sharp, J. D. Sims, J. C. Tinsley III, and J. I. Ziony, 1984, *Preliminary Slip-Rate Table and Map of Late-Quaternary Faults of California*, Open-file Report 84-0106, U.S. Geological Survey, 5 sheets, 1:1,000,000 scale, text, 13 pp.

Cleveland, J. M., T. F. Rees, and K. L. Nash, 1983a, "Neptunium and Americium Speciation in Selected Basalt, Granite, Shale and Tuff Ground Waters," *Science*, v. 221, n. 4607, pp. 271-273.

Cleveland, J. M., T. F. Rees, and K. L. Nash, 1983b, "Plutonium Speciation in Selected Basalt, Granite, Shale and Tuff Ground Waters," *Nucl. Tech.*, v. 62, n. 3, pp. 298-310.

Clifton, H. E., 1978, "Storms, Floods and the Coastal Sedimentary Record," (abstr.), Amer. Assoc. Petroleum Geologists Research Colloquium, Oklahoma City.

Clifton, P. M., and S. P. Neuman, 1982, "Effects of Kriging and Inverse Modeling on Conditional Simulation of the Avra Valley Aquifer in Southern Arizona," *Water Resources Research*, v. 18, pp. 1214-1234.

CLIMAP Project Members, 1976, "The Surface of the Ice-Age Earth," *Science*, v. 191, n. 4232, pp. 1131-1137.

CLIMAP Project Members, 1981, "Seasonal Reconstructions of the Earth's Surface at the Last Glacial Maximum," A. McIntyre (compiler), Geol. Soc. Amer. Map and Chart Series, MC-36, pp. 1-18.

Clough, R. W., and J. Penzien, 1975, *Dynamics of Structures*, McGraw-Hill, New York.

Cluff, L. S., A. S. Patwardhan, and K. J. Coppersmith, 1980, "Estimating the Probability of Occurrences of Surface Faulting Earthquakes on the Wasatch Fault Zone, Utah," *Bull. Seismol. Soc. Amer.*, v. 70, pp. 1463-1478.

Coats, D. W., and R. C. Murray, 1984, *Natural Phenomena Hazards Modeling Project: Seismic Hazard Models for DOE Sites*, UCRL-53582, Lawrence Livermore National Laboratories, Livermore, CA, 85 pp.

Coffman, J. L., and C. A. von Hake, *eds.*, 1973, *Earthquake History of the U.S.*, revised edition (through 1970), National Oceanic and Atmospheric Administration Publication 41-1, 208 pp.

COHMAP Members, 1988, "Climatic Changes of the Last 18,000 Years: Observations and Model Simulations," *Science*, v. 241, pp. 1043-1052.

Collins, David R., 1985, "Changes in Drilling Density and Discovery Rates Through Time," *Internat. Assoc. Math. Geol. Jour.*, v. 17, pp. 375-393.

Colman, S. M., and K. L. Pierce, 1979, *Preliminary Map Showing Quaternary Deposits and Their Dating Potential in the Conterminous United States*, U.S. Geological Survey Miscellaneous Field Studies Map MF-1052 (1 sheet, 1:7,500,000 side).

Colman, S. M., and K. L. Pierce, 1981, *Weathering Rinds on Andesitic and Basaltic*

Stones as a Quaternary Age Indicator, Western United States, Prof. Paper 1210, U.S. Geological Survey, 56 pp.

Colman, S. M., and K. Watson, 1983, "Ages Estimated from a Diffusion Equation Model for Scarp Degradation," *Science*, v. 221, n. 4607, pp. 263-265.

Comite Milieux Poreux, de l'association internationale de recherche hydraulique, 1985, *Symposium—The Stochastic Approach to Subsurface Flow, 1985*, Montvillargenne, France.

Committee on the Atmospheric Effects of Nuclear Explosions, National Research Council, 1985, *The Effects on the Atmosphere of a Major Nuclear Exchange*, National Academy Press, Washington, DC, 193 pp.

Cooley, R. L., 1977, "A Method for Estimating Parameters and Assessing Reliability for Models of Steady State Groundwater Flow, 1: Theory and Numerical Properties," *Water Resources Research*, v. 13, pp. 318-324.

Cooley, R. L., 1979, "A Method for Estimating Parameters and Assessing Reliability for Models of Steady State Groundwater Flow, 2: Applications of Statistical Analysis," *Water Resources Research*, v. 15, pp. 603-617.

Cooley, R. L., 1982, "Incorporation of Prior Information on Parameters into Nonlinear Regression Groundwater Flow Models, 1: Theory," *Water Resources Research*, v. 18, pp. 965-976.

Coppersmith, K. J., and R. Kulkarni, 1985, "Tectonic Stress Regime and Tectonic Framework of the Eastern United States," *in* EPRI, *Seismic Hazard Methodology for Nuclear Facilities in the United States, Vol. 1, Development of Methodology*, Research Project P101-29, Electric Power Research Institute, Palo Alto, CA, pp. 3-1-3-61.

Coppersmith, K. J., and D. R. Schwartz, 1984, "Special Section on Fault Behavior and the Earthquake Generation Process," *Jour. Geophys. Research*, v. 89, pp. 5669-5928.

Cording, E. J., A. J. Hendron, and D. U. Deere, 1971, "Rock Engineering for Underground Caverns," *Symp. Underground Rock Chambers, Proc.*, Amer. Soc. Civil Engineers, Phoenix, pp. 567-600.

Cornell, C. A., 1968, "Engineering Seismic Risk Analysis," *Bull. Seismol. Soc. Amer.*, v. 58, pp. 1583-1606.

Cornell, C. A., 1971, "Probabilistic Analysis of Damage to Structures under Seismic Loads," *in* D. A. Howells, I. P. Haigh, and C. Taylor, *eds., Dynamic Waves in Civil Engineering*, John Wiley and Sons, London, pp. 473-493.

Cornell, C. A., and S. R. Winterstein, 1986, *Applicability of the Poisson Earthquake Occurrence Model, Eastern U.S. Seismic Hazard Assessment Project*, Electric Power Research Institute, Palo Alto, CA.

Cornell, C. A., H. Banon, and A. F. Shakal, 1979, "Seismic Motion and Response Prediction Alternatives," *Earthquake Engineering and Structural Dynamics*, v. 7, pp. 295-315.

Corwin, P. F., and H. F. Morrison, 1977, "Self-Potential Variations Preceding Earthquakes in Central California," *Geophys. Research Let.*, v. 4, pp. 171-174.

Cox, J., and P. Talwani, 1984, "Three-Dimensional Description of an Earthquake-Induced Sandblow near Charleston, S. C.," (abstr.), *Eos, Trans. Amer. Geophys. Union*, v. 65, pp. 241.

Craig, R. G., 1987, "Modeling Future Climate Change Affecting Nuclear Waste Disposal: An Outline," *in* W. H. Berger and L. D. Labeyrie, *eds., Abrupt Climatic Change, Evidence and Implications*, D. Reidel, Dordrecht, pp. 341-356.

Crampin, S., R. Evans, and B. K. Atkinson, 1984, "Earthquake Prediction: A New Physical Basis," *Geophys. Jour. Royal Astronom. Soc.*, v. 76, pp. 147-156.

Crandall, S. H., and W. D. Mark, 1963, *Random Vibration in Mechanical Systems*, Academic Press, New York.

Cranwell, R. M., and F. A. Donath, 1980, "An Application of Geometric Probability to the Existence of Faults in Anisotropic Media," *in* C. J. M. Northrup, *ed., Scientific Bases for Nuclear Waste Management Vol. 2*, Proc. Mat. Res. Soc. Ann. Meeting, Plenum Press, New York, pp. 787-794.

Cranwell, R. M., R. V. Guzowski, J. E. Campbell, N. R. Ortiz, 1990, *Risk Methodology for Geologic Disposal of Radioactive Waste: Scenario Selection Procedure*, NUREG/CR-1667, SAND82-1429, U.S. Nuclear Regulatory Commission, Washington, DC.

Cranwell, R. M., J. E. Campbell, J. C. Helton, R. L. Iman, D. E. Longsine, N. R. Ortiz, G. E. Runkle, and M. J. Shortencarier, 1987, *Risk Methodology for Geologic Disposal of Radioactive Waste: Final Report*, NUREG/CR-2452, SAND81-2573, Sandia National Laboratories, Albuquerque, NM, 104 pp.

Crawford, A. M., 1982, "Rock Wedge Stability," *Proc. 23th U.S. Symp. Rock Mech.*, Berkeley, CA, pp. 1057-1064.

Crawford, A. M., and K. C. Lau, 1983, "Stochastic Analysis of Two-Dimensional Rock Wedge Stability," *Proc. 24th U.S. Symp. Rock Mech.*, pp. 71-78.

Cressie, N., and D. M. Hawkins, 1980, "Robust Estimation of the Variogram: I." *Internat. Assoc. Math. Geol. Jour.*, v. 12, pp. 115-125.

Crone, A. J., and S. T. Harding, 1984, "Relationship of Late Quaternary Fault Scarps to Subjacent Faults, Eastern Great Basin, Utah," *Geology*, v. 12, pp. 292-295.

Crone, A. J., and M. N. Machette, 1984, "Surface Faulting Accompanying the Borah Peak Earthquake, Central Idaho," *Geology*, v. 12, pp. 664-667.

Crouch, S. L., 1976, "Solution of Plane Elasticity Problems by the Displacement Discontinuity Method," *Internat. Jour. Numerical Methods in Engineering*, v. 10, n. 2, pp. 301-343.

Crouch, S. L., and A. M. Starfield, 1983, *Boundary Element Methods in Solid Mechanics*, George Allen and Unwin Ltd., London, 322 pp.

Crowe, B. M., 1980, *Disruptive Event Analysis: Volcanism and Igneous Intrusion*, PNL-2882, Battelle Pacific Northwest Laboratory, Richland, WA, 46 pp.

Crowe, B. M., 1981, "Assessment of Effectiveness of Geologic Isolation Systems," *WISAP Release Scenario Analysis Workshop, Sept. 13-15, 1979*, PNL-SA-8257, Battelle Pacific Northwest Laboratory, Richland, WA, pp. 90-92.

Crowe, B. M., and W. J. Carr, 1980, *Preliminary Assessment of the Risk of Volcanism at a Proposed Nuclear Waste Repository in the Southern Great Basin*, U.S. Geological Survey Open-file Report 80-357, 14 pp.

Crowe, B. M., M. E. Johnson, and R. J. Beckman, 1982, "Calculation of the Probability of Volcanic Disruption of a High-Level Radioactive Waste Repository Within Southern Nevada, USA," *Radioactive Waste Management*, v. 3, n. 2, pp. 167-190.

Crowe, B., R. Amos, F. Perry, S. Self, and D. Vaniman, 1982, *Aspects of Possible Magmatic Disruption of a High-Level Radioactive Waste Repository in Southern*

Nevada, Report 9326-MS, Los Alamos National Laboratory, Los Alamos, NM, 43 pp.

Crowe, B., S. Self, D. Vaniman, R. Amos, and F. Perry, 1983, "Aspects of Potential Magmatic Disruption of a High-Level Radioactive Waste Repository in Southern Nevada," *Jour. Geol.*, v. 91, pp. 259-276.

Cruden, D. M., 1977, "Describing the Size of Discontinuities," *Internat. Jour. Rock Mech. and Mining Sci.*, v. 14, pp. 133-137.

Cundall, P., 1971, "A Computer Model for Simulating Progressive Large Scale Movements in Blocky Rock Systems," *Proc. Internat. Symp. Rock Fractures*, Nancy, Topic II-8, 12 pp.

Curtis, D. J., 1976, "Discussion of 'The Circular Tunnel in Elastic Ground,' by A. M. Muir Wood," *Geotechnique*, v. 26, pp. 231-237.

Dacol, D. K., and H. Rabitz, 1983, "Arbitrary Order Functional Sensitivity Densities for Reaction-Diffusion Systems," *Jour. Chem. Phys.*, v. 78, pp. 4905-4914.

Daemen, J. J. K., and C. Fairhurst, 1972, "Rock Failure and Tunnel Support Loadings," *Proc. Internat. Symp. Underground Openings*, Lucerne, pp. 356-369.

Daemen, J. J. K., and C. Fairhurst, 1973, "Tunnel and Shafts in Rock--Part 2, Ground Support Interaction: Fundamentals and Design Implications," Contract No. DACW 45-72C-0018, Missouri River Div., U.S. Army Corps of Engineers, 149 pp.

Dagan, G., 1979, "Models of Groundwater Flow in Statistically Homogeneous Porous Formations," *Water Resources Research*, v. 15, pp. 47-63.

Dagan, G., 1982, "Stochastic Modeling of Groundwater Flow by Unconditional and Conditional Probabilities, 2: The Solute Transport," *Water Resources Research*, v. 18, pp. 835-848.

Dagan, G., 1984, "Solute Transport in Heterogeneous Porous Formations," *Jour. Fluid Mech.*, v. 145, pp. 151-177.

Dagan G., 1985a, "A Note on Higher-Order Corrections of the Head Covariances in Steady Aquifer Flow," *Water Resources Research*, v. 21, pp. 573-578.

Dagan, G., 1985b, "Stochastic Modeling of Groundwater Flow by Unconditional and Conditional Probabilities: The Inverse Problem," *Water Resources Research*, v. 21, pp. 65-72.

Dagan G., 1986, "Statistical Theory of Groundwater Flow: Pore-to-Laboratory, Laboratory-to-Formation, and Formation-to-Regional Scale," *Water Resources Research*, v. 22, pp. 120S-134S.

Dahlberg, E. L., 1975, "Effectiveness of Geologist and Computers in Mapping Potential Hydrocarbon Exploration Targets," *Internat. Assoc. Math. Geol.*, v. 7, pp. 373-394.

D'Alessandro, M., C. N. Murray, G. Bertozzi, and F. Girardi, 1980, "Probability Analysis of Geological Processes: A Useful Tool for the Safety Assessment of Radioactive Waste Disposal," *Radioactive Waste Management (Switzerland)*, v. 1, 25-42.

Dalkey, N. C., 1975, "Toward a Theory of Group Estimation," *in* H. A. Linstone and M. Turoff, *eds., The Delphi Method: Techniques and Applications*, Addison-Wesley, Reading, MA.

Dalkey, N. C., B. Brown, and S. Cochran, 1970, "The Use of Self Rankings to Improve Group Estimates," *Technological Forecasting*, v. 1, pp. 283-292.

Daus, A. D., and E. O. Frind, 1985, "An Alternating Galerkin Technique for Simulation of Contaminant Transport in Complex Groundwater Systems," *Water Resources Research*, v. 21, pp. 653-664.

David, M., 1977, *Geostatistical Ore Reserve Estimation*, Elsevier, Amsterdam, 364 pp.

Davis, E. H., and H. G. Poulos, 1974, *Elastic Solutions for Soil and Rock Mechanics*, John Wiley and Sons, Inc., New York, 411 pp.

Davis, John C., 1973, *Statistics and Data Analysis in Geology*, John Wiley and Sons, Inc., New York, 550 pp.

Davis, John C., and John W. Harbaugh, 1983, "Statistical Appraisal of Seismic Prospects in Louisiana-Texas Outer Continental Shelf," *Bull. Amer. Assoc. Petroleum Geologists*, v. 67, pp. 349-358.

Davis, John H., 1976, "Critical Values for a Measure of Contiguity of Anomalous Points in a Grid Survey," *Internat. Assoc. Math. Geol. Jour.*, v. 8, pp. 89-94.

Davis, S. N., and R. J. M. DeWiest, 1966, *Hydrogeology*, John Wiley and Sons, New York, 447 pp.

DeBra, D. B., 1979, "Impact of Technology on Geodesy," *in* National Academy of Sciences, *Studies in Geophysics: Impact of Technology on Geophysics*, Washington, DC, pp. 22-30.

Deere, D. U., A. J. Hendron, D. F. Patton, and E. J. Cording, 1967, "Design of Surface and Near Surface Construction in Rock," *Failure and Breakage in Rock*, Proc. 8th U.S. Symp. Rock Mech., AIME, pp. 237-302.

Deere, D. U., R. B. Peck, J. E. Monsees, and B. Schmidt, 1969a, "Design of Tunnel Liners and Support Systems," Report of Dept. of Civil Engineering, Univ. of Illinois, Urbana, to OHSGT-U.S. DOT, 419 pp.

Deere, D. U., R. B. Peck, H. W. Parker, J. B. Monsees, and B. Schmidt, 1969b, "Design of Tunnel Support Systems," *Highway Research Record No. 339*, pp. 26-33.

Deffeyes, K. S., B. D. Ripley, and G. S. Watson, 1982, "Stochastic Geometry in Petroleum Geology," *Internat. Assoc. Math. Geol. Jour.*, v. 14, pp. 419-432.

DeGeoffroy, J. G., and S. M. Wu, 1970, "A Statistical Study of Ore Occurrences in the Greenstone Belts of the Canadian Shield," *Econ. Geol.*, v. 65, pp. 496-504.

DeGroot, M. H., 1970, *Optimal Statistical Decisions*, McGraw Hill, New York, 489 pp.

Delany, J. M., S. R. Lundeen, 1989, *The LLNL Thermochemical Data Base and Revised Data and File Format for the EQ3/6 Package*, UCID-21658, Lawrence Livermore National Laboratory, Livermore, CA.

Delhomme, J. P., 1978, "Kriging in the Hydrosciences," *Adv. Water Resources*, v. 1, pp. 251-266.

Delhomme, J. P., 1979, "Spatial Variability and Uncertainty in Groundwater Flow Parameters: A Geostatistical Approach," *Water Resources Research*, v. 15, pp. 269-280.

DeMarsily, G., 1983, *Revue Critique des Methodes d'Analyse de Risques Utilisees Pour Estudier l'Enfouissement des Dechets Nucleaires en Formation Geologique*, Centre d'informatique Geologique, Fontainebleau, France, variously paginated.

DeMarsily, G., 1984, "Methodes et Domaines d'Application de la Mecanique des Fluides en Milieux Poreux et Fissures," *Annales des Mines*, v. 181, pp. 5-16.

DeMarsily, Ghislain, 1986, *Quantitative Hydrogeology*, Academic Press, Inc., New York.

de Montricher, G. F., R. A. Tapia, and J. R. Thompson, 1975, "Nonparametric Maximum Likelihood Estimation of Probability Densities by Penalized Function Methods," *Annals Stat.*, v. 3, pp. 1329-1348.

Denton, G. H., and T. J. Hughes, *eds.*, 1981, *The Last Great Ice Sheets*, Wiley-Interscience, New York, 484 pp.

der Kiureghian, A., and A. H.-S. Ang, 1977, "A Fault-Rupture Model for Seismic Risk Analysis," *Bull. Seismol. Soc. Amer.*, v. 67, pp. 1173-1194.

Dershowitz, W. S., 1979, "Rock Quality Indices and Rock Deformation Behavior in Foundations," M.S. Thesis, Department of Civil Engineering, Massachusetts Institute of Technology, Cambridge, MA, 125 pp.

Dershowitz, W. S., 1984, "Rock Joint Systems," Ph.D. Thesis, Massachusetts Institute of Technology, Cambridge, MA, 308 pp.

Dershowitz, W. S., H. H. Einstein, and G. B. Baecher, 1979, "Prediction of Rock Mass Deformability," *Proc. 4th Internat. Congress Internat. Soc. Rock Mech.*, v. 1, pp. 605-612.

Dershowitz, W. S., B. M. Gordon, J. C. Kafritsas, and H. Herda, 1985, "A New Three Dimensional Model for Flow in Fractured Rock," *Hydrogeology of Rocks of Low Permeability, Proc. 17th Internat. Congress Internat. Assoc. Hydro.*, Tucson, AZ, pp. 441-448.

Descoeudres, F., 1974, "Three-Dimensional Analysis of Tunnel Stability Near the Face in an Elasto-Plastic Rock," *Advances in Rock Mechanics*, Proc. Third Congress Internat. Soc. Rock Mech., Denver, vol. 2, Part B, pp. 1130-1135 (in French).

Detournay, E., 1983, "Two-Dimensional Elasto-Plastic Analysis of a Deep Cylindrical Tunnel Under Non-Hydrostatic Loading," Ph.D. Thesis, Univ. of Minnesota, 133 pp.

Deutsch, W. J., E. A. Jenne, and K. M. Krupka, 1982a, *Computed Solid Phases Limiting the Concentration of Dissolved Constituents in Basalt Aquifers in the Columbia Plateau in Eastern Washington*, PNL-4089, Pacific Northwest Laboratory, Richland, WA.

Deutsch, W. J., E. A. Jenne, and K. M. Krupka, 1982b, "Solubility Equilibria in Basalt Aquifers: The Columbia Plateau, Eastern Washington USA," *Chem. Geol.*, v. 36, pp. 15-34.

Devary, J., and P. G. Doctor, 1982, "Pore Velocity Estimation Uncertainties," *Water Resources Research*, v. 18, pp. 1157-1164.

DeWiest, R. J. M., *ed.*, 1969, *Flow Through Porous Media*, Academic Press, New York, 516 pp.

Dickinson, R. E., R. M. Errico, F. Giorgi, and G. T. Bates, 1989, "A Regional Climate Model for the Western United States," *Climatic Change*, v. 15, pp. 383-422.

Dieterich, J. H., 1972, "Time-Dependent Friction as a Possible Mechanism of Aftershocks," *Jour. Geophys. Research*, v. 77, pp. 3771-3781.

Dieterich, J. H., 1979, "Modeling of Rock Friction, 2, Simulation of Pre-Seismic Slip," *Jour. Geophys. Research*, v. 84, pp. 2169-2175.

Dmowska, R., 1977, "Electromechanical Phenomena Associated with Earthquakes," *Geophys. Surveys (Netherlands)*, v. 3, pp. 157-174.

Donath, Fred A., and Robert M. Cranwell, 1981, "Probabilistic Treatment of Faulting in Geologic Media," *Mechanical Behavior of Crustal Rocks*, Geophysical Monograph 24, American Geophysical Union, pp. 231-241.

Dorn, R. I., 1983, "Cation-Ratio Dating: A New Rock Varnish Age-Determination Technique," *Quaternary Research*, v. 20, pp. 49-73.

Doser, D. I., and R. B. Smith, 1982, "Seismic Moment Rates in the Utah Region," *Bull. Seismol. Soc. Amer.*, v. 72, pp. 525-551.

Dowding, C. H., 1977, "Seismic Stability of Underground Openings," *Proc. 1st Internat. Symp. "Rockstore,"* Stockholm, v. 2, pp. 231-238.

Dowding, C. H., 1978, "Future Challenges in Site Characterization," *in* C. H. Dowding, *ed., NSF Site Characterization Workshop*, Northwestern Univ.

Dowding, C. H., 1979, "Earthquake Stability of Rock Tunnels," *Tunnels and Tunneling*, v. 11, pp. 15-20.

Dowding, C. H., 1985, "Earthquake Response of Caverns: Empirical Correlations and Numerical Modeling," *Proc. 1985 Rapid Excavation and Tunneling Conf.*, Chicago, IL, pp. 71-83.

Dowding, C. H., and J. B. Miller, 1975, "Comparison of Predicted and Encountered Geology for 7 Colorado Tunnels," Report R75-6, Department of Civil Engineering, Massachusetts Institute of Technology, Cambridge, MA, 113 pp.

Dowding, C. H., and A. Rozen, 1978, "Damage to Rock Tunnels from Earthquake Shaking," *Jour. Geotechnical Div.*, Amer. Soc. Civil Engineers, v. 104, pp. 175-192.

Dowding, C. H., C. Ho, and T. Belytschko, 1983, "Earthquake Response of Caverns in Jointed Rock," *in* T. R. Howard, *ed., Seismic Design of Embankments and Caverns*, Amer. Soc. Civil Engineers, New York, pp. 142-156.

Drake, R. F., 1976, "Karst Topography in Florida," M.S. Thesis, Massachusetts Institute of Technology, Cambridge, MA.

Drever, James I., 1982, *The Geochemistry of Natural Waters*, Prentice-Hall, Inc., Englewood Cliffs, NJ.

Drew, Lawrence J., 1979, "Pattern Drilling Exploration: Optimum Pattern Types and Hole Spacings When Searching for Elliptical Shaped Targets," *Internat. Assoc. Math. Geol. Jour.*, v. 11, pp. 223-254.

Drew, L. J., J. H. Schuenemeyer, and W. J. Bawiec, 1979, "Petroleum Exhaustion Maps of the Cretaceous 'D-J' Sandstone Stratigraphic Interval of the Denver Basin," Misc. Invest. Ser., Map I-1138, U.S. Geological Survey.

Drew, L. J., J. H. Schuenemeyer, and D. H. Root, 1980, "Petroleum Resource Appraisal and Discovery-Rate Forecasting in Partially Explored Regions—An Application to the Denver Basin," U.S. Geological Survey Prof. Paper 1138A, pp. A1-A11.

Dubois, Jacques, and Jean Louis Cheminee, 1991, "Fractal Analysis of Eruptive Activity of Some Basaltic Volcanoes," *Jour. Volc. Geoth. Research*, v. 45, pp. 197-208.

Duddeck, H., and J. Erdmann, 1982, "Structural Design Models for Tunnels," *Tunneling 82*, The Institution of Mining and Metallurgy, London, pp. 83-91.

Duguid, J. O., and P. C. Y. Lee, 1977, "Flow in Fractured Porous Media," *Water Resources Research*, v. 13, pp. 558-566.

Duncan, R. A., J. Backman, and L. Peterson, 1989, "Reunion Hotspot Activity Through Tertiary Time: Initial Results from the Ocean Drilling Program, Leg 115," *Jour. Volc. Geoth. Research*, v. 36, pp. 193-198.

Ebel, J. E., 1984, "Statistical Aspects of New England Seismicity from 1975 to 1982 and Implications for Past and Future Earthquake Activity," *Bull. Seismol. Soc. Amer.*, v. 74, pp. 1311-1329.

Eckert, A., and G. M. Atkinson, 1985, "Ground Motion Data in Eastern North America," *Seismic Hazard Methodology for Nuclear Facilities in the Eastern United States, Vol. 3*, EPRI Research Project P101-29, Electric Power Research Institute, Palo Alto, CA.

Eckstand, O. R., 1977, "Mineral Resource Appraisal and Mineral Deposits Computer Files in the Geological Survey of Canada," *Internat. Assoc. Math. Geol. Jour.*, v. 9, pp. 235-243.

Efron, Bradley, and Carl Morris, 1973, "Stein's Estimation Rule and Its Competitions—An Empirical Bayes Approach," *Jour. Amer. Stat. Assoc.*, v. 68, pp. 117-130.

Egger, P., 1973, "Einfluss des Post Failure Verhaltens von Fels auf den Tunnelausbau," *Veroff des Inst. f. Bodenmech. und Felsmech.*, Univ. Karlsruhe, no. 57, 83 pp.

Einstein, H. H., 1977, "Observational Methods in Tunnel Design and Construction," *3rd Annual Conf. DOT Research and Development in Tunneling Technology*, 44 pp.

Einstein, H. H., and G. B. Baecher, 1983, "Probabilistic and Statistical Methods in Engineering Geology," Part I, Exploration, *Rock Mech. and Rock Engineering*, v. 16, pp. 39-72.

Einstein, H. H., and N. Bischoff, 1976, "Design of Tunnels in Swelling Rock," *Proc. 16th U.S. Symp. Rock Mech.*, pp. 185-196.

Einstein, H. H., and S. G. Vick, 1974, "Geologic Model for a Tunnel Cost Model," *Proc. Rapid Excavation and Tunneling Conf.*, Soc. of Min. Eng. of AIME, v. 2, pp. 1701-1720.

Einstein, H. H., G. B. Baecher, and D. Veneziano, 1979, "Risk Analysis for Rock Slopes in Open Pit Mines," BUMINES-OFR-46(8)-81, U.S. Bureau of Mines, Washington, DC, Exec. Summary, Vol. I-V, App. A-M, and App. UM.

Einstein, H. H., N. Bischoff, and E. Hofmann, 1979, "Verhalten von Stollen-sohlen in quellendem Mergel," *Proc. Internat. Symp. f. Untertagebaun*, pp. 296-319.

Einstein, H. H., W. Steiner, and A. S. Azzouz, 1980, *Improved Design of Tunnel Supports. Vol. IV: Tunneling Practices in Austria and Germany*, No. UMTA-MA-06-0100-80-7, 469 pp.

Einstein, H. H., A. S. Azzouz, A. F. McKown, and D. E. Thompson, 1983, "Evaluation of Design and Performance—Porter Square Transit Station Chamber Lining," *Proc. Rapid Excavation and Tunneling Conf.*, Chicago, IL, pp. 597-620.

Einstein, H. H., D. A. Labreche, M. J. Markow, and G. B. Baecher, 1978, "Decision Analysis Applied to Rock Tunnel Exploration," *Engineering Geol.*, v. 12, pp. 143-161.

Einstein, H. H., C. W. Schwartz, W. Steiner, M. M. Baligh, and R. E. Levitt, 1980, "Improved Design for Tunnel Supports," Final Report, Office of University Research, U.S. Department of Transportation, No. DOT-OS-60136, 466 pp.

Elderton, William Palen, and Norman Lloyd Johnson, 1969, *Systems of Frequency Curves*, Cambridge Univ. Press, Cambridge, MA.

Electric Power Research Institute, 1985, *Seismic Hazard Methodology for Nuclear Facilities in the Eastern United States, Vol. 1, Development of Methodology*, EPRI Research Project P101-29, Electric Power Research Institute, Palo Alto, CA, 235 pp.

Elsworth, Derek, and Richard E. Goodman, 1985, "Hydromechanical Modelling of Fractured Rock Masses Using Coupled Numerical Schemes," *in* C.-F. Tsang, *ed., International Symposium on Coupled Processes Affecting the Performance of a Nuclear Waste Repository*, Lawrence Berkeley Laboratory, Berkeley, CA, pp. 637-653.

Endo, H. K., J. C. S. Long, C. R. Wilson, and P. A. Witherspoon, 1984, "A Model for Investigating Mechanical Transport in Fracture Networks," *Water Resources Research*, v. 20, pp. 1390-1400.

Engdahl, E. R., and C. Kisslinger, 1977, "Seismological Precursors to a Magnitude 5 Earthquake in the Central Aleutian Islands," *Jour. Phys. Earth*, v. 25, Suppl., pp. S243-S250.

Engelder, P. Richard, 1979, "Application of the Unit Regional Value Concept to a Study of the Mineral Resources of Australia," Ph.D. Thesis, Pennsylvania State Univ., 357 pp.

Engineering Decision Analysis Company, 1979, *Probability Analysis of Surface Rupture Offset Beneath Reactor Building; General Electric Test Reactor*, EDAC-117-217.13, Engineering Decision Analysis Company, Palo Alto, CA.

Engle, J. H., 1957, "Use of Clustering in Mineralogical and Other Surveys," *Proc. 1st Internat. Conf. Operations Research*, Oxford.

Epstein, B., 1954, "Truncated Life Tests in the Exponential Case," *Annals Math. Stat.*, v. 25, p. 555 ff.

Epstein, B., and C. Lomnitz, 1966, "A Model for the Occurrence of Large Earthquakes," *Nature*, v. 211, n. 5052, pp. 954-956.

Erickson, K. L., M. S. Y. Chu, M. D. Siegel, and W. E. Beyeler, 1986, "Approximate Methods to Calculate Radionuclide Discharges for Performance Assessment of HLW Repositories in Fractured Rock," *in* R. G. Post, *ed., Waste Management 86, Vol. 2*, University of Arizona, Tucson, AZ.

Esteva, L., 1976, "Seismicity," *in* C. Lomnitz and E. Rosenblueth, *eds., Seismic Risk and Engineering Decisions*, Elsevier, Amsterdam, pp. 179-224.

Etheridge, M. A., V. J. Wall, S. F. Cox, and R. H. Vernon, 1984, "High Fluid Pressures During Regional Metamorphism and Deformation: Implications for Mass Transport and Deformation Mechanisms," *Jour. Geophys. Research B*, v. 89, pp. 4344-4358.

Evison, F. F., 1977a, "Fluctuations of Seismicity Before Major Earthquakes," *Nature*, v. 266, n. 5604, pp. 710-712.

Evison, F. F., 1977b, "The Precursory Earthquake Swarm," *Phys. Earth Planet. Inter.*, v. 15, pp. 19-23.

Ewing, R. E., and E. M. Raney, 1976, *The Three Dimensional Structural Analysis of Double-Entry and Single-Entry Coal Mines, Vol. II: Users' Guide for a Computer Program for Analytical Modeling of Rock/Structure Interaction*, Report No. U-7638-3-4183, BUMINES-OFR-16(2)-80, Agbabian Associates, El Segundo, CA, 291 pp.

Faccioli, E., 1986, "A Study of Strong Motions from Italy and Yugoslavia in Terms of Gross Source Properties," *in* S. Das, J. Boatright, and C. H. Scholz, *eds., 5th Maurice Ewing Symposium on Earthquake Source Mechanics, Geophysical Monograph 37*, Amer. Geophys. Union, Harriman, NY, pp. 297-309.

Fedotov, S. A., 1982, "Results of Research on Earthquake Prediction in Kamchatka," *Eos, Trans. Amer. Geophys. Union*, v. 63, pp. 1260.

Ferguson, T. S., 1967, *Mathematical Statistics: A Decision Theoretic Approach*, Academic Press, New York, 396 pp.

Fischer, C. M., 1985, "Geotechnical Measurements Around Joints and Faults in Higher

Fractural Rock During Excavation of Power Plant Caverns, *in* O. Stephansson, *ed., Proc. Internat. Symp. Fundamentals of Rock Joints*, Bjorkliden, pp. 525-534.

Fisher, R. A., 1925, "Theory of Statistical Estimation," *Proc. Camb. Phil. Soc.*, v. 22, pp. 700-725.

Fisher, R. A., 1934, "Two New Properties of Mathematical Likelihood," *Proc. Royal Soc.*, Ser. A, v. 144, pp. 285-307.

Fisher, R. A., 1953, "Dispersion on a Sphere," *Proc. Royal Soc. London*, Series A, v. 217, pp. 295-305.

Fisher, R. A., 1974, *Statistical Methods and Scientific Inference*, Oliver and Boyd, Edinburgh, 3rd ed.

Foerstner, U., G. Wittmann, F. Prosi, and J. H. Van Lierde, *eds.*, 1979, *Metal Pollution in the Aquatic Environment*, Springer-Verlag, New York.

Folayan, J. I., K. Hoeg, and J. R. Benjamin, 1970, "Decision Theory Applied to Settlement Predictions," *Jour. Soil Mech. and Foundations Div.*, Amer. Soc. Civil Engineers, v. 96, n. SM4, pp. 1127-1141.

Foley, M. G., G. M. Petrie, A. J. Baldwin, and R. G. Craig, 1982, *Assessment of Effectiveness of Geologic Isolation Systems. Geologic Simulation Model for a Hypothetical Site in the Columbia Plateau: Volume 2: Results*, PNL-3542-2, Battelle Pacific Northwest Laboratory, Richland, WA, 464 pp.

Forsyth, R. S., L. 0. Werme, and J. Bruno, 1986, "The Corrosion of Spent UO_2 Fuel in Synthetic Ground Water," *Jour. Nuclear Materials*, v. 138, pp. 1-15.

Fossum, A. F., G. D. Callahan, and D. K. Svalstad, 1989. *Documentation of SPECTROM-32: A Finite Element Thermomechanical Stress Analysis Program*, RSI-0269, RE/SPEC, Inc., Rapid City, SD.

Freeze, R. A., 1975, "A Stochastic-Conceptual Analysis of One-Dimensional Groundwater Flow in Non-Uniform Homogeneous Media," *Water Resources Research*, v. 11, pp. 725-741.

Freeze, R. A., and J. A. Cherry, 1979, *Groundwater*, Prentice Hall, Englewood Cliffs, NJ, 604 pp.

Friedman, Jerome H., and John W. Tukey, 1974, "A Projection Pursuit Algorithm for Exploratory Data Analysis," *IEEE Trans. Computers*, C-23, n. 9, pp. 881-890.

Friedman, Jerome H., Werner Stuetzle, and Anne Schrodder, 1984, "Projection Pursuit Density Estimation," *Jour. Amer. Stat. Assoc.*, v. 79, pp. 599-608.

Fritz, P., 1981, "RHEO-STAUB FE Program for Static Analysis of Slices with Rheologic Material Properties," ISETH Publication (in German), 105 pp.

Fugro, Inc., 1980, *Catalog of World-wide Accelerogram Data*, Prepared for Lawrence Livermore Laboratory, April, 1980.

Gable, D. J., and T. Hatton, 1980, *Vertical Crustal Movements in the Conterminous United States Over the Last 10 Million Years*, Open-file Report 80-180, U.S. Geological Survey, 9 pp. (extended bibliography, 29 pp., and 4 plates, 1:5,000,000 scale).

Gale, J. E., 1982, "Assessing the Permeability Characteristics of Fractured Rocks," *in T. N. Narasimhan, ed., Recent Trends in Hydrogeology*, Geol. Soc. Amer. Special Paper 189, pp. 163-181.

Gartung, E., P. Bauernfeind, and J. C. Bianchini, 1979, "Three-Dimensional Finite Element Method Study of a Subway Tunnel at Nuremberg," *Proc. 1979 Rapid Excavation and Tunneling Conf.*, Vol. 1, Sec. 9, pp. 773-789.

Garvin, David, Vivian B. Parker, and Donald D. Wagman, 1981, *Chemical Thermodynamic Data Banks*, NBSIR 81-2341, Office of Standard Reference Data, National Bureau of Standards, U. S. Department of Commerce, Washington, DC.

Garvin, D., V. B. Parker, D. D. Wagman, and W. H. Evans, 1976, *A Combined Least Sums and Least Squares Approach to the Evaluation of Thermodynamic Data Networks*, NBSIR 76-1147, Office of Standard Reference Data, National Bureau of Standards, U. S. Department of Commerce, Washington, DC.

Gates, W. L., 1976a, "The Numerical Simulation of Ice-Age Climate with a Global General Circulation Model," *Jour. Atmospheric Sci.*, v. 33, pp. 1844-1873.

Gates, W. L., 1976b, "Modeling the Ice-Age Climate," *Science*, v. 191, n. 4232, pp. 1138-1144.

Gates, W. L., 1981, "The Climate System and Its Portrayal by Climate Models: A Review of Basic Principals. I. Physical Basis for Climate," *in* A. Berger, *ed.*, *Climatic Variations and Variability: Facts and Theories*, D. Reidel Co., Dordrecht, Holland, pp. 3-20.

Gawronski, Wolfgang, and Ulrich Stadtmuller, 1980, "On Density Estimation by Means of Poisson's Distribution," *Scand. Jour. Stat.*, v. 7, pp. 90-94.

Gayer, K. H., and H. Leider, 1957, "The Solubility of Uranium (IV) Hydroxide in Solutions of Sodium Hydroxide and Perchloric Acid at 25°C," *Canadian Jour. Chem.*, v. 35, pp. 5-7.

Gelfand, I. M., Sh. A. Guberman, M. L. Izvekova, V. I. Keilis-Borok, and E. Ja. Ranzman, 1972, "Criteria of High Seismicity, Determined by Pattern Recognition," *Tectonophysics*, v. 13, pp. 415-422.

Gelhar, L. W., 1985, "Stochastic Subsurface Hydrology, from Theory to Application," *Internat. Symp. Stochastic Approach to Subsurface Flow*, Montevillargenne, France, A1-24.

Gelhar, L. W., and C. L. Axness, 1983, "Three Dimensional Stochastic Analysis of Macrodispersion in Aquifers," *Water Resources Research*, v. 19, pp. 161-180.

Gelhar, L. W., and A. L. Gutjahr, 1982, "Stochastic Solutions of the One-Dimensional Corrective Dispersion Equation," Hydrology Report H-11, New Mexico Institute of Mining and Technology, Socorro, NM, 20 pp.

Gelhar, L. W., A. L. Gutjahr, and R. L. Naff, 1979, "Stochastic Analysis of Macrodispersion in a Stratified Aquifer," *Water Resources Research*, v. 15, pp. 1387-1397.

Geller, R. J., 1976, "Scaling Relations for Earthquake Source Parameters and Magnitude," *Bull. Seismol. Soc. Amer.*, v. 66, pp. 1501-1523.

Geotechnical Engineers, Inc., 1976, *Report to Lawrence Livermore Laboratories on Site Characterization Technology for High Level Waste Disposal*, Geotechnical Engineers, Inc., Winchester, MA.

Geotechnical Engineers, Inc., 1980, *State-of-the-Art for Evaluating the Potential Impact of Tectonism and Volcanism on a Radioactive Waste Repository*, Geotechnical Engineers, Inc., Winchester, MA.

Ghil, M., 1981, "Internal Climatic Mechanisms Participating in Glaciation Cycles," *in* A. L. Berger, *ed.*, *Climate Variations and Variability: Facts and Theories*, D. Reidel, Dordrecht, Holland, pp. 539-557.

Gill, Dan, and John C. Griffiths, 1984, "Areal Value Assessment of the Mineral

Resources Endowment of Israel," *Internat. Assoc. Math. Geol. Jour.*, v. 16, pp. 37-89.

Gillham, R. W., and J. A. Cherry, 1982, "Contaminant Migration in Saturated Unconsolidated Geologic Deposits," *in* T. N. Narasimhan, *ed., Recent Trends in Hydrogeology*, Geol. Soc. Amer. Special Paper 189, pp. 31-62.

Glynn, E. F., 1979, "A Probabilistic Approach to the Stability of Rock Slopes," Ph.D. Thesis, Massachusetts Institute of Technology, Cambridge, MA, 463 pp.

Goel, S. C., and C. H. Page, 1982, "An Empirical Method for Predicting the Probability of Chimney Core Occurrence Over a Mining Area," *Internat. Jour. Rock Mech. and Mining Sci.*, v. 19, no. 6.

Gokhale, D. V., 1975, "Maximum Entropy Characterizations of Some Distributions," *in* G. P. Patel and others, *eds., Statistical Distributions in Scientific Work*, D. Reidel Publ. Co., Dordrecht, Holland, pp. 299-304.

Golder Associates, 1985, "Proposed Methodology for Completing of Scenario Analysis for the Basalt Waste Isolation Project," Report to Rockwell International, Richland, WA.

Good, I. J., and R. A. Gaskins, 1971, "Nonparametric Roughness Penalties for Probability Densities," *Biometrika*, v. 58, pp. 255-277.

Goodess, C. M., J. P. Palutikof, and T. D. Davies, 1990, "A First Approach to Assessing Future Climate States in the UK Over Very Long Timescales: Input to Studies of the Integrity of Radioactive Waste Repositories," *Climatic Change*, v. 16, pp. 115-140.

Goodman, R. E., and J. M. Duncan, 1971, "The Role of Structure and Solid Mechanics in the Design of Surface and Underground Excavations in Rock," *Proc. Conf. Structure, Solid Mech. and Engineering*, John Wiley and Sons, New York, pp. 1379.

Goodman, R. E., and G. H. Shi, 1985, *Block Theory and Its Application to Rock Engineering*, Prentice Hall, Englewood Cliffs, NJ, 338 pp.

Goodman, R. E., and J. C. St. John, 1977, "Static Finite Element Analysis of Jointed Rock," *in* J. T. Christian and C. S. Desai, *eds., Numerical Methods in Geotechnical Engineering*, McGraw Hill, New York, 783 pp.

Goodman, R. E., R. C. Taylor, and T. L. Brekke, 1968, "A Model for the Mechanics of Jointed Rock," *Amer. Soc. Civil Engineers Jour.*, v. 94, SM3, pp. 637-659.

Goudie, A. S., 1981, "Raised Shorelines," *in* A. S. Goudie, *ed., Geomorphological Techniques*, George Allen and Unwin, Ltd., Boston and Sydney, pp. 327-341.

Goudie, Andrew, 1983, *Environmental Change*, 2nd ed., Oxford University Press, Oxford, 258 pp.

Grandstaff, D. E., 1976, "A Kinetic Study of the Dissolution of Uraninite," *Econ. Geol.*, v. 71, pp. 1493-1506.

Grant, R., 1973, "A Probabilistic Approach to the Search for Solution Cavities in Limestone," M.S. Thesis, Massachusetts Institute of Technology, Cambridge, MA.

Grayson, C. J., Jr., 1960, *Decisions Under Uncertainty: Drilling Decisions by Oil and Gas Operators*, Graduate School of Business Admin., Harvard Univ., 402 pp.

Greenwood, J. Arthur, J. Maciunas Landwehr, N. C. Matalas, and J. R. Wallis, 1979, "Probability Weighted Moments: Definition and Relation to Parameters of Several Distributions Expressable in Inverse Form," *Water Resources Research*, v. 15, pp. 1049-1054.

Greenwood, M., and G. U. Yule, 1920, "An Inquiry into the Nature of Frequency Distributions of Multiple Happenings," *Jour. Royal Stat. Soc.*, v. 83, pp. 255 ff.

Gretener, P. E., 1967, "Significance of the Rare Event in Geology," *Bull. Amer. Assoc. Petroleum Geologists*, v. 51, pp. 2197-2206.

Griffiths, J. C., 1967, "Mathematical Exploration Strategy and Decision Making," *Proc. 7th World Petroleum Congress*, v. 2, Elsevier Pub. Co., London, pp. 599-604.

Griffiths, John C., 1978, "Mineral Resource Assessment Using the Unit Regional Value Concept," *Internat. Assoc. Math. Geol. Jour.*, v. 10, pp. 441-472.

Griffiths, J. C., A. T. Watson, and W. D. Menzie, 1980, "Relationship Between Mineral Resource and Geologic Diversity," *in* A. D. Miall, *ed., Facts and Principles of World Petroleum Occurrence*, Canadian Soc. of Petroleum Geologists, Mem. 6, pp. 329-340.

Grisak, G. E., and J. F. Pickens, 1980, "Solute Transport Through Fractured Media, 1: The Effect of Matrix Diffusion," *Water Resources Research*, v. 16, pp. 719-730.

Grisak, G. E., J. F. Pickens, and J. A. Cherry, 1980, "Solute Transport Through Fractured Media, 2: Column Study of Fractured Till," *Water Resources Research*, v. 16, pp. 731-739.

Grob, A., 1972, "Schwelldruck im Belchen Tunnel," *Proc. Internat. Symp. f. Untertagbau*, A. A. Balkema, Luzern, pp. 99-119.

Grocott, J., 1977, "The Relationships Between Precambrian Shear Belts and Modern Fault Systems," *Jour. Geol. Soc. London*, v. 133, pp. 257-262.

Gueven, O., F. J. Molz, and J. G. Melville, 1984, "An Analysis of Dispersion in a Stratified Aquifer," *Water Resources Research*, v. 20, pp. 1337-1354.

Gull, S. F., and J. Skilling, 1984, "The Maximum Entropy Method," *in* J. A. Roberts, *ed., Indirect Imaging*, Cambridge Univ. Press, Cambridge, pp. 267-279.

Gupta, I. N., 1975, "Premonitory Seismic-Wave Phenomena Before Earthquakes near Fairview Peak, Nevada," *Bull. Seismol. Soc. Amer.*, v. 65, pp. 425-437.

Gupta, I. N., K. L. McLaughlin, and R. A. Wagner, 1985, *Attenuation of Ground Motion in the Eastern United States*, Electric Power Research Institute, Palo Alto, CA.

Gureghian, A. B., and G. Jansen, 1985, "One-Dimensional Analytical Solutions for the Migration of a Three-Member Radionuclide Decay Chain in a Multi-Layered Geologic Medium," *Water Resources Research*, v. 21, pp. 733-742.

Gusev, A. A., 1983, "Descriptive Statistical Model of Earthquake Source Radiation and Its Application to an Estimation of Short-Period Strong Motion," *Geophys. Jour., Royal Astronom. Soc.*, v. 74, pp. 787-808.

Gutjahr, A. L., 1981, "Kriging in Stochastic Hydrology; Assessing the Worth of Data, *Amer. Geophys. Union Chapman Conference on Spatial Variability in Hydrologic Modeling*, poster session, Fort Collins, CO.

Gutjahr, A. L., 1984, "Stochastic Models of Subsurface Flow: Log-Linearized Gaussian Models Are 'Exact' for Covariances," *Water Resources Research*, v. 20, pp. 1909-1912.

Gutjahr, A. L., 1985, "Investigation of the Stochastic Solution of a One-Dimensional Solute Transport Equation," Statistical Research Report SRC-3, New Mexico Institute of Mining and Technology, Socorro, NM, 24 pp.

Gutjahr, A. L., and L. W. Gelhar, 1981, "Stochastic Models of Subsurface Flows: Infinite Versus Finite Domains and Stationarity," *Water Resources Research*, v. 17, pp. 337-350.

Gutjahr, A. L., and J. L. Wilson, 1985, "Co-Kriging and Stochastic Differential Equations," *Internat. Symp. Stochastic Approach to Subsurface Flow*, Montevillargenne, France, pp. 180-191.

Gutjahr, A. L., E. J. Bonano, and R. M. Cranwell, 1985, "Treatment of Parameter Uncertainties in Modeling Contaminant Transport in Geologic Media: Stochastic Models vs. Deterministic Models With Statistical Parameter Sampling," *Internat. Symp. Stochastic Approach to Subsurface Flow*, Montevillargenne, France, pp. 332-350.

Gutjahr, A. L., L. W. Gelhar, A. A. Bakr, and J. R. MacMillan, 1978, "Stochastic Analysis of Spatial Variability in Subsurface Flows, 2: Evaluation and Applications," *Water Resources Research*, v. 14, pp. 953-960.

Gysel, M., 1977, "A Contribution to the Design of a Tunnel Lining in Swelling Rock," *Rock Mech.*, v. 10, pp. 55-71.

Hadley, D. M., and D. V. Helmberger, 1980, "Simulation of Strong Ground Motions," *Bull. Seismol. Soc. Amer.*, v. 70, pp. 617-630.

Hadley, J. R., and J. F. Devine, 1974, *Seismotectonic Map of the Eastern U.S.*, Misc. Field Studies Map MF-620, U.S. Geological Survey.

Hagiwara, Y., 1974, "Probability of Earthquake Occurrence as Obtained from a Weibull Distribution Analysis of Crustal Strain," *Tectonophysics*, v. 23, pp. 313-318.

Hagiwara, Y., 1977, "Gravity Changes Associated with Seismic Activities," *Jour. Phys. Earth*, v. 25, Suppl., pp. S137-S146.

Hamblin, W. K., P. E. Damon, and W. B. Bull, 1981, "Estimates of Vertical Crustal Strain Rates Along the Western Margins of the Colorado Plateau," *Geology*, v. 9, pp. 293-298.

Hammer, C. U., H. B. Clausen, and W. Dansgaard, 1980, "Greenland Ice Sheet Evidence of Post-Glacial Volcanism and Its Climatic Impact," *Nature*, v. 288, pp. 230-235.

Handin, J., J. E. Russell, and N. L. Carter, 1984, *Transient Creep of Repository Rocks, Mechanistic Creep Laws for Rock Salt, Final Report*, BMI/ONWI-550, Battelle Memorial Institute, Columbus, OH, 173 pp.

Hanks, T. C., 1982, "f_{max}", *Bull. Seismol. Soc. Amer.*, v. 72, pp. 1867-1879.

Hanks, T. C., and Hiroo Kanamori, 1979, "A Moment Magnitude Scale," *Jour. Geophys. Research*, v. 84, pp. 2348-2350.

Hanks, T. C., and R. K. McGuire, 1981, "The Character of High-Frequency Strong Ground Motion," *Bull. Seismol. Soc. Amer.*, v. 71, pp. 2071-2095.

Hanks, T. C., R. C. Buckman, K. R. Lajoie, and R. E. Wallace, 1984, "Modification of Wave-Cut and Faulting-Controlled Landforms," *Jour. Geophys. Research*, v. 89, pp. 5771-5790.

Hansen, J., I. Fung, A. Lacis, D. Rind, S. Lebedeff, R. Rudey, and G. Russell, 1988, "Global Climate Changes As Forecast by Goddard Institute of Space Studies Three-Dimensional Model," *Jour. Geophys. Research*, v. 93, n. D8, pp. 9341-9364.

Hansen, J., A. Lacis, D. Rind, G. Russell, I. Fung, R. Rudey, and J. Lerner, 1984, "Climate Sensitivity: Analysis of Feedback Mechanisms," *in* James E. Hansen and Taro Takahashi, *eds., Climate Processes and Climate Sensitivity*, Geophysical Monograph 29, Amer. Geophys. Union, pp. 130-163.

Harbaugh, John W., and Michel Ducastaing, 1981, "Historical Changes in Oil-Field Populations as a Method of Forecasting Field Sizes of Undiscovered Populations:

A Comparison of Kansas, Wyoming, and California," Subsurface Geol. Ser. No. 5, Kansas Geological Survey, 56 pp.

Harbaugh, John W., John H. Doveton, and John C. Davis, 1977, *Probability Methods in Oil Exploration*, Wiley-Interscience, New York, 277 pp.

Hardee, H. C., 1986, "Replenishment Rates of Crustal Magma and Their Bearing on Potential Sources of Thermal Energy," *Jour. Volc. Geoth. Research*, v. 28, pp. 275-296.

Harden, J. W., 1982, "A Quantitative Index of Soil Development from Field Descriptions: Examples from a Chronosequence in Central California," *Geoderma*, v. 28, pp. 1-28.

Harding, E. F., and D. G. Kendall, *eds.*, 1974, *Stochastic Geometry*, John Wiley and Sons, New York, 465 pp.

Harr, M. E., 1977, *Mechanics of Particulate Media*, McGraw-Hill, New York, 543 pp.

Harris, D. P., 1984, *Mineral Resources Appraisal: Mineral Endowment, Resources and Potential Supply: Concepts, Methods and Cases*, Oxford Univ. Press, New York, 445 pp.

Harrison, J. M., 1963, "Nature and Significance of Geological Maps," *in* C. C. Albritton, Jr., *ed., The Fabric of Geology*, Addison-Wesley, Reading, MA, pp. 225-232.

Hart, Roger D., Peter A. Cundall, and Michael L. Cramer, 1985, "Analysis of a Loading Test on a Large Basalt Block," *in* Eileen Ashworth, ed., *Research & Engineering Applications in Rock Masses*, 26th U.S. Symposium on Rock Mechanics, Rapid City, SD, Vol. 2, A. A. Balkema, Boston, pp. 759-768.

Hartzell, S., 1985, "The Use of Small Earthquakes as Green's Functions," *in* R. E. Scholl and J. L. King, *eds., Strong Ground Motion Simulation and Earthquake Engineering Applications*, Publication No. 85-02, Earthquake Engineering Research Institute, Berkeley, CA.

Harwell, M. A., S. K. Gupta, A. Brandstetter, C. C. Harwell, G. L. Benson, B. A. Napier, J. R. Raymond, A. E. Reisenauer, D. J. Bradley, L. S. Prater, R. J. Serne, C. S. Simmons, J. K. Soldat, D. L. Strenge, C. R. Cole, J. F. Washburn, W. J. Deutsch, and J. T. Zellmer, 1982, *Reference Site Initial Assessment for a Salt Dome Repository, Volume 1*, PNL-2955 Vol. 1, Pacific Northwest Laboratory, Battelle Memorial Institute, Richland, WA.

Hasegawa, A., T. Hasegawa, and T. Hori, 1975, "Premonitory Variation in Seismic Velocity Related to the Southeastern Akita Earthquake of 1970," *Jour. Phys. Earth*, v. 23, pp. 189-203.

Haskell, N. A., 1966, "Total Energy and Energy Spectral Density of Elastic Wave Radiation from Propagating Faults: Part II. A Statistical Source Model," *Bull. Seismol. Soc. Amer.*, v. 56, pp. 125-140.

Haskell, N. A., 1969, "Elastic Displacements in the Near Field of a Propagating Fault," *Bull. Seismol. Soc. Amer.*, v. 59, pp. 865-908.

Hauksson, E., 1981, "Radon Content of Groundwater as an Earthquake Precursor: Evaluation of Worldwide Data and Physical Basis," *Jour. Geophys. Research*, v. 86, pp. 9397-9410.

Hays, J. D., J. Imbrie, and N. J. Shackleton, 1976, "Variations in the Earth's Orbit: Pacemaker of the Ice Ages," *Science*, v. 194, pp. 1121-1132.

Hays, W. W., 1980, *Procedures for Estimating Earthquake Ground Motions*, Prof. Paper 1114, U.S. Geological Survey.

Heathcote, R. L., 1983, *The Arid Lands: Their Use and Abuse*, Longman, New York, 321 pp.

Heaton, T. H., and D. V. Helmberger, 1978, "Predictability of Strong Ground Motion in the Imperial Valley: Modeling the M 4.9 November 4, 1976, Brawley Earthquake," *Bull. Seismol. Soc. Amer.*, v. 68, pp. 31-48.

Hecht, Alan D., 1985, "Paleoclimatology: A Retrospective of the Past 20 Years," *in* A. D. Hecht, *ed., Paleoclimate Analysis and Modeling*, John Wiley and Sons, New York, pp. 1-25.

Helton, Jon C., and Ronald L. Iman, 1982, "Sensitivity Analysis of a Model for the Environmental Movement of Radionuclides," *Health Physics*, v. 42, pp. 565-584.

Helton, J. C., and P. C. Kaestner, 1981, *Risk Methodology for Geologic Disposal of Radioactive Waste: Model Description and User Manual for Pathways Model*, SAND78-1711, NUREG/CR-1636, Vol. 1, Sandia National Laboratories, Albuquerque, NM, 154 pp.

Hendron, A., Jr., and G. Fernandez, 1983, "Dynamic and Static Considerations for Underground Chambers," *in* T. R. Howard, *ed., Seismic Design of Embankments and Caverns*, Amer. Soc. Civil Engineers, New York, pp. 157-197.

Hendron, A. J., Jr., and A. K. Aiyer, 1972, *Stress and Strains Around a Cylindrical Tunnel in an Elasto-Plastic Material with Dilatancy*, OMAHA-TR-10, Omaha District, U.S. Army Corps of Engineers, 83 pp.

Hendron, A. J., Jr., E. J. Cording, and A. K. Aiyer, 1971, "Analytical and Graphical Methods for the Analysis of Slopes in Rock Masses," *Technical Report No. 36*, Nuclear Cratering Group, U. S. Army Corps of Engineers, 162 pp.

Herrmann, R. B., 1985, "An Extension of Random Vibration Theory Estimates of Strong Ground Motion to Large Distances," *Bull. Seismol. Soc. Amer.*, v. 75, pp. 1447-1453.

Heuze, F. E., and R. E. Goodman, 1973, "Finite Element and Physical Model Studies of Tunnel Reinforcement in Rock," *Proc. 15th U.S. Symp. Rock Mech.*, "Applications of Rock Mechanics," South Dakota, pp. 37-67.

Hibbitt, Karlsson and Sorenson, Inc., 1981, *ABAQUS, Theory Manual.*

Hitchon, B., G. K. Billings, and J. E. Klovan, 1971, "Geochemistry and Origin of Formation Waters in the Western Canada Sedimentary Basin: III. Factors Controlling Chemical Composition," *Geochim. Cosmochim. Acta*, v. 35, pp. 567-598.

Ho, Chih-Hsiang, 1990, "Bayesian Analysis of Volcanic Eruptions," *Jour. Volc. Geoth. Res.*, v. 43, pp. 91-98.

Ho, C. H., and D. C. Doern, 1985, "The Sorption of Uranyl Species in a Hematite Solution," *Canadian Jour. Chem.*, v. 63, pp. 1100-1104.

Hocking, G., 1976, "Three-Dimensional Elastic Stress Distribution Around the Flat End of a Cylindrical Cavity," *Internat. Jour. Rock Mech., Mining Sci. and Geomech. Abstr.*, v. 13, n. 12, pp. 331-337.

Hocking, G., 1978, "Stress Analysis of Underground Excavations Incorporating Slip and Separation Along Discontinuities," *in* C. A. Brebbia, *ed., Recent Advances in Boundary Element Methods*, Pentech Press, London, pp. 195-214.

Hocking, G., E. T. Brown, and J. 0. Watson, 1976, "Three-Dimensional Elastic Stress Analysis of Underground Openings by the Boundary Integral Equation Method," *Proc. Third Symp. Engineering Applications of Solid Mech.*, Toronto, pp. 203-216.

Hocking, G., J. R. Williams, P. Boonlualohr, I. Mathews, and G. Mustoe, 1980, *Numerical Prediction of Basalt Response for Near-Surface Test Facility Heater Tests #1 and #2*, RHO-BWI-C-86, Rockwell Hanford Operations, Richland, WA.

Hoeg, K., 1968, "Stress Against Underground Structural Cylinders," *Jour. Soil Mech. and Foundation Engineering Div.*, Amer. Soc. Civil Engineers, v. 94, no. SM4, pp. 833-858.

Hoeg, K., and R. Murarka, 1974, "Probabilistic Analysis and Design of a Retaining Wall," *Jour. Geotechnical Div.*, Amer. Soc. Civil Engineers, v. 100, n. GT3, pp. 349-366.

Hoek, E., and J. W. Bray, 1981, *Rock Slope Engineering*, 3rd ed., The Institution of Mining and Metallurgy, London, 357 pp.

Hoek, E., and E. T. Brown, 1980, *Underground Excavations in Rock*, Institute of Mining and Metallurgy, London, 527 pp.

Hoeksema, R. J., and P. K. Kitanidis, 1984, "An Application of the Geostatistical Approach to the Inverse Problem in Two-Dimensional Groundwater Modeling," *Water Resources Research*, v. 20, pp. 1003-1020.

Hoeksema, R. J., and P. K. Kitanidis, 1985, "Analysis of the Spatial Structure of Properties of Selected Aquifers," *Water Resources Research*, v. 21, pp. 563-572.

Hofmann, R., 1981, *STEALTH—A Lagrange Explicit Finite Difference Code for Solids, Structural, and Thermohydraulic Analysis*, EPRI-NP-2080-CCM, Vols. 1A and 1B, Electric Power Research Institute, Palo Alto, CA.

Hogarth, R. M., 1975, "Cognitive Processes and the Assessment of Subjective Probability Distributions," *Jour. Amer. Stat. Assoc.*, v. 70, pp. 271-294.

Holdahl, S. R., and N. L. Morrison, 1974, "Regional Investigation of Vertical Crustal Movements in the U.S., Using Precise Relevelings and Mareograph Data," *Tectonophysics*, v. 23, pp. 373-390.

Holland, C. T., 1962, "Design of Pillars for Overburden Support," *Min. Congr. Jour.*, v. 48, pp. 24-28 and 66-71.

Holland, H. D., 1978, *The Chemistry of the Atmosphere and Oceans*, Wiley-Interscience, New York.

Honkura, Y., 1978, "Electrical Conductivity Anomalies in the Earth," *Geophys. Survey*, v. 3, pp. 225-253.

Houghton, John C., 1978, "The Incomplete Means Estimation Procedure Applied to Flood Frequency Analysis," *Water Resources Research*, v. 14, pp. 1111-1115.

Housner, G. W., 1969, "Engineering Estimates of Ground Shaking and Maximum Magnitude Earthquakes," *Proc. 4th World Conf. Earthquake Engineering*, Santiago, Chile.

Howard, K. A., J. M. Aaron, E. E. Brabb, M. R. Brock, H. D. Gower, S. J. Hunt, P. J. Milton, W. R. Muelberger, J. K. Nakata, G. Plafker, D. C. Prowell, R. E. Wallace, and I. J. Watkind, 1978, *Preliminary Map of Younger Faults in the United States as a Guide to Possible Fault Activity*, U.S. Geological Survey Misc. Field Studies Map MF-916, 2 sheets.

Howland, J. D., W. H. Roth, R. K. McGuire, and J. Sweet, 1983, "Probabilistic Analysis of a Geomechanical Prediction of Ground Surface Rupture," presented at Seismic Risk and Heavy Industrial Facilities Conference, San Francisco, CA, May, pp. 50-52.

Huang, L., J. McRaney, T. Teng, and M. Prebish, 1979, "A Preliminary Study on the Relationship Between Precipitation and Large Earthquakes in Southern California," *Pure Appl. Geophys.*, v. 117, pp. 1286-1300, Birkhauser Verlag, Basel.

Huber, N. K., 1981, *Amount and Timing of Late Cenozoic Uplift and Tilt of the Central Sierra Nevada, California—Evidence from the Upper San Joaquin River Basin*, Prof. Paper 1197, U.S. Geological Survey, 28 pp.

Hudson, D. E., 1976, *Strong Motion Earthquake Accelerograms—Index Volume*, Report No. EERL 76-02, Earthquake Engineering Research Laboratory, California Institute of Technology, Pasadena, CA.

Hufschmied, P., 1985, "Estimation of Three-Dimensional Anisotropic Hydraulic Conductivity Field by Means of Single Well Pumping Tests Combined With Flowmeter Measurements," *Internat. Symp. Stochastic Approach to Subsurface Flow*, Montevillargenne, France, pp. 110-116.

Huijbregts, C., and G. Matheron, 1977, "Universal Kriging—An Optimal Approach to Trend Surface Analysis," *Decision Making in the Mineral Industry*, Can. Inst. Min. Metall. Special Volume 12, pp. 159-169.

Hunter, Regina L., 1983, *Preliminary Scenarios for the Release of Radioactive Waste from a Hypothetical Repository in Basalt of the Columbia Plateau*, NUREG/CR-3353, SAND83-1342, Sandia National Laboratories, Albuquerque, NM, 85 pp.

Hunter, R. L., and C. J. Mann, 1989, *Techniques for Determining Probabilities of Events and Processes Affecting the Performance of Geologic Repositories, Literature Review*, NUREG/CR-3964, SAND86-0196, Vol. 1, U.S. Nuclear Regulatory Commission, Washington, DC, 187 pp.

Hunter, R. L., G. E. Barr, and F. W. Bingham, 1983, *Scenarios for Consequence Assessments of Radioactive-Waste Repositories at Yucca Mountain, Nevada Test Site*, SAND82-1277, Sandia National Laboratories, Albuquerque, NM, 137 pp.

Hunter, R. L., C. John Mann, and R. M. Cranwell, 1986, "Determining Probabilities of Geologic Events and Processes," *in* Harry C. Burkholder, *ed., High-Level Nuclear Waste Disposal*, Battelle Press, Columbus, OH, pp. 637-646.

Huyakorn, P. S., J. W. Mercer, and D. S. Ward, 1985, "Finite Element Matrix and Mass Balance Computational Schemes for Transport in Variably Saturated Porous Media," *Water Resources Research*, v. 21, pp. 346-358.

Hyde, W. T., and T. J. Crowley, K.-Y. Kim, and G. R. North, 1989, "Comparison of GCM and Energy Balance Model Simulations of Seasonal Temperature Changes Over the Past 18000 Years," *Jour. Climate*, v. 2, pp. 864-887.

Idiz, E. F., D. Carlisle, and I. R. Kaplan, 1986, "Interaction Between Organic Matter and Trace Metals in a Uranium Rich Bog, Kern County, California, U.S.A.," *Applied Geochem.*, v. 1, pp. 573-590.

Idriss, I. M., 1978, "Characteristics of Earthquake Ground Motions," *Specialty Conference on Earthquake Engineering and Soil Dynamics*, Geotech. Eng. Div., Amer. Soc. Civil Engineers, New York, v. 3, pp. 1151-1265.

Ikeya, M., T. Miki, and K. Tanaka, 1982, "Dating of a Fault by Electron Spin Resonance on Intrafault Materials," *Science*, v. 215, pp. 1392-1393.

Iman, Ronald L., and W. J. Conover, 1980, "Small Sample Sensitivity Analysis Techniques for Computer Models with an Application to Risk Assessment," *Comm. Stat.-Theor. Meth.*, v. A9(17), pp. 1749-1842.

Iman, R. L., and W. J. Conover, 1982, "A Distribution-Free Approach to Inducing Rank Correlation Among Input Variables," *Communications in Stat.*, v. B11(3), pp. 311-334.

Iman, R. L., and J. C. Helton, 1985, *A Comparison of Uncertainty and Sensitivity Analysis Techniques for Computer Models*, NUREG/CR-3904, SAND84-1461, Sandia National Laboratories, Albuquerque, NM, 116 pp.

Iman, R. L., and M. J. Shortencarier, 1984, *A FORTRAN 77 Program and Users Guide for the Generation of Latin Hypercube and Random Samples for Use With Computer Models*, NUREG/CR-3624, SAND83-2365, Sandia National Laboratories, Albuquerque, NM, 67 pp.

Iman, R. L., J. M. Davenport, D. K. Zeigler, 1980, *Latin Hypercube Sampling (Program User's Guide)*, SAND79-1473, Sandia National Laboratories, Albuquerque, NM, 77 pp.

Iman, R. L., J. C. Helton, and J. E. Campbell, 1978, *Risk Methodology for Geologic Disposal of Radioactive Waste: Sensitivity Analysis Techniques*, NUREG/CR-0394, SAND78-0912, Sandia National Laboratories, Albuquerque, NM, 107 pp.

Iman, R. L., H. P. Stephens, J. M. Davenport, R. K. Waddell, and D. I. Leap, 1979, "Sensitivity Study on the Parameters of the Regional Hydrology Model for the Nevada Nuclear Waste Storage Investigations," *DOE Statistical Symp.*, pp. 44-61.

Imbrie, John, 1985, "The Future of Paleoclimatology," *in* A. D. Hecht, *ed., Paleoclimate Analysis and Modeling*, John Wiley and Sons, New York, pp. 423-432.

Imbrie, John, and John Z. Imbrie, 1980, "Modeling the Climatic Response to Orbital Variations," *Science*, v. 207, n. 4434, pp. 943-953.

Intera Environmental Consultants, Inc., 1983, *Porosity, Permeability, and Their Relationship in Granite, Basalt and Tuff*, ONWI-458, Houston, TX, 96 pp.

International Symposium on Engineering Geology and Underground Construction, 1983, *Proc.*, Vols. 1 and 2, Soc. Portuguesa de Geotecnia and Lab. Nacional de Ensenharia Civil, Lisbon.

Ioannou, P. S., 1984, "The Economic Value of Exploration as a Risk Reduction Strategy in Underground Construction," Ph.D. Thesis, Massachusetts Institute of Technology, Cambridge, MA, 461 pp.

Irmay, S., 1955, "Flow of Liquids Through Cracked Media," *Bull. Resources Council*, Israel, v. 5A, pp. 84.

Isaaks, Edward H., and R. Mohan Srivastava, 1989, *Applied Geostatistics*, Oxford Universty Press, New York.

Ishida, M., and H. Kanamori, 1977, "The Spatio-Temporal Variation of Seismicity Before the 1971 San Fernando Earthquake, California," *Geophys. Research Let.*, v. 4, pp. 345-346.

Ishida, M., and H. Kanamori, 1978, "The Foreshock Activities of the 1971 San Fernando Earthquake, California," *Bull. Seismol. Soc. Amer.*, v. 68, pp. 1265-1279.

Iyer, H. M., 1984, "Geophysical Evidence for the Locations, Shapes, and Sizes, and Internal Structures of Magma Chambers Beneath Regions of Quaternary Volcanism," *Phil. Trans. Royal Soc. London*, v. A 310, pp. 473-510.

Jachens, R. C., W. Thatcher, C. W. Roberts, and R. S. Stein, 1983, "Correlation of Changes in Gravity, Elevation, and Strain in Southern California," *Science*, v. 219, pp. 1215-1217.

Jackson, D. D., W. B. Lee, and C. Liu, 1980, "Aseismic Uplift in Southern California: An Alternative Interpretation," *Science*, v. 210, pp. 534-536.

Jamison, D. B., and N. G. W. Cook, 1980, "Note on Measured Values for the State of Stress in the Earth's Crust," *Jour. Geophys. Research*, v. 85, pp. 1833-1838.

Jaynes, E. T., 1983, "Confidence Intervals vs. Bayesian Intervals," *in* R. D. Rosenkrantz, *ed., E. T. Jaynes: Papers on Probability, Statistics, and Statistical Physics*, D. Reidel Publishing Company, Dordrecht, Holland, 434 pp.

Jewell, W. S., 1973, "Credible Means Are Exact Bayesian for Exponential Families," Report 73-21, Operations Research Center, Univ. of California, Berkeley.

Johnpeer, G. D., D. Miller, and G. Goles, 1981, "Assessment of Potential Volcanic Hazards: Pasco Basin, Washington," RHO-BW-CR-130P, Rockwell International Corp., 35 pp.

Johnson, Norman L., and Samuel Kotz, 1969, *Discrete Distributions*, Houghton Mifflin Co., Boston, 323 pp.

Johnson, Norman L., and Samuel Kotz, 1970a, *Continuous Univariate Distributions*—1, Houghton Mifflin Co., Boston.

Johnson, Norman L., and Samuel Kotz, 1970b, *Continuous Univariate Distributions*—2, Houghton Mifflin Co., Boston, 301 pp.

Johnson, Norman L., and Samuel Kotz, 1972, *Distributions in Statistics; Continuous Multivariate Distributions*, John Wiley and Sons, Inc., New York, 308 pp.

Journel, A. G., 1973, "Geostatistics and Sequential Exploration," *Mining Eng.*, v. 25, pp. 44-48.

Journel, A. G., 1980, "Geostatistical Simulation Methods for Exploration and Mine Planning," *Geostatistics*, McGraw-Hill, New York, pp. 93-108.

Journel, A. G., 1983, "Nonparametric Estimation of Spatial Distributions," *Internat. Assoc. Math. Geol. Jour.*, v. 15, n. 3, pp. 445.

Journel, A. G., 1986, "Constrained Interpolation and Qualitative Information—The Soft Kriging Approach," *Internat. Assoc. Math. Geol. Jour.*, v. 18, n. 3, pp. 269.

Journel, Andre G., 1989, *Fundamentals of Geostatistics in Five Lessons*, Short Course in Geology: Volume 8, American Geophysical Union, Washington, DC.

Journel, A. G., and Ch. J. Huijbregts, 1978, *Mining Geostatistics*, Academic Press, New York, 600 pp.

Joyner, W. B., 1984, "A Scaling Law for the Spectra of Large Earthquakes," *Bull. Seismol. Soc. Amer.*, v. 74.

Joyner, W. B., and D. M. Boore, 1981, "Peak Horizontal Acceleration and Velocity from Strong Motion Records Including Records from the 1979 Imperial Valley, California, Earthquake," *Bull. Seismol. Soc. Amer.*, v. 71, pp. 2011-2038.

Joyner, W. B., R. E. Warrick, and T. E. Fumal, 1981, "The Effect of Quaternary Alluvium in Strong Ground Motion in the Coyote Lake, California, Earthquake of 1979," *Bull. Seismol. Soc. Amer.*, v. 71, pp. 1333-1349.

Jurina, L., G. Maier, and K. Podolak, 1977, "On Model Identification Problems in Rock Mechanics," *Proc. Internat. Symp. Geotechnics of Structurally Complex Formations*, Assoc. Geotechnica Italiana, Capri, Italy, pp. 287-295.

Kafritsas, J., M. Gencer, H. H. Einstein, 1984, "Coupled Flow Deformation Analysis by the Distinct Element Method," *Proc. 25th U.S. Symp. Rock Mech.*, Amer. Soc. Civil Engineers, pp. 239-247.

Kagan, Ya., and L. Knopoff, 1976, "Statistical Search for Non-Random Features of the Seismicity of Strong Earthquakes," *Phys. Earth Planet. Inter.*, v. 12, pp. 291-318.

Kagan, Ya., and L. Knopoff, 1977, "Earthquake Risk Prediction as a Stochastic Process," *Phys. Earth Planet. Inter.*, v. 14, pp. 97-108.

Kaiser, P. K., and M. E. Harr, 1984, "Evaluation of Two Rock Mass Classification Systems by Probabilistic Approach," *Proc. 3rd Seminar of Probabilistic Methods in Geotechnical Engineering*, Lausanne, 11 pp.

Kanamori, H., 1979, "A Semi-Empirical Approach to Prediction of Long-Period Ground Motions for Great Earthquakes," *Bull. Seismol. Soc. Amer.*, v. 69, pp. 1645-1670.

Kanamori, H., 1981, "The Nature of Seismic Patterns Before Large Earthquakes," *in* D. W. Simpson and P. G. Richards, *eds., Earthquake Prediction, an International Review*, Maurice Ewing Ser., Amer. Geophys. Union, Washington, DC, v. 4, pp. 1-19.

Kanamori, H., and D. L. Anderson, 1975, "Theoretical Basis of Some Empirical Relations in Seismology," *Bull. Seismol. Soc. Amer.*, v. 65, pp. 1073-1095.

Kanari, S., N. Fuji, and S. Horie, 1984, "The Paleo-Climatological Constituents of Paleotemperature in Lake Biwa," *in* A. Berger, J. Imbrie, J. Hays, G. Kukla, and B. Saltzman, *eds., Milankovitch and Climate—Understanding the Response to Astronomical Forcing*, Part 1, D. Reidel Co., Dordrecht, Holland, pp. 405-414.

Kane, M. F., 1977, "Correlation of Major Eastern Earthquake Centers with Mafic/Ultramafic Basement Masses," Prof. Paper 1028-O, U.S. Geological Survey, pp. 199-204.

Kates, R. W., 1978, *Risk Assessment of Environmental Hazard*, John Wiley, New York, 137 pp.

Kaufman, G. M., 1963, *Statistical Decision and Related Techniques in Oil and Gas Exploration*, Prentice-Hall, Englewood Cliffs, NJ, 307 pp.

Kaufman, G. M., and J. W. Wang, 1980, *Model Mis-Specification and the Princeton Study of Volume and Area of Oil Fields and Their Impact on Order of Discovery*, Massachusetts Institute of Technology Energy Lab. Working Paper No. MIT-EL 80-003WP, Cambridge.

Keefer, D. K., 1984, "Landslides Caused by Earthquakes," *Geol. Soc. Amer. Bull.*, v. 94, pp. 406-421.

Keller, E. A., M. S. Bonkowski, R. J. Korsch, and R. J. Shlemon, 1982, "Tectonic Geomorphology of the San Andreas Fault Zone in the Southern Indio Hills, Coachella Valley, California," *Geol. Soc. Amer. Bull.*, v. 93, pp. 46-56.

Kellogg, William W., and Robert Schware, 1981, *Climate Change and Society: Consequences of Increasing Atmospheric Carbon Dioxide*, Westview Press, Boulder, CO, 172 pp.

Kelmers, A. D., R. J. Clark, N. H. Cutshall, G. K. Jacobs, J. S. Johnson, J. H. Kessler, and R. E. Meyer, 1984, *Evaluation of Radionuclide Geochemical Information Developed by DOE High-Level Nuclear Waste Repository Site Projects, Annual Progress Report for the Period October 1982-September 1983,* NUREG/CR-3730, ORNL/TM-9109, U.S. Nuclear Regulatory Commision, Washington, DC.

Kelmers, A. D., J. H. Kessler, W. D. Arnold, R. E. Meyer, N. H. Cutshall, G. K. Jacobs, and S. Y. Lee, 1985, *Evaluation of Radionuclide Geochemical Information Developed by DOE High-Level Nuclear Waste Repository Site Projects, Annual Progress Report, October 1983 - September 1984*, NUREG/CR-3851, ORNL/TM-9191, Vol. 4, U.S. Nuclear Regulatory Commission, Washington, DC.

Kelmers, A. D., F. G. Seeley, W. D. Arnold, J. G. Blencoe, R. E. Meyer, G. K. Jacobs, and S. K. Whatley, 1985, *Progress in Evaluation of Radionuclide Geochemical Information Developed by DOE High-Level Nuclear Waste Repository Site Projects, Annual Progress Report for the Period October - December 1984*, NUREG/CR-4236, ORNL/TM-9614, Volume 1 - 4, U.S. Nuclear Regulatory Commission, Washington, DC.

Kendall, M. G., and P. A. Moran, 1963, *Geometrical Probability*, Griffin's Statistical Monographs and Sources Series, Vol. 10, Hafner, London, 125 pp.

Kendall, M. G., and A. Stuart, 1973, *The Advanced Theory of Statistics, Vol. 2*, Hafner, London, 723 pp.

Kendall, M. G., and A. Stuart, 1976, *The Advanced Theory of Statistics, Vol. 3*, Hafner, London, 585 pp.

Kendall, M. G., and A. Stuart, 1977, *The Advanced Theory of Statistics, Vol. 1*, Hafner, London, 472 pp.

Keneshige, O., T. Kawamoto, and H. Okamura, 1960, "On the Stress Concentration in the Horizontal Crossing Gallery," *Jour. Mining and Metallurgical Inst. of Japan*, v. 76, pp. 375-378, (in Japanese).

Kennedy, R. P., S. A. Short, K. L. Merz, F. J. Tokarz, I. M. Idriss, M. S. Power, and K. Sadigh, 1984, *Engineering Characterization of Ground Motion, Task I: Effects of Characteristics of Free Field Motions on Structural Response*, NUREG/CR-3805, U.S. Nuclear Regulatory Commission, 381 pp.

Kennedy, T. C., and H. E. Lindberg, 1978a, "Model Tests for Plastic Response of Lined Tunnels," *Jour. Engineering Mech. Div.*, Amer. Soc. Civil Engineers, v. 104, n. 2, pp. 399-420.

Kennedy, T. C., and H. E. Lindberg, 1978b, "Tunnel Closure for Nonlinear Mohr-Coulomb Functions," *Jour. Engineering Mech. Div.*, Amer. Soc. Civil Engineers, v. 104, n. 6, pp. 1313-1326.

Kennett, J. P., and R. C. Thunell, 1975, "Global Increase in Quaternary Explosive Volcanism," *Science*, v. 187, pp. 497-503.

Kennett, J. P., A. R. McBirney, and R. C. Thunell, 1977, "Episodes of Cenozoic Volcanism in the Circum-Pacific Region," *Jour. Volc. Geoth. Research*, v. 2, pp. 145-163.

Kerisel, J. M., *ed.*, 1979, 1980, "Symposium on Convergence Confinement Method," in *Tunnels et Ouvrages Souterrains*, no. 32, (in French, 1979), pp. 70-147, and *Underground Space*, v. 4, no. 6 (English summary, 1980), pp. 366-402.

Kerr, R. A., 1979, "Prospects for Earthquake Prediction Wane," *Science*, v. 206, pp. 542-545.

Kerrisk, J. F., 1984, "Solubility Limits on Radionuclide Dissolution," *in* C. M. Jantzen, J. A. Stone, and R. C. Ewing, *eds., Scientific Basis for Nuclear Waste Management VIII*, Proc. Materials Res. Soc. Symp., pp. 237-244.

Keys, W. S., E. Shuter, R. G. Wolff, J. D. Bredehoeft, and J. H. Healy, 1979, "In-Situ

Measurements Near the San Andreas Fault in Central California," *Jour. Geophys. Research*, v. 84, pp. 1583-1591.

Kier, S. N., 1984, "Earthquake Prediction from B-Values Discussion," *Nature*, v. 310, pp. 456.

Kim, Y. W., 1984, "Decision Support Framework in Adaptable Tunneling," Ph.D. Thesis, Massachusetts Institute of Technology, Cambridge, MA, 348 pp.

Kim, Yeong W., Herbert H. Einstein, and Robert D. Logcher, 1985, "Decision Support System for Tunneling," *Construction Research Applied to Practice*, ASCE, New York, pp. 68-85.

Kimura, T., and R. J. Mair, 1981, "Centrifugal Testing of Model Tunnels in Soft Clay," *Proc. Tenth Internat. Conf. Soil Mech. and Foundation Engineering*, Stockholm, A. A. Balkema, Rotterdam, Netherlands, v. 1, pp. 319-322.

King, C. Y., 1978, "Radon Emanation on San Andreas Fault," *Nature*, v. 271, pp. 516-519.

King, J. E., 1978, "New Evidence on the History of the St. Francis Sunk Lands, Northeastern Arkansas," *Geol. Soc. Amer. Bull.*, v. 89, pp. 1719-1722.

Kiraly, L., 1969, "Statistical Analysis of Fractures (Orientation Density)," *Geologische Rundschau*, v. 59, pp. 125-151.

Kirby, S. H., 1980, "Tectonic Stresses in the Lithosphere: Constraints Provided by the Experimental Deformation of Rocks," *Jour. Geophys. Research*, v. 85, pp. 6353-6363.

Kiremidjian, A. S., 1984, "Reliability of Structures Subjected to Differential Fault Slip," *Earthquake Engineering and Structural Dynamics*, v. 12, pp. 603-618.

Kiremidjian, A. S., and T. Anagnos, 1980, "A Homogeneous Alternating Markov Model for Earthquake Occurrences," *Proc. 7th World Conf. Earthquake Engineering*, v. 1, pp. 355-362, Istanbul, Turkey.

Kiremidjian, A. S., and T. Anagnos, 1984, "Stochastic Slip-Predictable Model for Earthquake Occurrences," *Bull. Seismol. Soc. Amer.*, v. 74, pp. 739-755.

Kitanidis, P. K., 1983, "Statistical Estimation of Polynominal Generalized Covariance Functions and Hydrologic Applications," *Water Resources Research*, v. 19, pp. 909-921.

Kitanidis, P. K., and E. G. Vomvoris, 1983, "A Geostatistical Approach to the Inverse Problem in Groundwater Modeling (Steady-State) and One-Dimensional Simulations," *Water Resources Research*, v. 19, pp. 677-690.

Klovan, J. E., 1975, "R- and Q-Mode Factor Analysis," *in* R. McCammon, *ed., Concepts in Geostatistics*, Springer-Verlag, New York, pp. 21-69.

Kobayashi, S., and N. Nishimura, 1982a, "Dynamic Analysis of Underground Structures by the Integral Equation Method," *in* Z. Eisenstein, *ed., Numerical Methods in Geomechanics 1982*, A. A. Balkema Press, Rotterdam, pp. 401-409.

Kobayashi, S., and N. Nishimura, 1982b, "Transient Stress Analysis of Tunnels and Caverns of Arbitrary Shape Due to Travelling Waves," *in* P. K. Banerjee and R. P. Shaw, *eds., Developments in Boundary Element Methods—2*, Applied Science Publishers, Ltd., London, pp. 177-210.

Kominz, Michelle A., and Nicklas G. Pisias, 1979, "Pleistocene Climate: Deterministic or Stochastic?" *Science*, v. 204, pp. 171-173.

Koopman, B. O., 1956a, "The Theory of Search 1: Kinematic Bases," *Operations Research*, v. 4, pp. 324-346.

Koopman, B. O., 1956b, "The Theory of Search 2: Target Detection," *Operations Research*, v. 4, pp. 503-531.

Koopman, B. O., 1957, "The Theory of Search 3: The Optimum Distribution of Searching Effort," *Operations Research*, v. 5, pp. 613-626.

Koplick, C. M., M. F. Kaplan, and B. Ross, 1982, "The Safety of Repositories for Highly Radioactive Wastes," *Rev. Modern Phys.*, v. 54, pp. 269-310.

Korbin, G. E., and T. L. Brekke, 1975, "A Model Study of Spiling Reinforcement in Underground Openings," MRD-TR-2-75, Missouri River Div., U.S. Army Corps of Engineers, 278 pp.

Kovach, R. L., A. Nur, R. L. Wesson, and R. Robinson, 1975, "Water Level Fluctuations and Earthquakes on the San Andreas Fault Zone," *Geology*, v. 3, pp. 437-440.

Kovari, K., *ed.*, 1983, "Field Measurements in Geomechanics," *Proc. Internat. Symp.*, Zurich, 2 vols., 1753 pp.

Kovari, K., and C. Amstad, 1979, "Decision Making and Field Measurements in Tunneling," 25th Oyo Anniversary Lecture, Tokyo.

Krizek, R. J., R. B. Corotis, and H. H. El-Moursi, 1977, "Probabilistic Analysis of Predicted and Measured Settlements," *Canadian Geotechnical Jour.*, v. 14, n. 1, pp. 17-33.

Kronenberg, A. K., and J. Tullis, 1984, "Flow Strengths of Quartz Aggregates: Grain Size and Pressure Effects due to Hydrolytic Weakening," *Jour. Geophys. Research*, v. 89, n. B6, pp. 4281-4297.

Krumbein, W. C., 1959, "Trend Surface Analysis of Contour-Type Maps with Irregular Control-Point Spacings," *Jour. Geophys. Res.*, v. 64, pp. 823-834.

Krumbein, W. C., and F. A. Graybill, 1965, *Introduction to Statistical Models in Geology*, McGraw-Hill, New York, 475 pp.

Krumbein, W. C., and L. L. Sloss, 1963, *Stratigraphy and Sedimentation*, 2nd ed., W. H. Freeman and Co., San Francisco, 660 pp.

Kukla, G. J., and R. K. Matthews, 1972, "When Will the Present Interglacial End?" *Science*, v. 178, pp. 190-191.

Kukla, G., A. Berger, R. Lotti, and J. Brown, 1981, "Orbital Signature of the Interglacials," *Nature*, v. 290, pp. 295-300.

Kutzbach, J. E., 1976, "The Nature of Climate and Climatic Variations," *Quaternary Research*, v. 6, pp. 471-480.

Kutzbach, J. E., 1981, "Monsoon Climate of the Early Holocene: Climate Experiment with the Earth's Orbital Parameters for 9000 Years Ago," *Science*, v. 214, pp. 59-61.

Kutzbach, J. E., and R. A. Bryson, 1974, "Variance Spectrum of Holocene Climate Fluctuations in the North Atlantic Sector," *Jour. Atmospheric Sci.*, v. 31, pp. 1958-1962.

Kutzbach, J. E., and P. J. Guetter, 1984, "The Sensitivity of Monsoon Climates to Orbital Parameter Changes for 9,000 Years BP: Experiments with the NCAR General Circulation Model," *in* A. Berger, J. Imbrie, J. Hays, G. Kukla, and B. Saltzman, *eds., Milankovitch and Climate—Understanding the Response to Astronomical Forcing*, Part 2, D. Reidel Co., Dordrecht, Holland, pp. 801-802.

Kutzbach, J. E., and P. J. Guetter, 1986, "The Influence of Changing Orbital Parameters and Surface Boundary Conditions on Climate Simulations for the Past 18000 Years," *Jour. Atmospheric Sci.*, v. 43, pp. 1726-1759.

Kutzbach, J. E., and B. L. Otto-Bliesner, 1982, "The Sensitivity of the African-Asian Monsoon to Orbital Parameter Changes 9000 Years BP in a Low-Resolution General Circulation Model," *Jour. Atmospheric Sci.*, v. 39, pp. 1177-1188.

Kutzbach, John E., and F. Alayne Street-Perrott, 1985, "Milankovitch Forcing of Fluctuations in the Level of Tropical Lakes from 18 to 0 Kyr BP," *Nature*, v. 317, pp. 130-134.

Kyburg, Henry E., Jr., and Howard E. Smokler, 1980, *Studies in Subjective Probability*, 2nd ed., Robert E. Krieger Publ. Co., Huntington, NY, 257 pp.

Laabmayr, F., and G. Swoboda, 1978, "The Importance of Shotcrete As Support Element of the NATM," *Proc. Engineering Foundation Conf. Shotcrete for Underground Support*, St. Anton, Austria.

Labovitz, Mark Lang, 1978, "Unit Regional Value of the Dominion of Canada," Ph.D. Thesis, Pennsylvania State Univ., 366 pp.

Labovitz, M. L., and J. C. Griffiths, 1982, "An Inventory of Undiscovered Canadian Mineral Resources," *Econ. Geol.*, v. 77, pp. 1642-1654.

Labovitz, M. L., W. D. Menzie, and J. C. Griffiths, 1977, "COMOD: A Program for Standardizing Mineral-Resource Commodity Data," *Comp. and Geosci.*, v. 3, pp. 497-537.

Labreche, D., 1983, "Damage Mechanisms in Tunnels Subjected to Explosive Loads," *Proc. Amer. Soc. Civil Engineers Symp. Seismic Design of Embankments and Caverns*, Philadelphia, PA, pp. 128-141.

Ladanyi, B., 1974, "Use of the Long-Term Strength Concept in the Determination of Ground Pressures on Tunnel Linings," *Adv. Rock Mech., Proc. Third Congress, Internat. Soc. Rock Mech.*, Denver, Vol. II, Part B, pp. 1150-1156.

Ladanyi, B., and D. E. Gill, 1984, "Tunnel Lining in Creeping Rock," *Proc. Symp. Design and Performance of Underground Excavations*, Cambridge, MA, pp. 19-26.

Lagedrost, John F., and Webster Capps, 1983, *Thermal Property and Density Measurements of Samples Taken from Drilling Cores from Potential Geologic Media*, BMI/ONWI-522, Battelle Memorial Institute, Columbus, OH.

Lamb, H. H., 1982, *Climate, History and the Modern World*, Methuen and Co. Ltd., London, 387 pp.

Lambie, F. W., J. W. Harbaugh, G. R. Kendall, and J. D. Juilland, 1983, "Classification of the California Desert for Geology-Energy-Mineral Resource Potential, A Geostatistical Classification," BLM-YA-PT-83-002-4335, BLM-TN-359, U.S. Bureau of Land Management, Riverside, CA, 183 pp.

Langbein, J. O., M. F. Linker, A. McGarr, R. C. Jachens, and L. E. Slater, 1983, "Two-Color Geodimeter Measurements of Crustal Strain Accumulation near Palmdale, California: Comparisons with Gravity and Meteorological Data," *Eos, Trans. Amer. Geophys. Union*, v. 64, p. 841.

Langer, F. Th., and K. Stockmann, 1985, "Stability Analysis of Tunnels Using the Program ADINA," *Nonlinear Finite Element Analysis and ADINA, Proc. 5th ADINA Conf.*, Massachusetts Institute of Technology, Cambridge, MA, pp. 341-351.

Langmuir, D., 1978, "Uranium Solution-Mineral Equilibria at Low Temperatures with Applications to Sedimentary Ore Deposits," *Geochim. Cosmochim. Acta*, v. 42, pp. 547-570.

LaPointe, P. R., 1980, "Analysis of the Spatial Variation in Rock Mass Properties Through Geostatistics," *Proc. 21st U.S. National Rock Mech. Symp.*, pp. 570-580.

Larsson, H., T. Olafsson, and O. Stephansson, 1985, "Reinforcement of Jointed Rock Mass—A Non-Linear Continuum Approach," *Proc. Internat. Symp. Fundamentals of Rock Joints*, Bjorkliden, pp. 567-577.

Latynina, L. A., and S. D. Rizaeva, 1976, "On Tidal-Strain Variations Before Earthquakes," *Tectonophysics*, v. 31, pp. 121-127.

Leckie, J. O., 1989, "Simulations of Distribution Coefficients Using Surface Complexation Models," *in* M. D. Siegel, *ed., Progress in Development of a Methodology for Geochemical Sensitivity Analysis for Performance Assessment: Volume 2, Speciation, Sorption, and Transport in Fractured Media*, Letter Report to the U.S. Nuclear Regulatory Commission, Washington, DC, Appendix C.

Ledezma, Guillermo Federico Salazar, 1983, "Stochastic and Economic Evaluation of Adaptability in Tunneling Design and Construction," Ph.D. Thesis, Massachusetts Institute of Technology, Cambridge, MA, 540 pp.

Lee, P. J., and P. C. C. Wang, 1983, "Conditional Analysis for Petroleum Resource Evaluations," *Internat. Assoc. Math. Geol. Jour.*, v. 15, pp. 349-361.

Lee, P. J., and P. C. C. Wang, 1985, "Probabilistic Formulation of a Method for the Evaluation of Petroleum Resources," *Internat. Assoc. Math. Geol. Jour.*, v. 15, pp. 163-182.

Lehman, Linda L., and Ellen J. Quinn, 1982, "Geochemical Interpretations of Groundwater Flow Systems in the Central Columbia Plateau," Internal Memo No. 3109.5/LL/82/02/02/0, U.S. Nuclear Regulatory Commission.

Lemire, R. J., and P. R. Tremaine, 1980, "Uranium and Plutonium Equilibria in Aqueous Solutions to 200°C," *Jour. Chem. and Eng. Data*, v. 25, pp. 361-370.

Lemos, J. V., R. D. Hart, and P. A. Cundall, 1985, "A Generalized Distinct Element Program for Modelling Jointed Rock Mass," *Proc. Internat. Symp. Fundamentals of Rock Joints*, Bjorkliden, pp. 335-343.

Lensen, G. J., 1964, "The General Case of Progressive Fault Displacement of Flights of Degradational Terraces," *New Zealand Jour. Geol. and Geophys.*, v. 7, pp. 864-870.

Lensen, G. J., 1968, "Analysis of Progressive Fault Displacement During Downcutting at the Branch River Terraces, South Island, New Zealand," *Geol. Soc. Amer. Bull.*, v. 79, pp. 545-555.

Lewins, J., and M. Becker, *eds.*, 1982, "Sensitivity and Uncertainty Analysis of Reactor Performance Parameters," *Adv. Nuclear Sci. and Technology*, 14, Plenum Press, New York.

Liedtke, L., and W. Bleich, 1985, "Convergence Calculations for Back-Filled Tunnels in Rock Salt," *Nonlinear Finite Element Analysis and ADINA, Proc. 5th ADINA Conf.*, Massachusetts Institute of Technology, Cambridge, MA, pp. 353-378.

Lindberg, H., 1982, "Tunnel Response in Modeled Jointed Rock," *Proc. 23rd U.S. Symp. Rock Mech.*, Berkeley, CA, pp. 824-836.

Lindh, A., G. Fuis, and C. Mantis, 1978, "Seismic Amplitude Measurements Suggest Foreshocks Have Different Focal Mechanism Than Aftershocks," *Science*, v. 201, pp. 56-59.

Linstone, Harold A., and Murray Turoff, 1975, *The Delphi Method Techniques and Applications*, Addison-Wesley Publ. Co., Reading, MA, 614 pp.

Lisitzin, E., 1974, *Sea-Level Changes*, Elsevier Oceanography Series, No. 8, Elsevier, Amsterdam, Holland, 286 pp.

Liu, H. S., 1980, "Mantle Convection and Subcrustal Stresses Under the United States," *Mod. Geol.*, v. 7, pp. 81-93.

Locke, W. W., J. T. Andrews, and P. J. Webber, 1979, *A Manual for Lichenometry*, Technical Bull. 26, British Geomorphological Research Group, Norwich, Univ. of East Anglia.

Loftsgaarden, D. O., and C. F. Quesenberry, 1965, "A Nonparametric Estimate of a Multivariate Density Function," *Annals Math. Stat.*, v. 36, pp. 1049-1051.

Logan, S. E., and M. C. Berbano, 1977, "Geologic Modeling in Risk Assessment Methodology for Radioactive Waste Management," *Risk Analysis and Geologic Modelling in Relation to the Disposal of Radioactive Wastes into Geologic Formations*, Proceedings of a Workshop Organized Jointly by the Nuclear Energy Agency, OECD, and the Commission of the European Communities.

Logan, S. E., and M. C. Berbano, 1978, *Development and Application of a Risk Assessment Method for Radioactive Waste Management*, EPA 520/6-78-005, U.S. Environmental Protection Agency.

Logan, S. E., R. L. Link, H. S. Ng, F. A. Rockenbach, and K. J. Hong, 1982, *Parametric Studies of Radiological Consequences of Basaltic Volcanism*, SAND81-2375, Sandia National Laboratories, Albuquerque, NM, 230 pp.

Loken, M. C., G. D. Callahan, D. K. Svalstad, and R. A. Wagner, 1987, *Thermomechanical Analyses of Conceptual Repository Designs for the Paradox and Permian Basins*, BMI/ONWI-664, Battelle Memorial Institute, Columbus, OH, 505 pp.

Lombardi, G., 1973, "Dimensioning of Tunnel Linings with Regard to Construction Procedure," *Tunnels and Tunneling*, v. 5, n. 4, pp. 340-351.

Long, J. L. S., J. S. Remer, C. R. Wilson, and P. A. Witherspoon, 1982, "Porous Media Equivalents of Networks of Discontinuous Fractures," *Water Resources Research*, v. 18, pp. 645-658.

Lorig, J. J., and B. G. H. Brady, 1984, "A Hybrid Computational Scheme for Excavation and Support Design in Jointed Rock Media," *Proc. ISRM Symp. Design and Performance of Underground Excavations*, British Geotechnical Society, London, pp. 105-112.

Louis, Thomas A., 1984, "Estimating a Population of Parameter Values Using Bayes and Empirical Bayes Methods," *Jour. Amer. Stat. Assoc.*, v. 79, pp. 393-398.

Lucchitta, I., 1979, "Late Cenozoic Uplift of the Southwestern Colorado Plateau and Adjacent Lower Colorado River Region," *Tectonophysics*, v. 61, pp. 63-95.

Luce, R. D., and H. Raiffa, 1957, *Games and Decisions: Introduction and Critical Survey*, John Wiley and Sons, New York, 509 pp.

Luce, R. D., and P. Suppes, 1965, "Preference, Utility and Subjective Probability," *in* R. D. Luce, R. R. Bush, and E. Galanter, *eds.*, *Handbook of Mathematical Psychology*, v. III, Ch. 19, John Wiley and Sons, New York, 537 pp.

Lumb, P., 1974, "Applications of Statistics in Soil Mechanics," *in* I. K. Lee, *ed.*, *Soil Mechanics—New Horizons*, American Elsevier, New York.

Lumb, P., 1975, "Spatial Variability of Soil Properties," *Proc 2nd Internat. Conf. Applications of Stat. and Probability in Soil and Structural Engineering*, Aachen, Germany.

Maasch, K. A., and B. Saltzman, 1990, "A Low-Order Dynamical Model of Global Climatic Variability Over the Full Pleistocene," *Jour. Geophys. Research*, v. 95, pp. 1955-1963.

Machette, M. N., 1978, "Dating Quaternary Faults in the Southwestern United States by Using Buried Calcic Paleosols," *U.S. Geological Survey Jour. Research*, v. 6, pp. 369-381.

Machette, M. N., 1982, *Soil Dating Techniques, Western Region (United States)*, Open-file Report OFR-82-840, U.S. Geological Survey, pp. 137-140.

Machette, M. N., and R. G. McGimsey, 1983, *Map of Quaternary and Pliocene Faults in the Socorro and Western Part of the Fort Sumner 1° x 2° Quadrangles, New Mexico*, U.S. Geological Survey Miscellaneous Field Studies Map MF-1465-A (1 sheet, scale 1:250,000, with explanatory text, 12 pp.).

Machette, M. N., and S. E. Personius, 1984, *Map of Quaternary and Pliocene Faults in the Eastern Part of the Aztec 1° x 2° Quadrangle and the Western Part of the Raton 1° x 2° Quadrangle, Northern New Mexico*, U.S. Geological Survey Miscellaneous Field Studies Map MF-1465-B (Scale 1:250,000, with text, 14 pp.).

MacMillan, J. R., and A. L. Gutjahr, 1986, "Geologic Controls on Spatial Variability for One-Dimensional Arrays of Porosity and Permeability Normal to Layering," *in* L. W. Lake and H. B. Carroll, Jr., *eds., Proc. of NIPER Conf. Reservoir Characterization*, Dallas, Academic Press, New York, pp. 265-292.

Magalhaes, M. C. F., J. D. Pedrosa de Jesus, and P. A. Williams, 1985, "The Chemistry of Uranium Dispersion in Groundwaters of the Pinhal Do Souto Mine, Portugal," *Inorg. Chim. Acta.*, v. 109, pp. 71-78.

Mahtab, M. A., D. D. Bolstad, J. R. Alldredge, and R. J. Shanley, 1972, *Analysis of Fracture Orientations for Input to Structural Models of Discontinuous Rock*, BUMINES-RI-7669, U.S. Bureau of Mines, Washington, DC, 82 pp.

Makurat, Axel, and Nick Barton, 1985, "Shear-Flow Coupling in Non-Planar Rock Joints," *Proc. Internat. Symp. on Coupled Processes Affecting the Performance of a Nuclear Waste Repository*, LBL-21850, Lawrence Berkeley Laboratory, Berkeley, CA, pp. 271-274.

Mallio, W. J., and J. H. Peck, 1981, "Practical Aspects of Geological Prediction," *Bull. Assoc. Engineering Geologists*, v. 18, pp. 369-374.

Manabe, S., and A. J. Broccoli, 1985, "The Influence of Continental Ice Sheets on the Climate of an Ice Age," *Jour. Geophys. Research*, v. 90, pp. 2167-2190.

Manabe, Syukuro, and Douglas G. Hahn, 1977, "Simulation of the Tropical Climate of an Ice Age," *Jour. Geophys. Research*, v. 82, pp. 3889-3911.

Mantoglou, A., and L. W. Gelhar, 1985, *Large Scale Models of Transient Unsaturated Flow and Contaminant Transport Using Stochastic Methods*, Ralph M. Parsons Lab Report 299, Massachusetts Institute of Technology, Cambridge, MA, 326 pp.

Mardia, K. V., 1972, *Statistics and Directional Data*, Academic Press, New York, 357 pp.

Marechal, A., and J. Serra, 1970, "Random Kriging," *in* D. F. Merriam, *ed., Geostatistics—A Colloquium*, Plenum Press, New York, pp. 91-112.

Mark, R. K., 1977, "Application of Linear Statistical Models of Earthquake Magnitude Versus Fault Length in Estimating Maximum Expectable Earthquakes," *Geology*, v. 5, pp. 464-466.

Mark, R. K., J. C. Tinsley, E. B. Newman, T. D. Gilmore, and R. O. Castle, 1981, "An Assessment of the Accuracy of the Geodetic Measurements that Led to the Recognition of the Southern California Uplift," *Jour. Geophys. Research*, v. 86, pp. 2783-2808.

Marle, C. M., 1981, *Multiphase Flow in Porous Media*, Gulf Publ. Co., Houston, TX, 254 pp.

Marsh, B. D., 1981, "On the Crystallinity, Probability of Occurrence, and Rheology of Lava and Magma," *Contrib. Min. Petrol.*, v. 78, pp. 85-98.

Marsh, B. D., 1984, "Mechanics and Energetics of Magma Formation and Ascension," *Explosive Volcanism: Inception, Evolution, and Hazards*, National Academy Press, Washington, DC, pp. 67-83.

Martin, R., and J. F. Davis, *eds.*, 1982, *Status of Volcanic Prediction and Emergency Response Capabilities in Volcanic Hazard Zones of California*, California Div. Mines and Geology, Sacramento, 275 pp.

Martz, H. F., and R. A. Waller, 1982, *Bayesian Reliability Analysis*, John Wiley and Sons, Inc., New York.

Martz, H. F., R. J. Beckman, K. Campbell, D. E. Whiteman, and J. M. Booker, 1983, *A Comparison of Methods for Uncertainty Analysis of Nuclear-Power-Plant Safety-System Fault-Tree Models (PWR; BWR)*, NUREG/CR-3263, LA-9729-MS, Los Alamos National Laboratory, Los Alamos, NM, 181 pp.

Matern, B., 1960, "Spatial Variation," *Medd. Fr. Statens Skogsfofshnings Institut*, v. 49, pp. 1-144.

Matheron, G., 1963, "Principles of Geostatistics," *Econ. Geol.*, v. 58, pp. 1246-1266.

Matheron, G., 1971, *The Theory of Regionalized Variables and Its Application*, Les Cahiers du Centre de Morphologie Mathematique de Fontainebleau, 5.

Matheron, G., and G. DeMarsily, 1980, "Is Transport in Porous Media Always Diffusive? A Counter Example," *Water Resources Research*, v. 16, pp. 901-917.

Matsuda, T., Y. Ota, M. Ando, and N. Yonekura, 1978, "Fault Mechanism and Recurrence Time of Major Earthquakes in Southern Kanto District, Japan, as Deduced from Coastal Terrace Data," *Geol. Soc. Amer. Bull.*, v. 89, pp. 1610-1618.

Matsuo, M., and K. Kawamura, 1983, "Reliability Based Design of Supporting System for NATM," *Proc. 4th Internat. Conf. Appl. Stat. and Probability in Soil and Structural Engineering*, Pitagora Editrice, Bologna, Italy.

Matti, J. C., J. C. Tinsley, D. M. Morton, and L. D. McFadden, 1982, "Holocene Faulting History as Recorded by Alluvial Stratigraphy Within the Cucamonga Fault Zone: A Preliminary View," *in* J. C. Tinsley, J. C. Matti, and L. D. McFadden, *eds.*, *Guidebook, Field Trip No. 12*, Geol. Soc. Amer., Cordilleran Section, pp. 29-44.

Maximus, Inc., 1980, "Handbook for the Calculation of Lower Statistical Confidence Bounds on System Reliability," unpublished manuscript.

Maxwell, Donald E., Barry W. Dial, and Michael B. Gross, 1984, *Joint Response of Queen Grayburg Siltstone Due to Excavation and Construction of a Repository Shaft*, SAI-83/1247, Sciences Applications, Inc., San Leandro, CA.

Maxwell, Donald E., Krishan K. Wahi, and Barry W. Dial, 1982, *Thermomechanical Analysis of Nuclear Waste Repositories with Horizontal Emplacement*, ONWI-210, Battelle Memorial Institute, Columbus, OH.

Mayer, Larry, 1979, "Evolution of the Mogollon Rim in Central Arizona," *Tectonophysics*, v. 61, pp. 49-62.

Mayer, Larry, 1982, *Quantitative Tectonic Geomorphology, with Applications to Neotectonics of Northwestern Arizona*, Ph.D. Thesis, Univ. of Arizona, Tucson, AZ, 512 pp.

Mayer, Larry, 1984, "Dating Quaternary Fault Scarps Formed in Alluvium Using Morphological Parameters," *Quaternary Research*, v. 22, pp. 300-313.

Mayer, Larry, M. Mergner-Keefer, and C. M. Wentworth, 1981, *Probability Models and Computer Simulation of Landscape Evolution*, U.S. Geological Survey Open-file Report OFR-81-656, 37 pp.

McCalpin, J. P., 1982, "Quaternary Geology and Neotectonics of the West Flank of the Northern Sangre de Cristo Mountains, South-Central Colorado," *Colorado School of Mines Quarterly*, v. 77, 97 pp.

McCann, M. W., 1980, *RMS Acceleration and Duration of Strong Ground Motion*, Rept. 46, J. A. Blume Earthquake Engineering Center, Stanford University, Stanford, CA.

McCann, W. R., S. P. Nishenko, L. R. Sykes, and J. Krause, 1978, "Seismic Gaps and Plate Tectonics: Seismic Potential for Major Plate Boundaries," *in* J. F. Evernden (Convener), B. L. Isacks, and G. Plafker (Organizers), *Proc. Conf. VI Methodology for Identifying Seismic Gaps*, Open-file Report 78-943, U.S. Geological Survey, pp. 441-584.

McFadden, L. D., and J. C. Tinsley, 1982, "Soil Profile Development in Xeric Climates: A Summary," *in* J. C. Tinsley, J. C. Matti, and L. D. McFadden, *eds., Guidebook, Field Trip No. 12*, Geol. Soc. Amer., Cordilleran Section, pp. 15-19.

McGarr, A., 1984, "Scaling of Ground Motion Parameters, State of Stress and Focal Depth," *Jour. Geophys. Research*, v. 89, pp. 6969-6979.

McGarr, A., and N. C. Gay, 1978, "State of Stress in the Earth's Crust," *Ann. Rev. Earth Planet. Sci.*, v. 6, pp. 405-436.

McGarr, A., R. W. E. Green, and S. M. Spottiswoode, 1981, "Strong Ground Motion of Mine Tremors, Some Implications for Near Source Ground Motion Parameters," *Bull. Seismol. Soc. Amer.*, v. 71, pp. 295-319.

McGarr, A., M. D. Zoback, and T. C. Hanks, 1982, "Implications of an Elastic Analysis of In-Situ Stress Measurements Near the San Andreas Fault," *Jour. Geophys. Research*, v. 87, pp. 7797-7806.

McGuire, R. K., 1977a, "Seismic Design Spectra and Mapping Procedures Using Hazard Analysis Based Directly on Oscillator Response," *Earthquake Eng. and Struct. Dyn.*, v. 5, pp. 211-234.

McGuire, R. K., 1977b, "Effects of Uncertainty in Seismicity on Estimates of Seismic Hazard on the East Coast of the United States," *Bull. Seismol. Soc. Amer.*, v. 67, pp. 827-848.

McGuire, R. K., 1978a, "A Simple Model for Estimating Fourier Amplitude Spectra of Horizontal Ground Acceleration," *Bull. Seismol. Soc. Amer.*, v. 68, pp. 803-822.

McGuire, R. K., 1978b, *FRISK—A Computer Program for Seismic Risk Analysis Using Faults as Earthquake Sources*, Open-file Report 78-1007, U.S. Geological Survey.

McGuire, R. K., 1979, "Adequacy of Simple Probability Models for Calculating Felt-Shaking Hazard, Using the Chinese Earthquake Catalog," *Bull. Seismol. Soc. Amer.*, v. 69, pp. 877-892.

McGuire, R. K., 1984, "Ground Motion Estimation in Regions with Few Data," *Proc. Eighth World Conf. Earthquake Engineering*, San Francisco, CA, Vol. II, pp. 327-334.

McGuire, R. K., and J. A. Barnhard, 1977, "Magnitude, Distance, and Intensity Data for C.I.T. Strong Motion Records," U.S. Geological Survey, *Jour. Research*, v. 5, pp. 437-443.

McGuire, R. K., and T. P. Barnhard, 1979, *Four Definitions of Strong-Motion Duration: Their Predictability and Utility for Seismic Hazard Analysis*, Open-file Report 79-1515, U.S. Geological Survey.

McGuire, R. K., and T. P. Barnhard, 1981, "Effects of Temporal Variations in Seismicity on Seismic Hazard," *Bull. Seismol. Soc. Amer.*, v. 71, pp. 321-334.

McGuire, R. K., and T. C. Hanks, 1980, "RMS Accelerations and Spectral Amplitudes of Strong Ground Motion during the San Fernando, California, Earthquake," *Bull. Seismol. Soc. Amer.*, v. 70, pp. 1907-1919.

McGuire, R. K., A. M. Becker, and N. C. Donovan, 1984, "Spectral Estimates of Seismic Shear Waves," *Bull. Seismol. Soc. Amer.*, v. 74, pp. 1427-1440.

McKay, M. D., R. J. Beckman, and W. J. Conover, 1979, "A Comparison of Three Methods for Selecting Values of Input Variables in Analysis of Output From a Computer Code," *Technometrics*, v. 21, pp. 239-245.

McLaughlin, R. J., K. R. Lajoie, D. H. Sorg, S. D. Morrison, and J. A. Wolfe, 1983, "Tectonic Uplift of a Middle Wisconsin Marine Platform near the Mendocino Triple Junction, California," *Geology*, v. 11, pp. 35-39.

McMahon, B., 1971, "A Statistical Method for the Design of Rock Slopes," *Dames and Moore Bulletin*, v. 39, pp. 1-9.

McNulty, E. Gregory, 1987, *Expected Near-Field Thermal Performance for Nuclear Waste Repositories at Potential Salt Sites: Technical Report*, BMI/ONWI-652, Battelle Memorial Institute, Columbus, OH, 134 pp.

Meehan, R., 1984, *The Atom and The Fault*, Massachusetts Institute of Technology Press, Cambridge, MA, 157 pp.

Meehl, Gerald A., 1984, "Modeling of the Earth's Climate," *Climatic Change*, v. 6, pp. 259-286.

Mehlhorn, G., J. Kollegger, M. Keuser, and W. Kolmar, 1985, "Nonlinear Contact Problems—A Finite Element Approach Implemented in ADINA," *Proc. 5th ADINA Conf.*, Mechanical Engineering Department, Massachusetts Institute of Technology, Cambridge, MA.

Meisner, J., and F. Demirmen, 1980, "The Creaming Method: A Bayesian Procedure to Forecast Future Oil and Gas Discoveries in Mature Exploration Provinces," *Jour. Royal Stat. Soc.*, v. 143, 17 pp.

Menzie, W. David, 1977, "The Unit Regional Value of the Republic of South Africa," Ph.D. Thesis, Pennsylvania State Univ., 236 pp.

Merifield, P. M., and D. L. Lamar, 1981, "Anomalous Water-Level Changes and Possible Relation with Earthquakes," *Eos, Trans. Amer. Geophys. Union*, v. 62, p. 1034.

Merriam, D. F., *ed., 1972, Mathematical Models of Sedimentary Processes*, Plenum Press, New York, 168 pp.

Merritt, A. H., 1972, "Geologic Predictions for Underground Excavations," *Proc. lst RETC*, v. 1, AIME, New York, pp. 115-132.

Merz, H. A., and C. A. Cornell, 1973, *Aftershocks in Engineering Risk Analysis*, Research Report R73-25, Department of Civil Engineering, Massachusetts Institute of Technology, Cambridge, MA.

Metcalfe, D. E., J. E. Campbell, B. S. RamaRao, and V. W. Harper, 1985, "Geostatistical and Adjoint Sensitivity Techniques Applied to a Conceptual Model of Groundwater Flow in the Paradox Basin, Utah," *in* R. G. Post, *ed., Waste Management '85*, Tucson, AZ, Vol. 1, pp. 219-225.

Meyer, Richard F., 1977, "Petroleum Resource Data Systems," *Internat. Assoc. Math. Geol. Jour.*, v. 9, pp. 281-299.

Meyer, Richard F., 1978, "The Volumetric Method for Petroleum Resource Estimation," *Internat. Assoc. Math. Geol. Jour.*, v. 10, pp. 501-518.

Mikumo, T., M. Kato, H. Doi, Y. Wada, and T. Tanaka, 1977, "Possibility of Temporal Variations in Earth Tidal Strain Amplitudes Associated with Major Earthquakes," *Jour. Phys. Earth*, v. 25, Suppl., pp. S123-S136.

Miles, R. E., 1969, "Poisson Flats in Euclidian Spaces, I," *Adv. Appl. Prob.*, v. 1, pp. 211-237.

Miles, R. E., 1971, "Poisson Flats in Euclidian Spaces, II," *Adv. Appl. Prob.*, v. 3, pp. 1-43.

Miller, S. M., 1975, "Determination of Spatial Variation in Fracture Set Characteristics by Geostatistical Methods," M.S. Thesis, Univ. of Arizona.

Miller, S. M., 1979, "Geostatistical Analysis for Evaluating Spatial Dependence in Fracture Set Characteristics," APCOM, Tucson.

Mitchell, J. Murray, Jr., 1976, "An Overview of Climatic Variability and Its Causal Mechanisms," *Quaternary Research*, v. 6, pp. 481-493.

Mizell, S. A., A. L. Gutjahr, and L. W. Gelhar, 1982, "Stochastic Analysis of Spatial Variability in Two-Dimensional Steady Groundwater Flow Assuming Stationary and Non-Stationary Heads," *Water Resources Research*, v. 19, pp. 1053-1067.

Moavenzadeh, F., M. J. Markow, R. D. Waytt, and C. H. Minott, 1974, "Tunnel Cost Model," *Proc. Rapid Excavation and Tunnel Conf.*, Vol. 2, Sec. 25, pp. 1721-1739, AIME Soc. of Min. Eng., New York.

Mogi, K., 1968, "Sequential Occurrences of Recent Great Earthquakes," *Jour. Phys. Earth*, v. 16, pp. 30-36.

Montazer, P., and W. E. Wilson, 1984, "Conceptual Hydrologic Model of Flow in the Unsaturated Zone, Yucca Mountain, Nevada," Water-Resources Investigations Report 84-4345, U.S. Geological Survey, Lakewood, CO, 55 pp.

Moore, I. D., and J. R. Brooker, 1982, "A Circular Boundary Element for the Analysis of Deep Underground Openings," *in* Z. Eisenstein, *ed., Numerical Methods in Geomechanics 1982*, A. A. Balkema Publishers, Rotterdam, pp. 53-60.

Morgan, H. D., 1961, "A Contribution to the Analysis of Stress in a Circular Tunnel," *Geotechnique*, v. 11, pp. 37-46.

Morgan, Harold S., Raymond D. Krieg, and Rudolph V. Matalucci, 1981, *Comparative Analysis of Nine Structural Codes Used in the Second WIPP Benchmark Problem*, SAND81-1389, Sandia National Laboratories, Albuquerque, NM.

Morgan, Harold S., Charles M. Stone, and Raymond D. Krieg, 1985, "The Use of Field Data to Evaluate and Improve Drift Response Models for the Waste Isolation Pilot Plant (WIPP)," *in* Eileen Ashworth, *ed., Research & Engineering Applications in*

Rock Masses, 26th U.S. Symposium on Rock Mechanics, Rapid City, SD, Vol. 2, A. A. Balkema, Boston, pp. 769-776.

Morgan, H. S., C. M. Stone, R. D. Krieg, and D. E. Munson, 1986, "Thermal/Structural Modeling of a Large Scale In Situ Overtest Experiment for Defense High Level Waste at the Waste Isolation Pilot Plant Facility," presented at The Second International Conference on Numerical Models in Geomechanics, Ghent, Belgium, March 31-April 14, 1986, 12 pp.

Morris, Carl N., 1983, "Parametric Empirical Bayes Inference: Theory and Application," *Jour. Amer. Stat. Assoc.*, v. 78, pp. 47-65.

Morris, P. A., 1974, "Decision Analysis Expert Use," *Management Sci.*, v. 20, n. 9, pp. 1233-1241.

Morrison, H. F., R. Fernandez, and R. F. Corwin, 1979, "Earth Resistivity, Self Potential Variations, and Earthquakes: A Negative Result for M = 4.0," *Geophys. Research Let.*, v. 6, pp. 139-142.

Morse, P. M., 1974, *Search Theory*, TR01-74, Operations Research Center, Massachusetts Institute of Technology, Cambridge, MA.

Moskvin, A. I., 1971, "Complex Formation of Neptunium (IV, V, VI) in Carbonate Solutions," *Sov. Radiochem.*, v. 13, pp. 694-699.

Moskvin, A. I., and A. D. Gel'man, 1958, "Determination of the Composition and Instability Constants of Oxalate and Carbonate Complexes of Plutonium (IV)," *Jour. Inorg. Chem. USSR*, v. 3, pp. 198-216.

Moyle, W. R., Jr., 1980, *Groundwater-Level Monitoring for Earthquake Prediction. A Progress Report Based on Data Collected in Southern California, 1976-1979*, Open-file Report 80-413, U.S. Geological Survey, 60 pp.

Mrugala, M., 1984, "Application of Statistical Methods to the Determination of Mechanical Model Parameters for Salt," *in* H. Reginald Hardy, Jr., and Michael Langer, *eds., The Mechanical Behavior of Salt, Proceedings of the First Conference*, Pennsylvania State University, University Park, PA, pp. 339-354.

Mucciardi, A. N., 1978, "Statistical Investigation of the Mechanics Controlling Radionuclide Sorption, Part II, Task 4," *Second Contractor Information Meeting Volume II*, PNL-SA-7352, Battelle Northwest Laboratory, pp. 333-425.

Mucciardi, A. N., and E. C. Orr, 1977, "Statistical Investigation of the Mechanics Controlling Radionuclide Sorption," *Waste Isolation Safety Assessment Program, Task 4, Contractor Information Meeting Proceedings*, PNL-SA-6957, Battelle Northwest Laboratory, pp. 151-188.

Muehlhaus, H. B., 1985, "Lower Bound Solutions for Circular Tunnels of Two and Three Dimensions," *Rock Mech. and Rock Engineering*, v. 18, pp. 37-52.

Mueller, L., 1963, *Der Felsbau*, v. 1, Enke Verlag, 624 pp.

Muhs, D. R., and B. J. Szabo, 1982, "Uranium-Series Age of the Eel Point Terrace, San Clemente Island, California," *Geology*, v. 10, pp. 23-26.

Muir Wood, A. M., 1975, "The Circular Tunnel in Elastic Ground," *Geotechnique*, v. 25, pp. 115-127.

Mulargia, F., S. Tinti, and E. Boschi, 1985, "A Statistical Analysis of Flank Eruptions on Etna Volcano," *Jour. Volc. Geoth. Research*, v. 23, pp. 263-272.

Mulargia, F., S. Tinti, and E. Boschi, 1986, "A Statistical Analysis of Flank Eruptions on Etna Volcano, Reply to the Comments of D. K. Chester," *Jour. Volc. Geoth. Research*, v. 28, pp. 385-395.

Muller, A. B., and L. E. Duda, 1982, "The Influence and Modeling of Waste Nuclide Aqueous Speciation on Geosphere Retardation Processes Relevant to Risk Assessment," *in* R. G. Post and M. E. Wacks, *eds., Waste Management '82*, vol. 1, Univ. of Arizona, Tucson, AZ.

Munguia, L., and J. N. Brune, 1985, "Simulation of Strong Ground Motion for Earthquakes in the Mexicali-Imperial Valley," *in* R. E. Scholl and J. L. King, *eds., Strong Ground Motion Simulation and Earthquake Engineering Applications*, Publication No. 85-02, Earthquake Engineering Research Institute, Berkeley, CA.

Munson, Darrell E., Rudolph V. Matalucci, and Teresa M. Torres, 1985, "Implementation of Thermal/Structural Interactions in In Situ Tests at the Waste Isolation Pilot Plant Facility," presented at Second International Conference on Mechanical Behavior of Salt, Hanover, Germany, Sept. 24-27, 1985, 53 pp.

Myer, L. R., T. L. Brekke, C. T. Dare, R. B. Dill, and G. E. Korbin, 1981, "An Investigation of Stand-Up Time of Tunnels in Squeezing Ground," *Proc. Rapid Excavation and Tunneling Conf.*, Vol. 2, Sec. 20, pp. 1415-1433, AIME Soc. Min. Eng., New York.

Namias, J., 1975, *Short Period Climatic Variations*, University of California, San Diego, 905 pp.

Nansen, G. C., 1980, "A Regional Trend to Meander Migration," *Jour. Geol.*, v. 88, pp. 100-108.

Narasimhan, T. N., 1982, "Multi-Dimensional Numerical Simulation of Fluid Flow in Fractured Porous Media," *Water Resources Research*, v. 18, pp. 1235-1247.

NASA, 1979, *Application of Space Technology to Crustal Dynamics and Earthquake Research*, NASA-TP-1464, National Aeronautics and Space Administration, Washington, DC, 274 pp.

Nash, D. B., 1980, "Morphologic Dating of Degraded Normal Fault Scarps," *Jour. Geol.*, v. 88, pp. 353-360.

Nash, D. B., 1984, "Morphologic Dating of Fluvial Terrace Scarps and Fault Scarps Near West Yellowstone, Montana," *Geol. Soc. Amer. Bull.*, v. 95, pp. 1413-1424.

Neftel, A., H. Oeschger, J. Schwander, B. Stauffer, and R. Zumbrunn, 1982, "Ice Core Sample Measurements Give Atmospheric CO_2 Content During the Past 40,000 Yr," *Nature*, v. 295, pp. 220-223.

Negmatullayev, S. K., K. M. Mirzoyev, O. V. Soboleva, A. S. Malamud, V. I. Starkov, R. S. Mihailova, M. V. Kulagina, and E. Y. Starkova, 1984, "Activation of the Seismic Processes in Regional Structures of the Lithosphere and a New Approach to the Problem of Long-Term Earthquake Prediction," *in* S. A. Fedotov, I. N. Galkin, A. V. Nikolayev, and E. N. Sedova, *eds., Earthquakes and Geological Hazard Prediction*, Internat. Geol. Congress, v. 27, pp. 34-44.

Nelson, D. P., and A. F. Fossum, 1985, "Salt Pressures on Defense High-Level Waste Packages," *in Eileen Ashworth, ed., Research & Engineering Applications in Rock Masses*, 26th U.S. Symposium on Rock Mechanics, Rapid City, SD, v. 2, A. A. Balkema, Boston, pp. 251-258.

Nelson, P. H., 1982, "Advances in Borehole Geophysics for Hydrology," *in* T. N. Narasimhan, *ed., Recent Trends in Hydrogeology*, Geol. Soc. Amer. Special Paper 189, pp. 207-219.

Neretnieks, I., 1980, "Diffusion in Rock Matrix: An Important Factor in Radionuclide Retardation?" *Jour. Geophys. Research*, v. 85, pp. 4379-4397.

Neretnieks, I., 1982, "The Influence of Microfissures on Crystalline Rock in Radionuclide Migration," *in* G. DeMarsily and D. F. Merriam, *eds., Predictive Geology*, Pergamon Press, Elmsford, NY, pp. 137-152.

Neretnieks, I., 1983, "A Note on Fracture Flow Dispersion Mechanisms in the Ground," *Water Resources Research*, v. 19, pp. 364-370.

Neretnieks, I., and A. Rasmuson, 1983, "An Integrated Approach to the Description of Radionuclide Release and Transport in the Geosphere," *in* G. L. McVay, *ed., Scientific Basis for Nuclear Waste Management VII*, Research Soc. Proc., v. 26, North-Holland, New York, pp. 269-278.

Neretnieks, I., T. Eriksen, and P. Taehtinen, 1982, "Tracer Movement in a Single Fissure in Granitic Rock: Some Experimental Results and Their Interpretation," *Water Resources Research*, v. 18, pp. 849-858.

Neuman, S. P., 1982, "Statistical Characterization of Aquifer Heterogeneities," *Recent Trends in Hydrogeology*, Geol. Soc. Amer. Special Paper 189, pp. 81-102.

Neuman, S. P., 1984, "Role of Geostatistics in Subsurface Hydrology," *in* G. Verly, M. David, A. G. Journel, and A. Marechal, *eds., Geostatistics for Natural Resources Characterization*, Part 2, pp. 787-816, D. Reidel Publ. Co., Boston, MA, 1072 pp.

Neuzil, C. E., and J. V. Tracy, 1981, "Flow Through Fractures," *Water Resources Research*, v. 17, pp. 191-199.

Newhall, C. G., 1981, "Probabilistic Hazard Assessment at Mount St. Helens," *in* M. Kent, *ed., New Horizons in Engineering Geology*, Assoc. Eng. Geologists, Sudbury, MA, pp. 47-48.

Newhall, C. G., 1982, *A Method for Estimating Intermediate- and Long-Term Risks from Volcanic Activity, with an Example from Mount St. Helens, Washington*, Open-file Report 82-396, U.S. Geological Survey, 58 pp.

Newmark, N. M., 1963, "Design of Structures for Dynamic Loads Including the Effects of Vibration on Ground Shock," *Symp. Scientific Problems of Protective Construction*, Swiss Federal Institute of Technology, Zurich.

Newmark, N. M., 1969, "Design of Rock Silo and Rock Cavity Linings," Technical Report to Space and Missile Systems Organization, Air Force Systems Command, Norton Air Force Base, Contract No. FO 4701-69-C-0155.

Newmark, N. M., and W. J. Hall, 1982, *Earthquake Spectra and Design*, Earthquake Engineering Research Institute, Berkeley, CA.

Newmark, N. M., and E. Rosenblueth, 1979, *Fundamentals of Earthquake Engineering*, Prentice-Hall, Englewood Cliffs, NJ.

Nguyen, V. U., and R. N. Chowdhury, 1984, "Probabilistic Study of Spoil Pile in Strip Coal Mines—Two Techniques Compared," *Internat. Jour. Rock Mech. Mining Sci.*, v. 26, n. 6, pp. 303-312.

Niazi, M., and C. P. Mortgat, 1984, "Application of Logistic Models to Earthquake Forecasting," *Earthquake Prediction Research*, v. 2, pp. 221-225.

Niccum, M., K. M. Cline, and M. Jackson, 1980, "Volcanic and Seismic Hazards on the Snake River Plain, Idaho," *18th Ann. Symp. Eng. Geol. and Soils Eng.*, Boise State Univ., Boise, Idaho, 81 pp.

Nicholas, D. E., and J. M. Marek, 1981, "The Feasibility Study—Selection of a Mining

Method Integrating Rock Mechanics and Mine Planning," *Proc. 1981 Rapid Excavation and Tunneling Conf.*, Vol. 2, Sec. 14, pp. 1018-1031, AIME Soc. of Min. Eng., New York.

Nie, N. H., 1975, *SPSS: Statistical Package for the Social Sciences*, 2nd ed., McGraw-Hill Book Co., New York, 343 pp.

Nigam, P. S., O. P. Jain, and M. B. Kanchi, 1976, "Three-Dimensional Analysis of Hydroelectric Power Station," *Jour. Power Div.*, Proc. Amer. Soc. Civil Engineers, v. 102, no. POl, pp. 35-52.

Niwa, Y., S. Kobayashi, and T. Fukui, 1977, "Application of Integral Equation Method to the Determination of Three Dimensional Stresses Around a Cavity," *Proc. Japanese Soc. Civil Engineers*, No. 266, pp. 25-37, (in Japanese).

NOAA, no date, "Digitized Strong Motion Data Tapes and Intensity File Tapes," World Data Center A, National Oceanic and Atmospheric Administration, Washington, DC.

Nord, G., and H. Stille, 1983, "Support Technique for a Large Silo in Poor Rock," *Rock Mech. and Rock Engineering*, v. 16, n. 4, pp. 233-252.

Noritomi, K., 1978, "Application of Precursory Geoelectric and Geomagnetic Phenomena to Earthquake Prediction in China," *Chin. Geophys.* (Amer. Geophys. Union), v. 1, pp. 377-391.

North, Gerald R., Robert F. Cahalan, and James A. Coakley, Jr., 1981, "Energy Balance Climate Models," *Rev. Geophys. Space Phys.*, v. 19, pp. 91-121.

North, G. R., J. G. Mengel, and D. A. Short, 1984, "A Two-Dimensional Climate Model Useful in Ice Age Applications," *in* A. Berger, J. Imbrie, J. Hays, G. Kukla, and B. Saltzman, *eds., Milankovitch and Climate—Understanding the Response to Astronomical Forcing*, Part 2, D. Reidel Co., Dordrecht, Holland, pp. 513-518.

Nucci, L. R., 1978, "An Analysis of the Rate of Error in Two Color Maps," M.S. Thesis, Massachusetts Institute of Technology, Cambridge, MA.

Nuttli, O. W., 1974, "Magnitude-Recurrence Relation for Central Mississippi Valley Earthquakes," *Bull. Seismol. Soc. Amer.*, v. 64, n. 4, pp. 1189-1207.

Nuttli, O. W., 1979, *State of the Art for Assessing Earthquake Hazards in the United States, Report 16, The Relation of Sustained Maximum Ground Acceleration and Velocity to Earthquake Intensity and Magnitude*, Misc. Paper S-73-1, U.S. Army Engineers Waterways Experimental Station, Vicksburg, MS.

Nuttli, O. W., 1983a, "Average Seismic Source—Parameter Relations for Mid-Plate Earthquakes," *Bull. Seismol. Soc. Amer.*, v. 73, n. 2, pp. 519-535.

Nuttli, O. W., 1983b, "Empirical Magnitude and Spectral Scaling Relations for Mid-Plate and Plate-Margin Earthquakes," *Tectonophysics*, v. 93, n. 3-4, pp. 207-223.

Nuttli, O. W., 1983c, "Letter of January 24, 1983," *in* D. L. Bernreuter, J. B. Savy, R. W. Mensing, J. C. Chen, and B. C. Davis, 1985, *Seismic Hazard Characterization of the Eastern United States*, UCID-20421, Vols. 1 and 2, Lawrence Livermore National Laboratory, Livermore, CA.

Nuttli, O. W., and R. B. Herrmann, 1982, "Earthquake Magnitude Scales," *Jour. Geotech. Engineering Div.*, Amer. Soc. Civil Engineers, pp. 783-786.

Obermeier, S. F., G. S. Gohn, R. E. Weems, R. L. Gelinkas, and M. Rubin, 1985, "Geologic Evidence for Recurrent Moderate to Large Earthquakes Near Charleston, South Carolina," *Science*, v. 227, pp. 408-411.

Obert, L., and W. I. Duvall, 1967, *Rock Mechanics and Design of Structures in Rock*, John Wiley and Sons, Inc., New York, 650 pp.

Obert, L., S. L. Windes, and W. I. Duvall, 1946, "Standardised Test for Determining the Physical Properties of Mine Rock," Department Investigation 3891, U.S. Bureau of Mines.

Oblow, E. M., 1983, *An Automated Procedure for Sensitivity Analysis Using Computer Calculus*, Oak Ridge National Lab, ORNL-TM-8776, Oak Ridge, TN, 35 pp.

O'Brien, T. J., and P. A. Williams, 1983, "The Aqueous Chemistry of Uranium Minerals 4, Schrockingerite, Grimselite, and Related Alkali Uranyl Carbonates," *Min. Mag.*, v. 47, n. 1, pp. 69-73.

Oerlemans, J., 1982, "Glacial Cycles and Ice-Sheet Modelling," *Climatic Change*, v. 4, n. 4, pp. 353-374.

Oerlemans, J., 1984, "On the Origin of the Ice Ages," *in* A. Berger, J. Imbrie, J. Hays, G. Kukla, and B. Saltzman, *eds., Milankovitch and Climate—Understanding the Response to Astronomical Forcing*, Part 2, D. Reidel Co., Dordrecht, Holland, pp. 607-611.

Ohnishi, Yuzo, Hideaki Shibata, and Akira Kobayashi, 1987, "Development of Finite Element Code for the Analysis of Coupled Thermo-Hydro-Mechanical Behaviors of Saturated-Unsaturated Medium," *International Symposium on Coupled Processes Affecting the Performance of a Nuclear Waste Repository*, Sept. 18-20, 1985, Lawrence Berkeley Laboratory, Berkeley, CA, Harcourt, Brace, Jovanovich, pp. 679-697.

Ohtake, M., 1976, "Search for Precursors of the 1974 Izu-Hanto-Oki Earthquake, Japan," *Pure Appl. Geophys.*, v. 114, pp. 1083-1093.

Ohtake, M., T. Matumoto, and G. V. Latham, 1977, "Seismicity Gap near Oaxaca Southern Mexico as a Probable Precursor to a Large Earthquake," *Pure Appl. Geophys.*, v. 115, n. 1-2, pp. 375-385.

Ohtake, M., T. Matumoto, and G. V. Latham, 1978, "Patterns of Seismicity Preceding Earthquakes in Central America, Mexico and California," *in* J. F. Evernden (Convener), B. L. Isacks, and G. Plafker (Organizers), *Proc. Conf. VI Methodology for Identifying Seismic Gaps*, Open-file Report 78-943, U.S. Geological Survey, pp. 585-610.

Oliver, J., and L. Murphy, 1971, "WWSSN: Seismology's Global Network of Observing Stations," (World Wide Standardized Seismograph Network) *Science*, v. 174, pp. 267-274.

Olofsson, U., and B. Allard, 1983, *Complexes of Actinides with Naturally Occurring Organic Substances, Literature Survey Technical*, KBS Report 83-09, Stockholm, Sweden.

Omre, Henning, 1983, "The Combination of Subjective Information and Drillhole Data for Estimating the Size of an Elliptical Target," *Internat. Assoc. Math. Geol. Jour.*, v. 15, n. 3, pp. 477-481.

O'Neill, J. R., and T. C. Hanks, 1980, "Geochemical Evidence for Water-Rock Interaction Along the San Andreas and Garlock Faults of California," *Jour. Geophys. Research B*, v. 85, n. 11, pp. 6286-6292.

Ontario Ministry of Transportation and Communication, 1976, *Tunnelling Technology*, 186 pp.

O'Riordan, N. J., K. W. Cole, D. J. Henkel, 1984, "Collapses of Abandoned Limestone Mines in the West Midlands of England," *Proc. ISRM Symp. Design and Performance of Underground Excavations*, British Geotechnical Society, London, pp. 401-408.

O'Rourke, John E., Pamela H. Rey, Ezio Alviti, and Clyde C. Capps, 1986, *Rock Index Properties for Geoengineering in the Paradox Basin*, BMI/ONWI-579, Battelle Memorial Institute, Columbus, OH.

Ota, Y., 1975, "Late Quaternary Vertical Movement in Japan Estimated From Deformed Shorelines," *Bull. Research Soc. New Zealand*, v. 13, pp. 231-239.

Otsuka, M., 1964, "Earthquake Magnitude and Surface Fault Formation," *Jour. Phys. Earth*, v. 12, pp. 19-24.

Ouchi, Shanji, 1985, "Response of Alluvial Rivers to Slow Active Tectonic Movement," *Geol. Soc. Amer. Bull.*, v. 96, n. 4, pp. 504-515.

Owen, G. N., and R. E. Scholl, 1985, "Earthquake Engineering of Tunnels—Revisited," *Proc. 1985 Rapid Excavation and Tunneling Conf.*, Vol. 1, Chicago, IL, pp. 63-70.

Owen, G. N., R. E. Scholl, and T. L. Brekke, 1979, "Earthquake Engineering of Tunnels," *Proc. 1979 Rapid Excavation and Tunneling Conf.*, Vol. 1, Sec. 9, pp. 709-721.

Page, R., 1970, "Dating Episodes of Faulting From Tree Rings: Effects of the 1958 Rupture of the Fairweather Fault on Tree Growth," *Geol. Soc. Amer. Bull.*, v. 81, n. 10, pp. 3085-3094.

Panek, L., 1962, "The Combined Effects of Friction and Suspension in Bolting Bedded Mine Roof," U.S. Bureau of Mines Report of Invest., No. 6139.

Panet, M., 1976, "Stability Analysis of a Tunnel Driven in a Rock Mass in Taking into Account the Post-Failure Behavior," *Rock Mech.*, v. 8, n. 4, pp. 209-223 (in French).

Panet, M., 1978, "Texte provisoire des recommendations pour une description des massifs rocheux utile a l'etude de la stabilite des ouvrages souterrains," *Tunnels et Ouvrages Souterrains*, no. 28, pp. 176-185.

Panet, M., and P. Guellec, 1974, "Contribution to the Problem of the Design of Tunnel Support Behind the Face," *Proc. Third Congress Internat. Soc. Rock Mech.*, Denver, Vol. 2, Part B, pp. 1163-1168 (in French).

Panet, M., and A. Guenot, 1982, "Analysis of Convergence Behind the Face of a Tunnel," *Tunneling 82*, Third Internat. Symp., Institution of Mining and Metallurgy, London, pp. 197-204.

Papageorgiou, A. S., and K. Aki, 1983a, "A Specific Barrier Model for the Quantitative Description of Inhomogeneous Faulting and the Prediction of Strong Ground Motion. I. Description of the Model," *Bull. Seismol. Soc. Amer.*, v. 73, n. 3, pp. 693-722.

Papageorgiou, A. S., and K. Aki, 1983b, "A Specific Barrier Model for the Quantitative Description of Inhomogeneous Faulting and the Prediction of Strong Ground Motion. II. Application of the Model," *Bull. Seismol. Soc. Amer.*, v. 73, n. 4, pp. 953-978.

Papageorgiou, A. S., and K. Aki, 1985, "Scaling Law of Farfield Spectra Based on Observed Parameters of the Specific Barrier Model," *Pure Appl. Geophys.*, v. 123, n. 3, pp. 353-374.

Papelis, C., K. F. Hayes, J. O. Leckie, 1988, *"HYDRAQL, A Program for the Computation of Chemical Equilibrium Composition of Aqueous Batch Systems Including Surface-*

Complexation Modeling of Ion Adsorption at the Oxide/Solution Interface, Tech. Rept. No. 306, Dept. Civil Engineering, Stanford Univ., Stanford, CA.

Parzen, E., 1962a, "On the Estimation of a Probability Density Function and the Mode," *Annals Math. Stat.*, v. 33, pp. 832-837.

Parzen, E., 1962b, *Stochastic Processes*, Holden Day, San Francisco, CA.

Passchier, C. W., 1982, "Pseudotachylyte and the Development of Ultramylorite Bands in the Saint-Barthelemy Massif, French Pyrenees," *Jour. Structural Geol.*, v. 4, n. 1, pp. 69-79.

Patel, Jagdish K., C. H. Kapakia, and D. B. Owen, 1976, *Handbook of Statistical Distributions*, Marcel Dekker, Inc., New York, 294 pp.

Patwardhan, A. S., R. B. Kulkarni, and D. Tocher, 1980, "A Semi-Markov Model for Characterizing Recurrence of Great Earthquakes," *Bull. Seismol. Soc. Amer.*, v. 70, n. 1, pp. 323-348.

Pearson, Karl, 1902a, "On the Systematic Fitting of Curves to Observations and Measurements," *Biometrika*, v. 1, pp. 265-303.

Pearson, Karl, 1902b, "On the Systematic Fitting of Curves to Observations and Measurements, Part II," *Biometrika*, v. 2, pp. 1-23.

Pearthree, P. A., and S. S. Calvo, 1982, *Late Quaternary Faulting West of the Santa Rita Mountains South of Tucson, Arizona*, M.S. Thesis, Univ. of Arizona, Tucson, AZ, 49 pp.

Pearthree, P. A., C. M. Menges, and Larry Mayer, 1983, *Distribution, Recurrence, and Possible Tectonic Implications of Late Quaternary Faulting in Arizona*, Open-file Report 83-20, Arizona Bureau of Geology and Mineral Technology, 51 pp.

Peck, R. B., 1969a, "Advantages and Limitations of the Observational Method in Applied Soil Mechanics," 9th Rankine Lecture, *Geotechnique*, v. 19, n. 2, pp. 171-187.

Peck, R. B., 1969b, "State-of-the-Art Report: Deep Excavations and Tunneling in Soft Ground," *Proc. Internat. Conf. Soil Mech. Found. Eng., 7th*, Mexico, State-of-the-Art Volume, pp. 225-290.

Pentz, D., 1985, Golder Associates, Seattle, personal communication.

Pepping, R. E., M. S. Y. Chu, and M. D. Siegel, 1983, *Technical Assistance for Regulatory Development: A Simplified Repository Analysis in a Hypothetical Bedded Salt Formation*, SAND82-0996, Sandia National Laboratories, Albuquerque, NM.

Petersen, Erik Lundtang, and Søren E. Larsen, 1978, "A Statistical Study of a Composite Isotopic Paleotemperature Series from the Last 700,000 Years," *Tellus*, v. 30, n. 3, pp. 193-200.

Peterson, G. M., T. Webb III, J. E. Kutzbach, T. van der Hammen, T. A. Wijmstra, and F. A. Streets, 1979, "The Continental Record of Environmental Conditions at 18,000 Yr B.P.: An Initial Evaluation," *Quaternary Research*, v. 12, n. 1, pp. 47-82.

Petrie, G. M., J. T. Zellmer, J. W. Lindberg, and M. G. Foley, 1981, *Geologic Simulation Model for a Hypothetical Site in the Columbia Plateau (AEGIS)*, PNL-3542, Batelle Pacific Northwest Laboratories, 603 pp.

Pfeifle, Tom W., Kirby D. Mellegard, and Paul E. Senseny, 1983, *Preliminary Constitutive Properties for Salt and Nonsalt Rocks from Four Potential Repository Sites*, ONWI-450, Battelle Memorial Institute, Columbus, OH.

Phillips, S. L., F. V. Hale, L. F. Silvester, and M. D. Siegel, 1988, *Thermodynamic Tables for Nuclear Waste Isolation: Aqueous Solutions Data Base*, NUREG/CR-4864, LBL-22860, SAND87-0323, Lawrence Berkeley Laboratory, Berkeley, CA.

Pickens, J. F., and G. E. Grisak, 1981, "Scale-Dependent Dispersion in a Stratified Granular Aquifer," *Water Resources Research*, v. 17, n. 4, pp. 1191-1209.

Pielke, Roger A., 1984, *Mesoscale Meteorological Modeling, An Introductory Survey*, Academic Press, Orlando, FL, 612 pp.

Pillans, B., 1983, "Upper Quaternary Marine Terrace Chronology and Deformation, South Taranaki, New Zealand," *Geology*, v. 11, n. 5, pp. 292-297.

Pisias, N. G., and T. C. Moore, Jr., 1981, "The Evolution of Pleistocene Climate: A Time Series Approach," *Earth and Planet Sci. Let.*, v. 52, n. 2, pp. 450-458.

Plafker, G., and M. Rubin, 1978, *Uplift History and Earthquake Recurrence as Deduced from Marine Terraces on Middleton Island, Alaska*, Open-file Report 78-9434, U.S. Geological Survey, pp. 687-721.

Pollard, David, 1983, "Ice-Age Simulations with a Calving Ice-Sheet Model," *Quaternary Research*, v. 20, n. 1, pp. 30-48.

Poppe, B. B., 1979, "Historical Survey of U.S. Seismograph Stations," Prof. Paper 1096, U.S. Geological Survey, 389 pp.

Powers, D. W., S. J. Lambert, S. E. Shaffer, L. R. Hill, and W. D. Weart, *eds.*, 1978, *Geological Characterization Report, Waste Isolation Pilot Plant (WIPP) Site, Southeastern New Mexico*, SAND78-1596, Sandia National Laboratories, Albuquerque, NM, 2 vols.

Pratt, H. R., D. E. Stephenson, G. Zandt, M. Bouchon, and W. A. Hustrulid, 1979, "Earthquake Damage to Underground Facilities," *Proc. 1979 Rapid Excavation and Tunneling Conf.*, Vol. 1, Sec. 1, pp. 19-51.

Prescott, W. H., J. C. Savage, and W. T. Kinoshita, 1979, "Strain Accumulation Rates in the Western U.S. Between 1970 and 1978," *Jour. Geophys. Research*, v. 84, B10, pp. 5423-5435.

Priest, S. D., 1985, *Hemispherical Projection Methods in Rock Mechanics*, George Allen & Unwin, London, 124 pp.

Priest, S. D., and E. T. Brown, 1983, "Probabilistic Stability Analysis of Variable Rock Slopes," *Trans. Inst. Min. Metall.*, v. 92, pp. A1-A12.

Priest, S. D., and J. A. Hudson, 1976, "Discontinuity Spacings in Rock," *Internat. Jour. Rock Mech. and Mining Sci.*, v. 13, n. 5, pp. 135-148.

Priestley, M. B., 1981, *Spectral Analysis and Time Series*, v. 1, Academic Press, London, 653 pp.

Purcarn, G., and H. Berkhemer, 1982, "Quantitative Relations of Seismic Source Parameters and a Classification of Earthquakes," *Tectonophysics*, v. 84, pp. 57-128.

Puri, M. L., and P. K. Sen, 1971, *Nonparametric Methods in Multivariate Analysis*, John Wiley and Sons, New York, 440 pp.

Rabcewicz, L. V., 1969, "The Stability of Tunnels Under Rock Load," *Water Power*, Part 1, pp. 225-229; Part 2, pp. 266-273; Part 3, pp. 297-302.

Raiffa, Howard, and Robert Schlaifer, 1961, *Applied Statistical Decision Theory*, Massachusetts Institute of Technology Press, Cambridge, MA, 356 pp.

Raleigh, B., G. Bennet, H. Craig, T. Hanks, P. Molnar, A. Nur, J. Savage, C. Scholz, R. Turner, and F. Wu, 1977, "Prediction of the Haicheng Earthquake," *Eos, Trans. Amer. Geophys. Union*, v. 58, n. 5, pp. 236-272.

Ranken, R. E., and J. Ghaboussi, 1975, *Tunnel Design Considerations: Analysis of Stresses and Deformations Around Advancing Tunnels*, Report No. FRA/ORD-75-84,

Federal Railroad Administration, U.S. Department of Transportation, 169 pp.

Ranken, R. E., J. Ghaboussi, and A. J. Hendron, Jr., 1978, *Analysis of Ground-Liner Interaction for Tunnels*, Report No. UMTA-IL-06-0043-78-3, Urban Mass Transit Administration, U.S. Department of Transportation, 441 pp.

Rantz, S. E., 1964, "Surface-Water Hydrology of Coastal Basins of Northern California," U.S. Geological Survey Water Supply Paper 1758, 77 pp.

Ratcliffe, N. M., 1983, "Fault Reactivation Models for the Origin of Eastern United States Seismicity: Does the Solution to Charleston Earthquake Reside at Charleston?" *in* W. W. Hays and P. L. Gori, *eds., A Workshop on The 1886 Charleston, South Carolina, Earthquake and Its Implications for Today*, Open-file Report 83-0843, U.S. Geological Survey, pp. 137-141.

Ratigan, J. L., 1977, *Ground Water Movements Around a Repository, Rock Mechanics Analysis*, KBS-TR-54-04, Kaernbraenslesaekerhet, Stockholm, Sweden, 50 pp.

Razak, M., 1985, "Approche Probabiliste de l'etude en Subsurface de la Geometrie des Reservoirs Fissures—Effet de l'echelle d'investigation," *Internat. Symp. Stochastic Approach to Subsurface Flow*, Montevillargenne, France, pp. 487-502.

Reches, Z., 1983, "Faulting of Rocks in Three-Dimensional Strain Field. II. Theoretical Analysis," *Tectonophysics*, v. 95, pp. 133-156.

Reddy, D. P., *ed.*, 1983, *Seismic Design Technology for Breeder Reactor Structures, Vol. 1. Special Topics in Earthquake Ground Motion*, DOE/SF/01011-T25-Vol. 1, Agbabian Associates, El Segundo, CA, 275 pp.

Reed, J. C., Jr., 1981, "Disequilibrium Profile of the Potomac River Near Washington, D.C.—A Result of Lowered Base Level or Quaternary Tectonics Along the Fall Line?," *Geology*, v. 9, n. 10, pp. 445-450.

Reeder, S. W., B. Hitchon, and A. A. Levinson, 1972, "Hydrogeochemistry of the Surface Waters of the Mackenzie River Drainage Basin, Canada: I. Factors Controlling Inorganic Composition," *Geochim. Cosmochim. Acta*, v. 36, n. 8, pp. 825-865.

Reeves, M., and R. M. Cranwell, 1981, *User's Manual for the Sandia Waste Isolation Flow and Transport Model (SWIFT), Release 4.81*, NUREG/CR-2324, SAND81-2516, Sandia National Laboratories, Albuquerque, NM, 143 pp.

Rehbinder, G., 1984, "Strains and Stresses in the Rock Around an Unlined Hot Water Cavern," *Rock Mech. and Rock Engineering*, v. 17, n. 3, pp. 129-145.

Reilinger, R., M. Bevis, and G. Jurkowski, 1984, "Tilt from Releveling: An Overview of the U.S. Data Base," *Tectonophysics*, v. 107, n. 3-4, pp. 315-330.

Remson, I., S. J. Dreiss, and A. G. Journel, 1982, "Radioactive Waste Disposal—An Application of Predictive Geology," *in* G. DeMarsily and D. F. Merriam, *eds., Predictive Geology*, Pergamon Press, Elmsford, NY, pp. 25-32.

Rescher, O. J., K. H. Abraham, F. Braeutigam, and A. Pahl, 1973, "Ein Kavernenbau mit Ankerung und Spritzbeton unter Berucksichtigung der geomechanischen Bedingungen," *Rock Mech.*, Suppl. 2, pp. 313-354.

Rhoades, D. A., and F. F. Evison, 1979, "Long-Term Earthquake Forecasting Based on a Single Predictor," *Geophys. Jour. Royal Astronom. Soc.*, v. 59, n. 1, pp. 43-56.

Richardson, R. M., and S. C. Solomon, 1979, "Tectonic Stress in the Plates," *Rev. Geophys. Space Phys.*, v. 17, n. 5, pp. 981-1019.

Richter, C. F., 1958, *Elementary Seismology*, W. H. Freeman and Co., San Francisco, CA.

Rickertsen, L. D., J. G. Arbital, and H. C. Claiborne, 1982, *Expected Near-Field Thermal Environments in a Sequentially Loaded Spent-Fuel or High-Level Waste Repository in Salt*, ORNL/TM-8083, Oak Ridge National Laboratory, Oak Ridge, TN, 63 pp.

Riecker, R. E., *ed.*, 1979, *Rio Grande Rift—Tectonics and Magmatism*, Internat. Symp. on the Rio Grande Rift, Santa Fe, NM, Oct. 1978, Amer. Geophys. Union., Washington, DC, 438 pp.

Rikitake, T., 1974, "Probability of Earthquake Occurrence as Estimated from Crustal Strain," *Tectonophysics*, v. 23, n. 3, pp. 299-312.

Rikitake, T., 1975a, "Earthquake Precursors," *Bull. Seismol. Soc. Amer.*, v. 65, n. 5, pp. 1133-1162.

Rikitake, T., 1975b, "Statistics of Ultimate Strain of the Earth's Crust and Probability of Earthquake Occurrence," *Tectonophysics*, v. 26, no. 1-2, pp. 1-21.

Rikitake, T., 1976, *Earthquake Prediction*, Elsevier, Amsterdam, 357 pp.

Rikitake, T., 1979a, "Changes in the Direction of Magnetic Vector of Short-Period Geomagnetic Variations Before the 1972 Sitka, Alaska, Earthquake," *Jour. Geomag. Geoelectr.*, v. 31, n. 4, pp. 441-448.

Rikitake, T., 1979b, "Classification of Earthquake Precursors," *Tectonophysics*, v. 54, n. 3-4, pp. 293-309.

Rikitake, T., 1982, *Earthquake Forecasting and Warning*, Center for Academic Publications, Tokyo, D. Reidel Publ. Co., Dordrecht, Holland, 424 pp.

Rikitake, T., 1984, "Active Fault and Crustal Strain," *Earthquake Prediction Research*, v. 2, n. 3, pp. 167-189.

Rind, D., D. Peteet, W. Broeker, A. McIntyre, and W. Ruddiman, 1986, "The Impact of Cold North Atlantic Sea Surface Temperatures on Climate: Implications for the Younger Dryas Cooling (11-10k)," *Climate Dynamics*, v. 1, pp. 3-33.

Ripley, B. D., 1981, *Spatial Statistics*, John Wiley and Sons, New York, 252 p.

Roberds, W. J., 1979, "Numerical Modelling of Jointed Rock," Ph.D. Thesis, Massachusetts Institute of Technology, Cambridge, MA, 748 pp.

Robertson, A. M., 1970, "The Interpretation of Geologic Factors for Use in Slope Stability," *Proc. Symp. Theoretical Background to the Planning of Open Pit Mines*, Johannesberg, South Africa, pp. 55-71.

Robinson, C. S., and F. T. Lee, 1964, "Geological Research at the Straight Creek Tunnel Site, Colorado," *Highway Research Record*, 57, Publ. 1241, National Academy of Science--National Research Council.

Rockwell, T. K., E. A. Keller, M. N. Clark, and D. L. Johnson, 1984, "Chronology and Rates of Faulting of Ventura River Terraces, California," *Geol. Soc. Amer. Bull.*, v. 95, n. 12, pp. 1466-1474.

Rodriguez-Iturbe, I., and J. M. Mejia, 1974, "The Design of Rainfall Networks in Time and Space," R74-6, R. M. Parsons Laboratory, Research Department, Massachusetts Institute of Technology, Cambridge, MA.

Rogers, A., 1974, *Statistical Analysis of Spatial Dispersion*, Pion Monograph, London, 164 pp.

Rogers, A. M., D. M. Perkins, and F. A. McKeown, 1977, "A Preliminary Assessment of the Seismic Hazard of the Nevada Test Site Region," *Bull. Seismol. Soc. Amer.*, v. 67, n. 6, pp. 1587-1606.

Rollin, F., 1979, "The Updating of Information During Tunnel Construction," M.S. Thesis, Massachusetts Institute of Technology, Cambridge, MA, 111 pp.

Root, D. H., and J. H. Schuenemeyer, 1980, "Petroleum Resource Appraisal and Discover-Rate Forecasting in Partially Explored Regions—Mathematical Foundations," U.S. Geological Survey Prof. Paper 1138B, pp. B1-B9.

Rosenblatt, Murray, 1956, "Remarks on Some Nonparametric Estimates of a Density Function," *Annals Math. Stat.*, v. 27, pp. 832-837.

Ross, S., 1985, *Introduction to Probability Models*, Academic Press, 3rd ed., New York, 502 pp.

Rossotti, H. S., 1974, "Design and Publication of Work on Stability Constants," *Talanta*, vol. 21, pp. 809-829.

Rothe, J. P., 1969, *The Seismicity of the Earth (1953-1965)*, UNESCO, New York, 336 pp.

Rothfuchs, T., D. Lubker, A. J. Coyle, and H. N. Kalia, 1984, *Nuclear Waste Repository Simulation Experiments, Asse Salt Mine, Federal Republic of Germany: Annual Report 1983*, BMI/ONWI-539, Battelle Memorial Institute, Columbus, OH, 159 pp.

Rouleau, Alain, 1984, "Statistical Characterization and Numerical Simulation of a Fracture System: Application to Groundwater Flow in the Stripa Granite," Ph.D. Thesis, Univ. of Waterloo, Canada.

Rouleau, A., and J. E. Gale, 1985, "Fracture Size, Interconnectivity and Groundwater Flow: A Site Specific Stochastic Analysis," *Internat. Symp. Stochastic Approach to Subsurface Flow*, Montevillargenne, France, pp. 514-522.

Royle, A. G., 1980, "Why Geostatistics?", *in* A. G. Royle, *ed., Geostatistics*, McGraw-Hill, New York, pp. 1-16.

Rozen, A., 1976, "Response of Rock Tunnels to Earthquake Shaking," M.S. Thesis, Massachusetts Institute of Technology, Cambridge, MA, 238 pp.

Ruddiman, W. F., and M. E. Raymo, 1988, "Northern Hemisphere Climate Regimes During the Past 3 Ma: Possible Tectonic Connections," *Phil. Trans. Royal Soc. London B*, v. 318, pp. 411-430.

Russ, D. P., 1982, "Style and Significance of Surface Deformation in the Vicinity of New Madrid, Mo.," *in* F. A. McKeown and L. C. Pakiser, *eds., Investigation of the New Madrid, Mo., Earthquake Region*, Prof. Paper 1286, U.S. Geological Survey, pp. 95-114.

Rutledge, J. C., and R. L. Preston, 1978, "New Zealand Experience with Engineering Classifications of Rock for the Prediction of Tunnel Support," *Internat. Tunnel Symp.*, Tokyo, pp. A-1.3-7.

Ryall, A., D. B. Slemmons, and L. D. Gedney, 1966, "Seismicity, Tectonics, and Surface Faulting in the Western United States During Historic Time," *Bull. Seismol. Soc. Amer.*, v. 56, n. 5, pp. 1105-1135.

Ryan, J. L., and D. Rai, 1983, "The Solubility of Uranium (IV) Hydrous Oxide in Sodium Hydroxide Solutions under Reducing Conditions," *Polyhedron*, v. 2, pp. 947-952.

Sadler, P. M., 1981, "Sediment Accumulation Rates and the Completeness of Stratigraphic Sections," *Jour. Geol.*, v. 89, n. 5, pp. 569-584.

Sagar, B., and P. M. Clifton, 1985, "Stochastic Groundwater Flow Modeling Using the Second Moment Method," *Hydrogeology of Rocks of Low Permeability, Proc. 17th Internat. Congress Internat. Assoc. Hydro.*, pp. 498-512, Tucson, AZ.

Sagar, B., and A. Runchal, 1982, "Permeability of Fractured Rock: Effect of Fracture Size and Data Uncertainties," *Water Resources Research*, v. 18, n. 2, pp. 266-274.

Sakurai, S., and M. Shinji, 1984, "A Monitoring System for the Excavation of Underground Openings Based on Microcomputers," *Proc. ISRM Symp. Design and Performance of Underground Excavations*, British Geotechnical Society, London, pp. 471-476.

Salamon, M. D. G., 1974, "Rock Mechanics of Underground Excavations," *Advances in Rock Mechanics, Vol. I, Part B*, Internat. Soc. Rock Mech., Congr., Proc. 3, National Academy of Sciences, Washington, DC, pp. 951-1359.

Salamon, M. D. G., and A. H. Munro, 1967, "Coal Mine Pillars in South Africa," *Jour. South African Institute of Mining and Metallurgy*, v. 68, pp. 55-67.

Saltzman, B., 1978, "A Survey of Statistical-Dynamical Models of the Terrestrial Climate," *Adv. Geophys.*, v. 20, pp. 183-304.

Saltzman, B., 1983, "Climatic Systems Analysis," *in* Barry Saltzman, *ed., Theory of Climate, Adv. Geophys.*, v. 25, pp. 173-233.

Saltzman, Barry, 1984, "On the Role of Equilibrium Atmospheric Climate Models in the Theory of Long-Period Glacial Variations," *Jour. Atmospheric Sci.*, v. 41, pp. 2263-2266.

Saltzman, Barry, 1985, "Paleoclimatic Modeling," *in* A. D. Hecht, *ed., Paleoclimate Analysis and Modelling*, Wiley-Interscience, Inc., New York, pp. 341-396.

Saltzman, B., and K. A. Maasch, 1988, "Carbon Cycle Instability as a Cause of the Late Pleistocene Ice Age Oscillations: Modelling the Asymmetric Response," *Global Biogeochemical Cycles*, v. 2, pp. 177-185.

Saltzman, Barry, and Alfonso Sutera, 1984, "A Model of the Internal Feedback System Involved in Late Quaternary Climatic Variations," *Jour. Atmospheric Sci.*, v. 41, n. 5, pp. 736-745.

Saltzman, B., A. R. Hansen, and K. Maasch, 1984, "The Late Quaternary Glaciations as a Response of a Three Component Feedback System to Earth-Orbital Forcing," *Jour. Atmospheric Sci.*, v. 41, pp. 3380-3389.

Samaniego, J. A., and S. D. Priest, 1984, "The Prediction of Water Flows Through Discontinuity Networks into Underground Excavations," *Proc. Symp. Design and Performance of Underground Excavations*, Cambridge, MA, pp. 157-164.

Sandefur, R. L., and D. C. Grant, 1980, "Applying Geostatistics to Roll Front Uranium in Wyoming," *in* A. G. Royle, *ed., Geostatistics*, McGraw-Hill, New York, pp. 127-143.

Santa Fe Southern Pacific Corp., 1987, *Minerals for Industry, Volume III, Southern California, Summary of Geological Survey of 1955-1961*, Special Publication 95, California Division of Mines and Geology, 242 pp.

Sato, H., 1977, "Some Precursors Prior to Recent Great Earthquakes Along the Nankai Trough," *Jour. Phys. Earth*, v. 25, Suppl., pp. S115-S121.

Sauber, J., K. C. McNally, J. C. Pechmann, and H. Kanamori, 1983, "Seismicity Near Palmdale, California, and Its Relation to Strain Changes," *Jour. Geophys. Research B*, v. 88, n. 3, pp. 2213-2219.

Saul, Richard B., H. Clifton, Jr., and James R. Evans, 1971, *Mines and Mineral Resources of Riverside County, California*, California Div. of Mines and Geol., Sacramento, California, unpublished report, 1016 pp.

Savage, J. C., 1972, "Relation of Corner Frequency to Fault Dimensions," *Jour. Geophys. Research*, v. 77, n. 20, pp. 3788-3795.

Savage, J. C., and L. M. Hastie, 1966, "Surface Deformation Associated with Dip-Slip Faulting," *Jour. Geophys. Research*, v. 71, n. 20, p. 4897-4904.

Savin, G. N., 1970, "Stress Distribution Around Holes," NASA-TT-F-607, National Aeronautical and Space Administration, 1005 pp.

Savinskii, I. D., 1965, *Probability Tables for Locating Elliptical Underground Masses with a Rectangular Grid*, Consultants Bureau, New York, 110 p.

Savy, J. B., 1980, *A Geophysical Model of Strong Motion for Large Ensemble Generation*, Research Report R81-6, Department of Civil Engineering, Massachusetts Institute of Technology, Cambridge, MA.

Savy, J. B., H. C. Shah, and D. Boore, 1980, "Nonstationary Risk Model with Geophysical Input," *Jour. Structural Div.*, Amer. Soc. Civil Engineers, v. 106, pp. 145-163.

Schlesinger, M. E., 1984, "Climate Model Simulations of CO_2-Induced Climate Change," *Adv. Geophys.*, v. 26, pp. 141-235.

Schlesinger, M. E., and J. F. B. Mitchell, 1987, "Climate Model Simulations of the Equilibrium Climatic Response to Increased Carbon Dioxide," *Rev. Geophys.*, v. 25, pp. 760-798.

Schneider, K. J., and A. M. Platt, *eds.*, 1974, *High-Level Radioactive Waste Management Alternatives*, BNWL-1900, Battelle Pacific Northwest Laboratory, Richland, WA.

Schneider, Stephen H., 1984, "On the Empirical Verification of Model-Predicted CO_2-Induced Climate Effects," *in* James E. Hansen and Taro Takahashi, *eds., Climate Processes and Climate Sensitivity*, Geophysical Monograph 29, pp. 187-207.

Scholl, R. E., and J. L. King, *eds.*, 1985, *Strong Ground Motion Simulation and Earthquake Engineering Applications; a Technological Assessment*, Publication No. 85-02, Earthquake Engineering Research Institute, Berkeley, CA.

Schramke, J. A., S. V. Mattigod, Dhanpat Rai, 1987, *Salt Repository Project, Available Thermochemical Data and Data Requirements for a Salt Repository for Thorium, Plutonium, Americium, and Curium*, BMI/ONWI/C--8, Pacific Northwest Laboratory, Richland, WA.

Schuenemeyer, J. H., W. J. Bawiec, and L. J. Drew, 1980, "Computational Methods for a Three-Dimensional Model of the Petroleum-Discovery Process," *Comp. and Geosci.*, v. 6, n. 4, pp. 323-360.

Schuenemeyer, J. H., L. J. Drew, and W. J. Bawiec, 1980, "A Three-Dimensional Model to Predict Future Oil Discoveries in Spatially Connected Multiple Plays," *Internat. Assoc. Math. Geol. Jour.*, v. 12, n. 5, pp. 459-472.

Schwartz, C. W., and H. H. Einstein, 1980, "Simplified Analysis for Ground-Structure Interactions in Tunneling," Report No. UMTA-MA-05-0100-80-4, Urban Mass Transit Administration, U.S. Department of Transportation, 450 pp.

Schwartz, C. W., A. S. Azzouz, and H. H. Einstein, 1980, "Aspects of Yielding in Ground-Structure Interaction," Report No. UMTA-MA-05-0100-80-5, Urban Mass Transit Administration, U.S. Department of Transportation, 75 pp.

Schwartz, D. P., and K. J. Coppersmith, 1984, "Fault Behavior and Characteristic Earthquakes: Examples From the Wasatch and San Andreas Fault Zones," *Jour. Geophys. Research B*, v. 89, n. 7, pp. 5681-5698.

Schwartz, F. W., and F. A. Donath, 1982, "Influence of Faulting on Groundwater Flow and

Contaminant Transport," *in* G. DeMarsily and D. F. Merriam, *eds., Predictive Geology*, Pergamon Press, Elmsford, NY, pp. 117-136.

Schwartz, F. W., and L. Smith, 1981, "An Evaluation of the Ability to Predict Radionuclide Transport in Groundwater Flow Systems," *in* J. G. Moore, *ed., Proc. Third Internat. Symp. Scientific Basis for Nuclear Waste Management*, v. 3, Plenum Press, New York, pp. 599-605.

Schwartz, F. W., L. Smith, and A. S. Crowe, 1983, "A Stochastic Analysis of Macroscopic Dispersion in Fractured Media," *Water Resources Research*, v. 19, n. 5, pp. 1253-1265.

Schwartz, S. C., 1967, "Estimation of a Probability Density by an Orthogonal Series," *Annals Math. Stat.*, v. 38, pp. 1261-1265.

Scott, W. E., 1978, *Hazardous Volcanic Events and Assessment of Long-Term Volcanic Hazards*, Open-file Report, U.S. Geological Survey, pp. 447-498.

Segall, P., 1984, "Rate-Dependent Extensional Deformation Resulting from Crack Growth in Rock," *Jour. Geophys. Research B,* v. 89, n. 6, pp. 4185-4195.

Segall, P., and D. D. Pollard, 1980, "Mechanics of Discontinuous Faults," *Jour. Geophys. Research B*, v. 85, n. 8, pp. 4337-4350.

Segall, P., and D. D. Pollard, 1983, "Joint Formation in Granitic Rock of the Sierra Nevada," *Geol. Soc. Amer. Bull.*, v. 94, n. 5, pp. 563-575.

Sekiya, H., 1977, "Anomalous Seismic Activity and Earthquake Prediction," *Jour. Phys. Earth*, v. 25, Suppl., pp. S85-S93.

Self, S., and R. S. J. Sparks, *eds.*, 1981, "Tephra Studies," *Proc. NATO Advanced Studies Institute, Tephra Studies as a Tool in Quaternary Research*, D. Reidel Publ. Co., Dordrecht, Holland.

Sellers, W. D., 1969, "A Climate Model Based on the Energy Balance of the Earth-Atmosphere System," *Jour. Appl. Meteorology*, v. 8, pp. 392-400.

Semenov, A. N., 1969, "Variations in the Travel Time of Transverse and Longitudinal Wave Before Violent Earthquakes," *Izv. Acad. Sci. USSR (Phys. Solid Earth)* (English edition), v. 4, pp. 245-248.

Semple, R. M., A. T. Hendron, and G. Mesri, 1973, "The Effects of Time-Dependent Properties of Altered Rock on Tunnel Support Requirements," FRA-ORD/D-74-30, Federal Railroad Administration, U.S. Department of Transportation, 217 pp.

Semple, R. M., A. T. Hendron, and G. Mesri, 1974, "The Effects of Time-Dependent Properties of Altered Rock on Tunnel Support Requirements," *Proc. Rapid Excavation and Tunneling Conf.*, Vol. 2, Sec. 20, pp. 1371-1377.

Senseny, P. E., 1983, *Review of Constitutive Laws Used to Describe the Creep of Salt*, ONWI-295, Battelle Memorial Institute, Columbus, OH, 55 pp.

Serne, R. J., and A. B. Muller, 1987, "A Perspective on Adsorption of Radionuclides onto Geologic Media," *in* D. G. Brookins, *ed., The Geological Disposal of High Level Radioactive Waste*, Theophrastus Publications, S. A., Athens, pp. 407-444.

Sexton, J. L., W. J. Hinze, R. R. B. von Frese, and L. W. Braile, 1982, "Long-Wavelength Aeromagnetic Anomaly Map of the Conterminous United States," *Geology*, v. 10, n. 7, pp. 364-369.

Shackleton, N. J., and N. G. Pisias, 1985, "Atmospheric Carbon Dioxide, Orbital Forcing, and Climate," *in* E. T. Sundquist and W. S. Broecker, *eds., The Carbon Cycle and Atmospheric* CO_2*: Natural Variations Archean to Present*, Geophysical Monograph 32, Amer. Geophys. Union, pp. 303-317.

Shackleton, N. J., M. A. Hall, J. Line, and Cang Shuxi, 1983, "Carbon Isotope Data in Core V19-30 Confirm Reduced Carbon Dioxide Concentration in the Ice Age Atmosphere," *Nature*, v. 306, n. 5941, pp. 319-322.

Shakal, A. F., and D. E. Willis, 1972, "Estimated Earthquake Probabilities in the North Circum-Pacific Area," *Bull. Seismol. Soc. Amer.*, v. 62, n. 6, pp. 1397-1410.

Shannon & Wilson/Agbabian Associates, 1976, *Geotechnical and Strong Motion Earthquake Data from U.S. Accelerograph Stations*, NUREG-0029, U.S. Nuclear Regulatory Commission, Washington, DC, 2 vols.

Sharp, R. V., 1979, "Implications of Surficial Strike-Slip Fault Patterns for Simplification and Widening with Depth," *Proc. Conf. VIII, Analysis of Actual Fault Zones in Bedrock*, Open-file Report 79-1239, U.S. Geological Survey, pp. 66-78.

Shaw, H. R., and A. E. Gartner, 1984, *Empirical Laws of Order Among Rivers, Faults, and Earthquakes*, Open-file Report 84-0356, U.S. Geological Survey, 55 pp.

Sheets, P. D., and D. K. Grayson, *eds.*, 1979, *Volcanic Activity and Human Ecology*, Academic Press, New York.

Shimamura, H., 1982, "Microtemperature Measurements of Groundwater in Earthquake Impending Areas in Japan," *Eos, Trans. Amer. Geophys. Union*, v. 63, n. 51, p. 1260.

Shimazaki, K., and T. Nakata, 1980, "Time-Predictable Recurrence Model for Large Earthquakes," *Geophys. Research Let.*, v. 7, n. 4, pp. 279-282.

Shore, John E., and Rodney W. Johnson, 1980, "Axiomatic Derivation of the Principle of Maximum Entropy and the Principle of Minimum Cross-Entropy," *IEEE Trans. Information Theory*, v. IT-26, n. 1, pp. 26-37.

Shurygin, A. M., 1976a, "Discovery of Deposits of Given Size in Boreholes with Preselected Probability," *Internat. Assoc. Math. Geol. Jour.*, v. 8, n. 1, pp. 85-88.

Shurygin, A. M., 1976b, "The Probability of Finding Deposits and Some Optimal Search Grids," *Internat. Assoc. Math. Geol. Jour.*, v. 8, n. 3, pp. 323-330.

Sibson, R. H., 1977, "Fault Rocks and Fault Mechanisms," *Jour. Geol. Soc. London*, v. 133, n. 3, pp. 191-213.

Sibson, R. H., 1981, "Fluid Flow Accompanying Faulting: Field Evidence and Models," *in* D. W. Simpson and P. G. Richards, *eds.*, *Earthquake Prediction: An International Review*, M. Ewing Series, v. 4, Amer. Geophys. Union, pp. 593-603.

Sibson, R. H., 1982, "Fault Zone Models, Heat Flow, and the Depth Distribution of Earthquakes in the Continental Crust of the United States," *Bull. Seismol. Soc. Amer.*, v. 72, n. 1, pp. 151-163.

Sibson, R. H., 1983, "Continental Fault Structure and the Shallow Earthquake Source," *Jour. Geol. Soc. London*, v. 140, n. 5, pp. 741-767.

Sibson, R. H., 1984, "Roughness at the Base of the Seismographic Zone: Contributing Factors," *Jour. Geophys. Research*, v. 89, pp. 5794-5799.

Siegel, M. D., and C. D. Leigh, 1989, *Progress in Development of a Methodology for Geochemical Sensitivity Analysis for Performance Assessment, Parametric Calculations, Preliminary Databases, and Computer Code Evaluation*, SAND85-1644, NUREG/CR-5085, U.S. Nuclear Regulatory Commission, Washington, DC.

Siegel, M. D., M. S. Chu, and R. E. Pepping, 1983, "Compliance Assessments of Hypothetical Geological Nuclear Waste Isolation Systems With the Draft EPA Standard," *in* D. G. Brookins, *ed.*, *Scientific Basis for Nuclear Waste Management*

VI, Mat. Research Soc. Symp. Proc., 15, Elsevier Science Publishing Co., Inc., Amsterdam.

Siegel, M. D., K. L. Robinson, and J. Myers, 1990, "Solute Relationships in Groundwaters from the Culebra Member of the Rustler Formation Near the WIPP Site, Southeastern New Mexico," *in* M. D. Siegel, S. J. Lambert, and K. L. Robinson, *eds., Hydrogeochemical Studies of the Rustler Formation and Related Rocks in the WIPP Area, Southeastern New Mexico*, SAND88-0196, Sandia National Laboratories, Albuquerque, NM.

Siegel, M. D., S. L. Phillips, J. O. Leckie, and W. R. Kelly, 1989, "Development of a Methodology of Geochemical Sensitivity Analysis for Performance Assessment," *in* B. E. Buxton, *ed., Proc. Conference on Geostatistical Sensitivity, and Uncertainty Methods for Ground-Water Flow and Radionuclide Transport Modeling*, Battelle Press.

Siegel, M. D., J. O. Leckie, S. W. Park, S. L. Phillips, and T. Sewards, 1990, "Studies of Radionuclide Sorption by Clays in the Culebra Dolomite at the WIPP Site, Southeastern New Mexico," *in* Roy G. Post, *ed., Waste Management 90, Waste Processing, Transportation, Storage and Disposal, Technical Programs and Public Education, Vol. 2*, University of Arizona, Tucson, AZ.

Sieh, K. E., 1978a, "Prehistoric Large Earthquakes Produced by Slip on the San Andreas Fault at Pallett Creek, California," *Jour. Geophys. Research*, v. 83, n. B8, pp. 3907-3939.

Sieh, K. E., 1978b, "Slip Along the San Andreas Fault Associated with the Great 1857 Earthquake," *Bull. Seismol. Soc. Amer.*, v. 68, n. 5, pp. 1421-1447.

Sieh, K. E., 1984, "Lateral Offsets and Revised Dates of Prehistoric Earthquakes at Pallett Creek, Southern California," *Jour. Geophys. Research*, v. 89, n. B9, pp. 7641-7670.

Sieh, K. E., and R. H. Jahns, 1984, "Holocene Activity of the San Andreas Fault at Wallace Creek, California," *Geol. Soc. Amer. Bull.*, v. 95, n. 8, pp. 883-896.

Silva, R. J., 1985, "Carbonate Complexation of Pu(IV) in Aqueous Solutions," Paper No. 44, Div. of Nucl. Chem. and Tech., 189th Amer. Chem. Soc. Mtg., Miami Beach, FL, April 28-May 3, 1985.

Simkin, T., and L. Siebert, 1982, "Large Historic Explosive Eruptions: How Much Warning?" *Eos, Trans. Amer. Geophys. Union*, v. 63, n. 18, p. 459.

Simkin, T., and L. Siebert, 1984, "Explosive Eruptions in Space and Time: Durations, Intervals, and a Comparison of the World's Active Volcanic Belt," *Explosive Volcanism: Inception, Evolution, and Hazards*, National Academy of Science National Research Council, pp. 110-121.

Simkin, T., L. Siebert, L. McClelland, D. Bridge, C. Newhall, and J. H. Latter, 1981, *Volcanoes of the World*, Smithsonian Institution, Hutchinson Ross Publishing Co., Stroudsburg, PA, 232 pp.

Simmons, C. S., 1982, "A Stochastic-Convective Transport Representation of Dispersion in One-Dimensional Porous Media Systems," *Water Resources Research*, v. 18, n. 4, pp. 1193-1214.

Simpson, D. W., and P. G. Richards, *eds.*, 1981, *Earthquake Prediction: An International Review*, M. Ewing Series, v. 4, Amer. Geophys. Union, 680 p.

Simpson, George Gaylord, 1952, "Probabilities of Dispersal Through Time," *Bull. Amer. Museum Natural History*, v. 99, pp. 163-176.

Singer, Donald A., 1971, "Multivariate Statistical Analysis of the Unit Regional Value of Mineral Resources," Ph.D. Thesis, Pennsylvania State Univ., 210 pp.

Singer, D. A., 1975, "Relative Efficiencies of Square and Triangular Grids in the Search for Elliptically Shaped Resource Targets," *U.S. Geological Survey Jour. of Research*, v. 3, n. 2, pp. 163-167.

Singer, Donald A., 1976, "RESIN, a FORTRAN IV Program for Determining the Area of Influence of Samples or Drill Holes in Resource Target Search," *Comp. and Geosci.*, v. 2, n. 2, pp. 249-260.

Singer, D. A., and Lawrence J. Drew, 1976, "The Area of Influence of an Exploratory Hole," *Econ. Geol.*, v. 71, n. 3, pp. 642-647.

Singer, Donald A., and Frans E. Wickman, 1969, "Probability Tables for Locating Elliptical Targets with Square, Rectangular, and Hexagonal Point Nets," *Penn. State Univ. Min. Sci. Exp. Sta. Spec. Pub. 1-69*, 100 pp.

Singh, R. S., 1977, "Improvement on Some Known Nonparametric Uniformly Consistent Estimators of Derivatives of a Density," *Annals Stat.*, v. 5, pp. 394-399.

Singh, S. K., L. Astiz, and J. Havskov, 1981, "Seismic Gaps and Recurrence Periods of Large Earthquakes along the Mexican Subduction Zone: A Re-examination," *Bull. Seismol. Soc. Amer.*, v. 71, n. 3, pp. 827-843.

Slemmons, D. B., 1977, *State-of-the-Art for Assessing Earthquake Hazards in the United States, Report 6, Faults and Earthquake Magnitude*, WES-MP-S-73-1, U.S. Army Engineers Waterways Experimental Station, Vicksburg, MS, 170 pp.

Slemmons, D. B., 1982, "Determination of Design Earthquake Magnitudes for Microzonation," *Proc. Third Internat. Earthquake Microzonation Conf.*, v. 1, pp. 119-130.

Slichter, L. B., 1955, "Geophysics Applied to Prospecting for Ores," *in* A. M. Bateman, *ed., Econ. Geol.*, Part 2, 50th Anniversary Volume, pp. 885-969.

Smith, B. E., and M. J. S. Johnston, 1976, "A Tectonomagnetic Effect Observed Before a Magnitude 5.2 Earthquake Near Hollister, California," *Jour. Geophys. Research*, v. 81, n. 20, pp. 3556-3560.

Smith, James L., and Geoffrey L. Ward, 1981, "Maximum Likelihood Estimate of the Size Distribution of North Sea Oil Fields," *Internat. Assoc. Math. Geol. Jour.*, v. 13, n. 5, pp. 399-413.

Smith, L., and R. A. Freeze, 1979a, "Stochastic Analysis of Steady State Groundwater Flow in a Bounded Domain, 1: One-Dimensional Simulations," *Water Resources Research*, v. 15, n. 3, pp. 521-528.

Smith, L., and R. A. Freeze, 1979b, "Stochastic Analysis of Steady State Groundwater Flow in a Bounded Domain, 2: Two-Dimensional Simulations," *Water Resources Research*, v. 15, n. 6, pp. 1543-1559.

Smith, L., and F. W. Schwartz, 1980, "Mass Transport, 1: Stochastic Analysis of Macrodispersion," *Water Resources Research*, v. 16, pp. 303-313.

Smith, L., and F. W. Schwartz, 1984, "An Analysis of the Influence of Fracture Geometry on Mass Transport in Fractured Media," *Water Resources Research*, v. 20, n. 9, pp. 1241-1252.

Smith, L., C. W. Mase, and F. W. Schwartz, 1985, "A Stochastic Model for Transport in Networks of Planar Fractures," *Internat. Symp. Stochastic Approach to Subsurface Flow*, Montevillargenne, France, pp. 523-536.

Smith, R. B., 1977, "Intraplate Tectonics of the Western North American Plate," *Tectonophysics*, v. 37, n. 4, pp. 323-336.

Smith, R. B., and R. L. Bruhn, 1984, "Intraplate Extensional Tectonics of the Eastern Basin-Range: Inferences on Structural Style from Seismic Reflection Data, Regional Tectonics, and Thermal-Mechanical Models of Brittle-Ductile Deformation," *Jour. Geophys. Research B*, v. 89, n. 7, pp. 5733-5762.

Smith, R. L., and R. G. Luedke, 1984, "Potentially Active Volcanic Lineaments and Loci in Western Conterminous United States," *in* National Research Council, Geophysics Study Committee, *Explosive Volcanism: Inception, Evolution, and Hazards*, National Academy of Science Press, Washington, DC, pp. 47-65.

Snow, D. T., 1968a, "Rock Fracture Spacings, Openings, and Porosities," *Proc. Amer. Soc. Civil Engineers, Jour. Soil Mech. Foundations Div.*, v. 94, n. SM 1, pp. 73-91.

Snow, D. T., 1968b, "Fracture Deformation and Changes of Permeability and Storage Upon Changes of Fluid Pressure," in "Geophysical and Geological Studies of the Relationships Between the Denver Earthquake and the Rocky Mountain Arsenal Well, Part A," *Colorado School of Mines Quarterly*, v. 63, n. 1, pp. 201-244.

Snow, D. T., 1969, "Anisotropic Permeability of Fractured Media," *Water Resources Research*, v. 5, n. 6, pp. 1272-1289.

Snow, D. T., 1970, "The Frequency and Apertures of Fractures in Rock," *Internat. Jour. Rock. Mech. Min. Sci.*, v. 7, n. 1, pp. 23-40.

Sokolnikoff, I. S., 1956, *Mathematical Theory of Elasticity*, McGraw-Hill Book Company, New York.

Sommerville, P., J. McLaren, R. Burger, V. LeFevre, and D. Helmberger, 1985, *Source Scaling Relations of Eastern North American Earthquakes*, Electric Power Research Institute, Palo Alto, CA.

Sorenson, Harold W., 1980, *Parameter Estimation, Principles and Problems*, Marcel Dekker, Inc., New York, 375 pp.

Speed, R., 1979, "Extension Faulting in the Great Basin: Kinematics and Possible Changes with Depth," *in* R. C. Speed and R. V. Sharp, *eds., Proc. Conf. VIII, Analysis of Actual Fault Zones in Bedrock*, Open-file Report 79-1239, U.S. Geological Survey, pp. 36-65.

Spencer, F. W., and R. G. Easterling, 1984, *Lower Confidence Bounds on System Reliability Using Component Data: The Maximus Methodology*, presented at Internat. Conf. on Reliability and Quality Control, Columbia, MO, June 4, 1984, SAND84-1199C, Sandia National Laboratories, Albuquerque, NM, 24 pp.

Starfield, A. M., and W. C. McClain, 1973, "Project Salt Vault: A Case Study in Rock Mechanics," *Internat. Jour. Rock Mech. Min. Sci. and Geomech. Abstr.*, v. 10, pp. 641-657.

Steiner, W., and H. H. Einstein, 1980, *Improved Design of Tunnel Supports: Volume 5—Empirical Methods in Rock Tunneling—Review and Recommendations*, U.S. Department of Transportation, Report No. UMTA-MA-06-0100-80-8, 541 pp.

Stepp, J. C., 1972, "Analysis of Completeness of the Earthquake Sample in the Puget Sound Area and Its Effect on Statistical Estimates of Earthquake Hazard," *Microzonation Conf.*, v. 2, pp. 897-909, Univ. of Seattle, WA.

Stepp, J. C., W. A. Rinehart, and S. T. Algermissen, 1965, *Earthquakes in the United*

States 1963-64 and an Evaluation of the Detection Capability of the U.S. Seismograph Stations, U.S. Coast and Geodetic Survey, Washington, DC, 2 pp.

Stevens, A. E., 1980, "History of Some Canadian and Adjacent American Seismograph Stations," *Bull. Seismol. Soc. Amer.*, v. 70, n. 4, pp. 1381-1393.

St. John, C. M., 1985, *Nevada Nuclear Waste Storage Investigations Project: Thermal Analysis of Spent Fuel Disposal in Vertical Emplacement Boreholes in a Welded Tuff Repository*, SAND84-7207, Sandia National Laboratories, Albuquerque, NM, 60 pp.

St. John, C. M., and D. E. Van Dillen, 1983, "Rockbolts: A New Numerical Representation and Its Application in Tunnel Design," *in* C. C. Mathewson, *ed., Proc. 24th U.S. Symp. Rock Mech.; Theory, Experiment, Practice*, June 20-23, 1983, College Station, TX, pp. 13-25.

St. John, C. M., and T. F. Zahrah, 1985, "Applicable Analytical Tools for a Seismic Design of Underground Excavations," *in* C. D. Mann and M. N. Kelley, *eds., Proc. 1985 Rapid Excavation and Tunneling Conf.*, v. 1, June 16-20, 1985, New York, pp. 89-103.

Stone, Charles M., Raymond D. Krieg, and Zelma E. Beisinger, 1985, *SANCHO—A Finite Element Computer Program for the Quasistatic, Large Deformation, Inelastic Response of Two-Dimensional Solids*, SAND84-2618, Sandia National Laboratories, Albuquerque, NM, 102 pp.

Stone, L. D., 1972, *Theory of Optimal Search*, Academic Press, New York, 260 pp.

Street, R. L., R. B. Herrmann, and O. W. Nuttli, 1975, "Spectral Characteristics of the Lg Wave Generated by Central United States Earthquakes," *Geophys. Jour., Royal Astron. Soc.*, v. 41, n. 1, pp. 51-63.

Street-Perrott, F. A., Max Beran, and Robert Ratcliffe, *eds.*, 1983, *Variations in the Global Water Budget Symposium*, 1981, Oxford, UK, D. Reidel Publishing Co., Dordrecht, 532 pp.

Stuart, W. D., 1985, "Instability Model for Recurring Large and Great Earthquake in Southern California," *Pure Appl. Geophys. (Switzerland)*, v. 122, n. 6, pp. 793-811.

Sudicky, E. A., 1985, "Spatial Variability of Hydraulic Conductivity at the Borden Tracer Test Site," *Internat. Symp. Stochastic Approach to Subsurface Flow*, Montevillargenne, France, pp. 150-169.

Sudicky, E. A., and E. O. Frind, 1982, "Contaminant Transport in Fractured Porous Media; Analytical Solutions for a System of Parallel Fractures," *Water Resources Research*, v. 18, n. 6, pp. 1634-1642.

Sudicky, E. A., and E. O. Frind, 1984, "Contaminant Transport in Fractured Porous Media; Analytical Solution for a Two-Member Decay Chain in a Single Fracture," *Water Resources Research*, v. 20, n. 7, pp. 1021-1029.

Suggate, R. P., 1960, "The Interpretation of Progressive Fault Displacement of Flights of Terraces," *New Zealand Jour. Geol. and Geophys.*, v. 3, n. 3, pp. 364-374.

Sundberg, Wayne D., and Roger R. Eaton, 1982, *Three-Dimensional Thermal Analysis for a Conceptual High-Level Waste Repository in Welded Tuff*, SAND81-0215, Sandia National Laboratories, Albuquerque, NM, 30 pp.

Sundquist, E. T., 1985, "Geological Perspective on Carbon Dioxide and the Carbon Cycle," *in* E. T. Sundquist and W. S. Broecker, *eds., The Carbon Cycle and Atmospheric* CO_2*: Natural Variations Archean to Present*, Geophys. Monograph 32, Amer. Geophys. Union, Washington, DC, pp. 5-59.

Swan, F. H., III, D. P. Schwartz, and L. S. Cluff, 1980, "Recurrence of Moderate to Large Magnitude Earthquakes Produced by Surface Faulting on the Wasatch Fault Zone, Utah," *Bull. Seismol. Soc. Amer.*, v. 70, n. 5, pp. 1431-1462.

Swan, G., 1983, "Determination of Stiffness and Other Joint Properties from Roughness Measurements," *Rock Mech.*, v. 16, n. 1, pp. 19-38.

Swanberg, C. A., and P. Morgan, 1978, "The Linear Relationship Between Temperatures Based on the Silica Content of Groundwater and Regional Heat Flow: A New Heat Flow Map of the United States," *Pure Appl. Geophys.*, v. 117, n. 1-2, pp. 227-241.

Swanger, H. J., S. M. Day, J. R. Murphy, and R. Guzman, 1981, *State-of-the-Art Study Concerning Near-Field Earthquake Ground Motion*, NUREG/CR-1978, U.S. Nuclear Regulatory Commission, 141 pp.

Swanger, H. J., J. R. Murphy, T. J. Bennett, and R. Guzman, 1980, *State-of-the-Art Study Concerning Near-Field Earthquake Ground Motion*, NUREG/CR-1340, U.S. Nuclear Regulatory Commission, 139 pp.

Swanson, P. L., 1984, "Subcritical Crack Growth and Other Time- and Environment-Dependent Behavior in Crustal Rocks," *Jour. Geophys. Research B*, v. 89, n. 6, pp. 4137-4152.

Sweeney, N. F., and H. J. Hovland, 1981, "A Large Underground Opening—Measured Versus Predicted Rock Deformations," *Proc. 1981 Rapid Excavation and Tunneling Conf.*, Vol. 2., Sec. 21, pp. 1473-1489, AIME Soc. of Min. Eng., New York.

Switzer, P., 1967, "Reconstructing Patterns from Sample Data," *Annals Math. Stat.*, v. 38, pp. 138-154.

Switzer, P., 1969, "Mapping a Geographically Correlated Environment," Technical Report 145, Department of Statistics, Stanford Univ., Stanford, CA.

Switzer, P., 1975, "Estimation of the Accuracy of Qualitative Maps," *in* J. C. Davis and M. J. McCullagh, *eds., Display and Analysis of Spatial Data*, John Wiley and Sons, New York.

Sykes, J. F., J. L. Wilson, and R. W. Andrews, 1985, "Sensitivity Analysis for Steady State Groundwater Flow Using Adjoint Operators," *Water Resources Research*, v. 21, n. 3, pp. 359-371.

Sykes, L. R., 1971, "Aftershock Zones of Great Earthquakes, Seismicity Gaps, and Earthquake Prediction for Alaska and the Aleutians," *Jour. Geophys. Research*, v. 76, n. 32, pp. 8021-8041.

Sykes, L. R., 1978, "Intraplate Seismicity, Reactivation of Preexisting Zones of Weakness, Alkaline Magmatism, and Other Tectonism Postdating Continental Fragmentation," *Rev. Geophys. Space Phys.*, v. 16, n. 4, pp. 621-688.

Sykes, L. R., and S. P. Nishenko, 1984, "Probabilities of Occurrence of Large Plate Rupturing Earthquakes for the San Andreas, San Jacinto, and Imperial Faults, California, 1983-2003," *Jour. Geophys. Research B*, v. 89, n. 7, pp. 5905-5927.

Sykes, L. R., and R. C. Quittmeyer, 1981, "Repeat Times of Great Earthquakes Along Simple Plate Boundaries," *in* D. W. Simpson and P. G. Richards, *eds., Earthquake Prediction; An International Review*, Maurice Ewing Series 4, Amer. Geophys. Union, pp. 217-247.

Szechy, K., 1967, *The Art of Tunneling*, Akademiai Kiado, Budapest, 1092 pp.

Tagutschi, Y., M. Takahashi, S. Nagai, and N. Tono, 1983, *The Interpretation of Changes in Ground Water Level Caused by Earthquakes*, Open-file Report 80-108, U.S. Geological Survey.

Tammemagi, H. Y., and J. D. Chieslar, 1985, *Interim Rock Mass Properties and Conditions for Analyses of a Repository in Crystalline Rock*, BMI/OCRD-18, Battelle Memorial Institute, Columbus, OH, 84 pp.

Tang, W. H., 1971, "A Bayesian Evaluation of Information for Foundation Engineering Design," *Proc. 1st Internat. Conf. Applications of Stat. and Probability in Soil and Structural Engineering*, Hong Kong.

Tanimoto, C., 1978, "Discussion on 'Analysis of Tunnel Stability by the Convergence-Confinement Method,'" *Underground Space*, v. 4, pp. 389-398.

Tapia, Richard A., and James R. Thompson, 1978, *Nonparametric Probability Density Estimation*, Johns Hopkins Univ. Press, Baltimore, 171 pp.

Tarantola, A., A. Nercessian, and E. Trygvasson, 1982, "La prevision volcanique ou sismique traitee comme un probleme inverse," *Bull. Programme Interdisciplinaire de Recherche sur la Prevision et la Surveillance des Eruptions Volcanique*, No. 61, 20 pp.

Tazieff, H., and J. C. Sabroux, *eds.*, 1983, *Forecasting Volcanic Events*, Elsevier Scientific Publishing Co., 635 pp.

Teague, L. S., 1980, *Secondary Minerals Found in Cores DC2 A1 and DC2 A2 Taken from the Grande Ronde Basalt Formation, Pasco Basin, Washington*, LBL-10387, Lawrence Berkeley Laboratory, 29 pp.

Telles, J. C. F., and C. A. Brebbia, 1981, "New Developments in Elastoplastic Analysis," *in* C. A. Brebbia, *ed., Boundary Element Methods*, Springer-Verlag, Berlin, pp. 350-370.

Terzaghi, Karl Charles, 1929, "Effect of Minor Geologic Details on the Safety of Dams," *IMME, Tech. Publ. 215*, pp. 31-44.

Terzaghi, K., 1946, "Introduction to Tunnel Geology; Rock Defects and Loads on Tunnel Supports," *in* R. V. Proctor and T. L. White, *eds., Rock Tunneling with Steel Supports*, Youngstown, Ohio, pp. 15-99.

Terzaghi, K., 1961, "Past and Future of Applied Soil Mechanics," *Jour. Boston Soc. Civil Engineers*, v. 48, n. 2, pp. 110-139.

Terzaghi, K., 1962, "Stability of Steep Slopes on Hard Unweathered Rock," *Geotechnique*, v. 12, n. 4, pp. 251-270.

Terzaghi, K., 1964, "Sources of Errors in Joint Surveys," *Geotechnique*, v. 15, pp. 287-303.

Tien, Pei-Lin, M. D. Siegel, C. D. Updegraff, K. K. Wahi, and R. V. Guzowski, 1985, *Repository Site Data Report for Unsaturated Tuff, Yucca Mountain, Nevada*, NUREG/CR-4110, SAND84-2668, U.S. Nuclear Regulatory Commission, Washington, DC, 400 pp.

Tilling, R. I., and R. A. Bailey, 1984, "Volcano Hazards Program in the USA," *in* S. A. Fedetov, I. N. Galkin, A. V. Nikolayev, and E. N. Sedova, *eds., Earthquakes and Geological Hazard Prediction*, 27th Internat. Geol. Congress, August 4-14, 1984, Moscow, v. 27, pp. 106-118.

Timoshenko, S., and N. Goodier, 1951, *Theory of Elasticity*, McGraw-Hill, New York, 567 pp.

Tocher, Don, 1958, "Earthquake Energy and Ground Breakage," *Bull. Seismol. Soc. Amer.*, v. 48, n. 2, pp. 147-153.

Toro, G. R., and R. K. McGuire, 1986, "Random-Vibration Model of Ground Motions in the Eastern United States," *Methods for Earthquake Ground-Motion Prediction for the Eastern United States*, Electric Power Research Institute, Project RP2556-16, Palo Alto, CA.

Townley, L. R., and J. L. Wilson, 1983, "Conditional Second Moment Analysis of Groundwater Flow: The Cumulative Effects of Transmissivity and Head Measurements," *Papers Internat. Conf. Groundwater and Man*, v. 1, December 5-9, 1983, Sydney, Australia, 10 pp.

Townley, Lloyd R., and John L. Wilson, 1985, "Estimation of Boundary Values and Identification of Boundary Type for Numerical Values of Aquifer Flow," *Internat. Symp. Stochastic Approach to Subsurface Flow*, Montevillargenne, France, pp. 211-224.

Trask, N. J., 1982, *Performance Assessment for Radioactive Waste Repositories: the Rate of Movement of Faults*, Open-file Report 82-972, U.S. Geological Survey, 25 pp.

Tremaine, P. R., J. D. Chen, G. J. Wallace, and W. A. Boivin, 1981, "Solubility of Uranium (IV) Oxide in Alkaline Aqueous Solutions to 300°C," *Jour. Solution Chem.*, v. 10, pp. 221-230.

Trifunac, M. D., 1976, "Preliminary Empirical Model for Scaling Fourier Amplitude Spectra of Strong Ground Acceleration in Terms of Earthquake Magnitude, Source-to-Station Distance, and Recording Site Conditions," *Bull. Seismol. Soc. Amer.*, v. 66, n. 4, pp. 1343-1373.

Trifunac, M. D., and A. G. Brady, 1975, "A Study on the Duration of Strong Earthquake Ground Motion," *Bull. Seismol. Soc. Amer.*, v. 65, n. 3, pp. 581-626.

Tripathi, Vijay Shankar, 1984, "Uranium(VI) Transport Modeling: Geochemical Data and Submodels," Ph.D. dissertation, Stanford University, Stanford, CA.

Tsang, Chin-Fu, and Donald C. Mangold, *eds.*, 1984, *Panel Report on Coupled Thermo-Mechanical-Hydro-Chemical Processes Associated with a Nuclear Waste Repository*, LBL-18250, Lawrence Berkeley Laboratory, Berkeley, CA.

Tsui, K. K., and A. Tsai, 1985, *Thermal Analyses for Different Options of Nuclear Fuel Waste Emplacement*, AECL-7823, Whiteshell Nuclear Research Establishment, Pinawa, Manitoba.

Turcotte, D. L., and G. Schubert, 1982, *Geodynamics: Applications of Continuum Physics to Geological Problems*, John Wiley and Sons, New York, 450 pp.

Tyler, L. D., 1979, "Evaluation of Tuff as a Waste Isolation Medium," *in* M. E. Wacks, coordinator, R. E. Post, *ed., Proceedings, State of Waste Disposal Technology and the Social and Political Implications*, Univ. of Arizona, pp. 199-215.

Ubbes, William F., and James O. Duguid, *eds.*, 1985, *Geotechnical Assessment and Instrumentation Needs for Isolation of Nuclear Waste in Crystalline Rocks: Symposium Proceedings*, BMI/OCRD-24, Battelle Memorial Institute, Columbus, OH.

Uffink, G. J. M., 1985, "A Random Walk Method for the Simulation of Macrodispersion in a Stratified Aquifer," *Proc. Hamburg Symp. Relation of Groundwater Quantity and Quality*, Internat. Assoc. Hydro. Publ. 146, pp. 103-114.

Uhler, R. S., and P. G. Bradley, 1970, "A Stochastic Model for Petroleum Exploration Over Large Regions," *Jour. Amer. Stat.*, v. 65, pp. 623-630.

U.S. Bureau of Land Management, 1979, *Occurrence of Industrial Minerals in the CDCA*, Unpublished map, Desert Plan Staff, U.S. Bureau of Land Management, Riverside, CA.

U.S. Bureau of Mines, 1977, *Mineral Industry Location System*, Western Field Operations Center.

U.S. Committee for the Global Atmospheric Research Program, 1975, *Understanding Climate Change—A Program for Action*, National Research Council, National Academy of Sciences, Washington, DC, 239 pp.

U.S. Department of Energy, 1979, *Technology for Commercial Radioactive Waste Management*, DOE/ET-0028, Vols. 1-5, U.S. Department of Energy, Washington, DC.

U.S. Department of Energy, 1984, *Draft Environmental Assessment, Reference Repository Location, Hanford Site, Washington*, DOE/RW-0017, U.S. Department of Energy, Washington, DC.

U.S. Department of Energy, 1986a, *Environmental Assessment, Yucca Mountain Site, Nevada Research and Development Area, Nevada, Volume II*, DOE/RW-0073, U.S. Department of Energy, Washington, DC.

U.S. Department of Energy, 1986b, *Environmental Assessment, Reference Repository Location, Hanford Site, Washington, Volume II*, DOE/RW-0070, U.S. Department of Energy, Washington, DC.

U.S. Department of Energy, 1986c, *Environmental Assessment, Richton Dome Site, Mississippi, Volumes I-III*, DOE/RW-0072, U.S. Department of Energy, Washington, DC.

U.S. Department of Energy, 1988, *Site Characterization Plan, Yucca Mountain Site, Nevada Research and Development Area, Nevada*, DOE/RW-0199, U.S. Department of Energy, Office of Civilian Radioactive Waste Management, Washington, DC.

U.S. Environmental Protection Agency, 1985, "Environmental Standards for the Management and Disposal of Spent Nuclear Fuel, High-Level and Transuranic Radioactive Wastes, Final Rule," 40 CFR Part 191, *Federal Register*, v. 50, pp. 38066-38089.

U.S. Geological Survey, 1975, "Estimation of Undiscovered Resources in the US," Circular 725.

U.S. Geological Survey, 1980, *Proceedings of Conference X, Earthquake Hazards Along the Wasatch and Sierra-Nevada Frontal Fault Zones*, Open-File Report 80-801, J. F. Evernden, *compiler*, U.S. Geological Survey, Menlo Park, CA, 688 pp.

U.S. National Committee on Rock Mechanics, 1981, *Rock-Mechanics Research Requirements for Resource Recovery, Construction and Earthquake-Hazard Reduction*, National Academy Press, Washington, DC, 222 pp.

U.S. Nuclear Regulatory Commission, 1975, *Reactor Safety Study: An Assessment of Accident Risks in U.S. Commercial Nuclear Power Plants*, NUREG-75/014, WASH-1400, U.S. Nuclear Regulatory Commission, Washington, DC.

Uppuluri, V. R. R., 1983, *Expert Opinion and Ranking Methods*, NUREG/CR-3115, ORNL/CSD/TM-201, Oak Ridge National Laboratory, Oak Ridge, TN, 20 pp.

Ushijima, R. S., 1982, "Three-Dimensional Analysis of Advancing Tunnels with Ground-Structure Interaction," M.S. Thesis, Massachusetts Institute of Technology, Cambridge, MA, 287 pp.

Ushijima, R. S., 1983, "A Finite Element/Boundary Element Procedure for Multiple Step Static Analysis of Three Dimensional Underground Structures," Eng. Thesis, Massachusetts Institute of Technology, Cambridge, MA, 536 pp.

Ushijima, R. S., and H. H. Einstein, 1985, "Application of Coupled Finite Element-Boundary Element Method," *Proc. Behavior of Rocks in Engineering Applications*, Amer. Soc. Civil Engineers, pp. 21-37.

Van Dillen, D. E., 1978, *The Three-Dimensional Structural Analysis of Double-Entry and Single-Entry Coal Mines, Vol. III: Three-Dimensional Finite Element Analyses of Single- and Double-Entry Portions of Sunnyside Mine No. 1*, BUMINES-OFR-16(3)-80, Report No. R-7638-4534, Agbabian Associates, El Segundo, CA.

Vaniman, D., and B. Crowe, 1981, *Geology and Petrology of the Basalts of Crater Flat: Applications to Volcanic Risk Assessment for the Nevada Nuclear Waste Storage Investigations*, LA-8845-MS, Los Alamos National Laboratory, Los Alamos, NM, 68 pp.

Vanmarcke, E. H., and S. P. Lai, 1980, "Strong-Motion Duration and RMS Amplitude of Earthquake Records," *Bull. Seismol. Soc. Amer.*, v. 70, n. 4, pp. 1293-1307.

Van Sint, M. L., and E. J. Cording, 1984, "Ground and Lining Behavior of Shallow Underground Rock Chambers for the Washington DC Subway," *Proc. ISRM Design and Performance of Underground Excavation*, British Geotechnical Society, London, pp. 415-422.

Varotsos, P., and K. Alexopoulos, 1984, "Physical Properties of the Variations of the Electrical Field of the Earth Preceeding Earthquakes; I," *Tectonophysics*, v. 110, n. 1-2, pp. 73-98.

Veneziano, D., 1974, "Statistical Estimation and Precision in Probabilistic Models, with Application to Structural Reliability," Ph.D. Thesis, Massachusetts Institute of Technology, Cambridge, MA.

Veneziano, D., 1978, "Probabilistic Model of Joints in Rock," unpublished manuscript, Massachusetts Institute of Technology, Cambridge, MA.

Veneziano, D., 1979, "Statistical Estimation and Data Collection: A Review of Procedures for Civil Engineers," *Third Internat. Conf. Applications of Stat. and Probability in Soil and Structural Engineering*, Vol. 3, Univ. New South Wales, pp. 8-39.

Veneziano, D., 1985, "Statistical Estimation of Ground Motion Using MM Intensities," *in* Electric Power Research Institute, 1985, *Seismic Hazard Methodology for Nuclear Facilities in the Eastern United States, Vol. 3*, EPRI Research Project P101-29, Electric Power Research Institute, Palo Alto, CA.

Veneziano, D., and K. Coppersmith, 1985, "Estimation of Seismicity Parameters," *in* Electric Power Research Institute, *Seismic Hazard Methodology for Nuclear Facilities in the Eastern United States, Vol. 1, Development of Methodology*, Research Project P101-29, Electric Power Research Institute, Palo Alto, CA, pp. 5-1-5-37.

Veneziano, D., and C. A. Cornell, 1974, *Earthquake Models with Spatial and Temporal Memory for Engineering Seismic Risk Analysis*, Research Report R74-78, Department of Civil Engineering, Massachusetts Institute of Technology, Cambridge, MA.

Veneziano, D., and E. Faccioli, 1975, "Bayesian Design of Optimal Experiments for the Estimation of Soil Properties," *Proc. Second Internat. Conf. Applications of Stat. and Probability in Soil and Structural Engineering*, Deutsche Ges. fur Erd- und Grundbau, Essen, Germany, v. 2, pp. 191-213.

Veneziano, D., and M. Heidari, 1986, "Statistical Analysis of Attenuation in the Eastern

United States," *Methods for Earthquake Ground-Motion Prediction for the Eastern United States*, EPRI Project RP2556-16, Electric Power Research Institute, Palo Alto, CA.

Veneziano, D., and J. M. Van Dyke, 1985, *Seismic Hazard Methodology for Nuclear Facilities in the Eastern United States*, EPRI Research Project Report P101-29, Electric Power Research Institute, Palo Alto, CA.

Veneziano, D., C. A. Cornell, and T. O'Hara, 1983, "Historical Seismic Hazard Analysis," *Proc. Seismic Risk and Heavy Industrial Facilities Conf.*, San Francisco, May 11, 1983.

Vere-Jones, D., 1978, "Earthquake Prediction--A Statistician's View," *Jour. Phys. Earth (Tokyo)*, v. 26, n. 2, pp. 129-146.

Vistelius, A. B., 1961, "Sedimentation Time Trend Functions and Their Application for Correlation of Sedimentary Deposits," *Jour. Geol.*, v. 69, n. 6, pp. 703-728.

Voegele, M. D., 1979, "A Rational Design of Tunnel Supports: An Interactive Graphics Based Analysis of the Support Requirements of Excavations in Jointed Rock Masses," WES/TR/GL-79-15, U.S. Army Corps of Engineers Waterways Experimental Station, Vicksburg, MS, 522 pp.

Vomvoris, S. G., and L. W. Gelhar, 1985, "Stochastic Analysis of Concentration Variability in Transport in Aquifers," *Internat. Symp. Stochastic Approach to Subsurface Flow*, Montevillargenne, France, pp. 314-321.

Von Seggern, D. H., 1982, *Investigation of Seismic Precursors Before Major Earthquakes and the State of Stress on Fault Planes*, Ph.D. Thesis, Pennsylvania State Univ., University Park, PA, 352 pp.

Wagner, H., and B. J. Madden, 1984, "15 Years' Experience with the Design of Coal Pillars in Shallow South African Collieries: An Evaluation of the Performance of the Design Procedures and Recent Improvements," *Proc. ISRM Symp. Design and Performance Underground Excavations*, British Geotechnical Society, London, pp. 391-399.

Wagner, Ralph A., 1985, *Comparison and Evaluation of Field and Numerical Results from the Site A Heater Test at Avery Island*, ONWI-300, Battelle Memorial Institute, Columbus, OH, 71 pp.

Wahba, Grace, 1971, "A Polynomial Algorithm for Density Estimation," *Annals Math. Stat.*, v. 42, pp. 1870-1886.

Wahba, Grace, 1975, "Interpolating Spline Methods for Density Estimation, I: Equispaced Knots," *Annals Stat.*, v. 3, pp. 30-48.

Wahi, K. K., and B. C. Trent, 1982, "Analysis of Seismic Response of Repository Tunnels," *Proc. Fourth Internat. Conf. on Numerical Methods in Geomechanics*, Edmonton, UK, May 31-June 4, 1982, pp. 625-633.

Wahi, K. K., B. Baltes, and W. Salz, 1980, "A Numerical Simulation of a Thermally-Loaded Rock Mass Surrounding a Borehole," ASME paper #80-HT-40, *9th National Heat Transfer Conference*, Orlando, FL.

Wahi, Krishan K., Donald E. Maxwell, and Ronald Hofmann, 1977, *Simulation of the Thermomechanical Response of Project Salt Vault*, Y/OWI/Sub-77/16519/1, Office of Waste Isolation, Oak Ridge, TN, 84 pp.

Wahi, Krishan K., Donald E. Maxwell, and Ronald Hofmann, 1978, *The Thermomechanical Response of a One-Level Spent Fuel Nuclear Waste Repository*, Y/OWI/Sub-78/16549/1, Office of Waste Isolation, Oak Ridge, TN.

Wahi, K. K., B. C. Trent, D. E. Maxwell, R. M. Pyke, C. Young, and D. M. Ross-Brown, 1980, *Numerical Simulations of Earthquake Effects on Tunnels for Generic Nuclear Waste Repositories*, DP-1579, Savannah River Laboratory, Aiken, SC.

Wahlstrom, E. E., 1964, "The Validity of Geological Projection—A Case History," *Econ. Geol.*, v. 59, n. 3, pp. 465-474.

Walker, G. P. L., 1982, "Volcanic Hazards," *Interdisciplinary Sci. Rev.*, v. 7, n. 2, pp. 148-157.

Wallace, R. E., 1977, "Profiles and Ages of Young Fault Scarps, North-Central Nevada," *Geol. Soc. Amer. Bull.*, v. 88, n. 9, pp. 1267-1281.

Wallace, R. E., 1980, "Discussion--Nomographs for Estimating Components of Fault Displacement from Measured Height of Fault Scarp," *Bull. Assoc. Engineering Geologists*, v. 17, pp. 39-45.

Wallace, R. E., 1981, "Active Faults, Paleoseismology, and Earthquake Hazards in the Western United States," *in* D. W. Simpson and P. G. Richards, *eds., Earthquake Prediction--An International Review*, Maurice Ewing Series 4, Amer. Geophys. Union, pp. 209-216.

Wallace, R. E., and H. T. Morris, 1979, "Characteristics of Faults and Shear Zones as Seen in Mines at Depths as Much as 2.5 km Below the Surface," *Proc. Conf. VIII, Analysis of Actual Fault Zones in Bedrock*, Open-file Report 79-1239, U.S. Geological Survey, Denver, CO, pp. 79-100.

Walsh, Deborah A., 1979, "An Assessment of the Mineral Resources of the United Kingdom and Republic of Ireland," M.S. Thesis, Pennsylvania State Univ., University Park, PA, 142 pp.

Wang, J. S. Y., and T. N. Narasimhan, 1984, *Hydrologic Mechanisms Governing Fluid Flow in Partially Saturated, Fractured, Porous Tuff at Yucca Mountain*, LBL-18473, Lawrence Berkeley Laboratory, Berkeley, CA, 47 pp.

Wang, J. S. Y., D. C. Mangold, R. K. Spencer, C.-F. Tsang, 1983, *Thermal Impact of Waste Emplacement and Surface Cooling Associated with Geologic Disposal of Nuclear Waste*, NUREG/CR-2910, LBL-13341, U.S. Nuclear Regulatory Commission, Washington, DC, 241 pp.

Warburton, P. M., 1981, "Vector Stability Analysis of an Arbitrary Polyhedral Rock Block with Any Number of Free Faces," *Internat. Jour. Rock Mech. and Mining Sci.*, v. 18, n. 5, pp. 415-427.

Ward, P. L., 1978, "Earthquake Prediction," *in* National Academy of Sciences, *Geophysical Predictions*, National Academy of Sciences, Washington, DC, pp. 37-46.

Ward, W. T., and J. Chappell, 1975, "Geology of Coral Terraces, Huon Peninsula, New Guinea: A Study of Quaternary Tectonic Movements and Sea-Level Changes," *Geol. Soc. Amer. Bull.*, v. 86, pp. 1482-1486.

Washington, W. M., and C. L. Parkinson, 1986, *An Introduction to Three-Dimensional Climate Modeling*, University Science Books, Mill Valley, CA, 422 pp.

Waste Isolation Systems Panel, 1983, *A Study of the Isolation System for Geologic Disposal of Radioactive Wastes*, U.S. National Research Council, National Academy Press, Washington, DC, 345 pp.

Watkins, N. D., and A. K. Baksi, 1974, "Magnetostratigraphy and Oroclinal Folding of the Columbia River, Steens and Owyhee Basalts in Oregon, Washington, and Idaho," *Amer. Jour. Sci.*, v. 274, n. 2, pp. 148-189; erratum, v. 274, n. 7, p. 832.

Watson, Alicia Tyler, 1977, "An Appraisal of the Mineral Resources of New Zealand," M.S. Thesis, Pennsylvania State Univ., University Park, PA, 129 pp.

Watson, G. S., 1972, *Trend Surface Analysis and Spatial Correlations*, Geol. Soc. Amer. Special Paper 146, pp. 39-46.

Watts, M. J., and G. D. Williams, 1979, "Fault Rocks as Indicators of Progressive Shear Deformation in the Guingamp Region, Britanny," *Jour. Structural Geology*, v. 1, n. 4, pp. 323-332.

Wawersik, Wolfgang R., and David H. Zeuch, 1984, *Creep and Creep Modeling of Three Domal Salts--A Comprehensive Update*, SAND84-0568, Sandia National Laboratories, Albuquerque, NM.

Webb, T., III, and T. M. L. Wigley, 1985, "What Past Climates Can Indicate About a Warmer World," *in* M. C. MacCracken and F. J. Luther, *eds., Projecting the Climatic Effects of Increasing Carbon Dioxide*, DOE/ER-0237, U.S. Department of Energy, pp. 237-257.

Webb, T., III, P. J. Bartlein, and J. E. Kutzbach, 1987, "Climatic Change in Eastern North America During the Past 18,000 Years; Comparisons of Pollen Data with Model Resutls, in North America and Adjacent Oceans During the Last Deglaciation," *in* W. F. Ruddiman and H. E. Wright, Jr., *eds., Geology of North America*, v. K-3, Geological Society of America, Boulder, CO, pp. 447-462.

Webb, Thompson, III, John Kutzbach, and F. Alayne Street-Perrott, 1985, "20,000 Years of Global Climatic Change: Paleoclimatic Research Plan," *in* T. F. Malone and J. G. Roederer, *eds., Global Change*, ICSU Press, Miami, FL, pp. 182-218.

Wegman, E. J., 1972, "Nonparametric Density Estimation, I. A Summary of Available Methods," *Technometrics*, v. 14, pp. 533-546.

Weldon, Ray J., II, 1991, "Active Tectonic Studies in the United States, 1987-1990," *U.S. National Report to International Union of Geodesy and Geophysics, Reviews of Geophysics, Supplement*, American Geophysical Union, Washington, DC, pp. 890-906.

Weldon, R. J., II, and K. E. Sieh, 1985, "Holocene Rate of Slip and Tentative Recurrence Interval for Large Earthquakes on the San Andreas Fault, Cajon Pass, Southern California," *Geol. Soc. Amer. Bull.*, v. 96, n. 6, pp. 793-812.

Wentworth, C. M., and M. Mergner-Keefer, 1983, "Regenerate Faults of Small Cenozoic Offset—Probable Earthquake Sources in the Southeastern United States," *in* G. S. Gohn, *ed., Studies Related to the Charleston, South Carolina, Earthquake of 1886; Tectonics and Seismicity*, Prof. Paper 1313, U.S. Geological Survey, pp. S1-S20.

Wesson, R. L., 1981, "Interpretation of Changes in Water Level Accompanying Fault Creep and Implications for Earthquake Prediction," *Jour. Geophys. Research B*, v. 86, n. 10, pp. 9259-9267.

West, Graham, 1983, "Comparisons Between Real and Predicted Geology in Tunnels: Examples from Recent Cases," *Q. Jour. Eng. Geol. (London)*, v. 16, n. 2, pp. 113-126.

Westercamp, D., and H. Traineau, 1983, "The Past 5,000 Years of Volcanic Activity at Mt. Pelee, Martinique (F. W. I.): Implications for Assessment of Volcanic Hazards," *Jour. Volc. Geoth. Research*, v. 17, n. 1-4, pp. 159-185.

White, David A., and Harry M. Gehman, 1979, "Methods of Estimating Oil and Gas Resources," *Bull. Amer. Assoc. Petroleum Geologists*, v. 63, n. 2, pp. 2183-2192.

Whitman, R. V., 1984, "Evaluating Calculated Risk in Geotechnical Engineering," *Amer. Soc. Civil Engineers, Jour. Geotechnical Eng.*, v. 110, n. 2, pp. 143-188.

Whitten, E. H. T., 1977, "Stochastic Models in Geology," *Jour. Geology*, v. 85, n. 3, pp. 321-330.

Whittle, P., 1963, *Prediction and Regulation by Linear Least-Square Methods*, English Univ. Press, London, 147 pp.

Wickman, F. E., 1966, "Repose-Period Patterns of Volcanoes," *Arkiv for Mineralogi och Geologi*, v. 4, n. 4, pp. 291-352.

Wight, L. H., 1979, "Analysis of the Seismic Hazard to an Underground Waste Repository," *in* B. L. Scott, G. L. Benson, R. A. Craig, and M. A. Harwell, *eds., Assessment of Effectiveness of Geologic Isolation Systems: A Summary of FY-1978 Consultant Input for Scenario Methodology Development*, PNL-2851, Pacific Northwest Laboratories, Richland, WA, pp. III-1-III-21.

Wight, L. H., 1980, *Assessment of Effectiveness of Geologic Isolation Systems: The Feasibility for Computer Interrogation of Experts for WISAP*, PNL-2862, Battelle Pacific Northwest Laboratory, Richland, WA, 101 pp.

Wigley, T. M. L., and S. C. B. Raper, 1990, "Natural Variability of the Climate System and Detection of the Greenhouse Effect," *Nature*, v. 344, pp. 324-327.

Williams and Associates, Inc., 1983, *Statistical Evaluation of Hydrochemical Data from the Saddle Mountains, Wanapum and Grande Ronde Basalts - Basalt Waste Isolation Project—Hanford Site*, Report to the U.S. Nuclear Regulatory Commission under Contract No. NRC-02-82-044.

Williams, H., and A. R. McBirney, 1979, *Volcanology*, Freeman, Cooper, and Co., San Francisco, CA, 397 p.

Williams, Jill, R. G. Barry, and W. M. Washington, 1974, "Simulation of the Atmospheric Circulation Using the NCAR Global Circulation Model with the Ice Age Boundary Conditions," *Jour. Atmospheric Sci.*, v. 13, pp. 305-317.

Williams, T. R., 1982, *Late Pleistocene Lake Level Maxima and Shoreline Deformation in the Basin and Range Province, Western United States*, M.S. Thesis, Colorado State Univ., Fort Collins, CO, 52 pp. (1 plate, 1:2,500,000 scale).

Wilson, J., and M. D. Dettinger, 1974, "Stochastic Modeling of Groundwater Systems," Report No. 189, Parsons Laboratory, Massachusetts Institute of Technology, Cambridge, MA.

Wilson, L., and J. W. Head, III, 1981, "Ascent and Eruption of Basaltic Magma on the Earth and Moon," *Jour. Geophys. Research B*, v. 86, n. 4, pp. 2971-3001.

Wilson, R., S. D. Colome, J. D. Spengler, and D. G. Wilson, 1980, *Health Effects of Fossil Fuel Burning: Assessment and Mitigation* , Ballinger Publishing Company, Cambridge, MA.

Winkler, R. A., 1968, "The Consensus of Subjective Probability Distribution," *Management Sci.*, v. 15.

Winkler, R. L., 1985, "Bayesian Inference," *in* A. H. Murphy and R. W. Katz, *eds., Probability Statistics and Decision Making in Atmospheric Sciences*, pp. 461-491, Westview Press, Boulder, CO, 545 pp.

Wittke, W., and B. Pierau, 1976, "3D Stability Analysis of Tunnels in Jointed Rock," *in* C. S. Desai, *ed., Numerical Methods in Geomechanics*, Amer. Soc. Civil Engineers, Vol. III, pp. 1401-1418.

Wittke, W., and P. Rissler, 1976, "Dimensioning of the Lining of Underground Openings in Swelling Rock Applying the Finite Element Method," Publ. No. 2 of the Institute for Foundation Engineering, Soil Mech., Rock Mech., and Water Ways Construction, Univ. of Aachen, pp. 7-48.

Wolfowitz, J., 1957, "The Minimum Distance Method," *Annals Math. Stat.*, v. 28, pp. 75-88.

Woodward, Wayne A., William C. Parr, William R. Schucany, and Hildegard Lindsey, 1984, "A Comparison of Minimum Distance and Maximum Likelihood Estimation of a Mixture Proportion," *Jour. Amer. Stat. Assoc.*, v. 79, pp. 590-598.

Worotniki, G., M. Wold, J. Enever, and R. Walton, 1976, "The Application of Physical Modelling Techniques to Problems in Tunneling," *Design and Construction of Tunnels and Shafts*, 2nd Australian Conf. Tunneling, The Australasian Institute of Mining and Metallurgy, Parkville, Victoria, Australia, pp. 25-40.

Wu, C. L., S. W. Chen, and R. D. Wise, 1982, "Seismic Evaluation of Underground Facilities for Nuclear Waste Storage," *in* R. G. Post and M. E. Wacks, *eds., Waste Management '82*, Univ. of Arizona, Tucson, AZ, pp. 113-129.

Wu, T. H., 1974, "Uncertainty, Safety, and Decision in Soil Engineering," *Amer. Soc. Civil Engineers Proc., Jour. Geotechnical Engineering Div.*, v. 100, n. GT3, pp. 329-348.

Wu, T. H., and L. M. Kraft, 1967, "Probability of Foundation Safety," *Jour. Soil Mech. and Foundation Div.*, Amer. Soc. Civil Engineers, v. 93.

Wu, T. H., W. B. Thayer, and S. S. Lin, 1975, "Stability of Embankment on Clay," *Amer. Soc. Civil Engineers, Jour. Geotechnical Eng. Div.*, v. 101, no. GT9, pp. 913-932.

Wyss, M., 1975, "A Search for Precursors to the Sitka, 1972, Earthquake: Sea Level, Magnetic Field, and P-Residuals," *Pure Appl. Geophys.*, v. 113, pp. 297-309.

Wyss, M., 1976a, "Local Changes of Sea Level Before Large Earthquakes in South America," *Bull. Seismol. Soc. Amer.*, v. 66, n. 3, pp. 903-914.

Wyss, M., 1976b, "Local Sea Level Changes Before and After the Hyuganada, Japan, Earthquakes of 1961 and 1968," *Jour. Geophys. Research*, v. 81, n. 29, pp. 5315-5321.

Wyss, M., 1977, "The Appearance Rate of Premonitory Uplift," *Bull. Seismol. Soc. Amer.*, v. 67, n. 4, pp. 1091-1098.

Wyss, Max, 1979, "Estimating Maximum Expectable Magnitude of Earthquakes from Fault Dimensions," *Geology*, v. 7, n. 7, pp. 336-340.

Wyss, M., 1981, "Recent Earthquake Prediction Research in the United States," *in* T. Rikitake, *ed., Current Research in Earthquake Prediction I*, Center for Academic Publications Japan, D. Reidel Publ. Co., Tokyo, pp. 81-128.

Yamazaki, Y., 1975, "Precursory and Coseismic Resistivity Changes," *Pure Appl. Geophys.*, v. 113, pp. 219-227.

Yamazaki, Y., 1977, "Tectonoelectricity," *Geophysical Surveys (Netherlands)*, v. 3, n. 2, pp. 123-142.

Yankee Atomic Electric Co., 1985, *Seismic Source Method of Seismic Hazard Analysis and EQZONE Computer Code Manual*, EPRI NP-4349-CCM, Electric Power Research Institute, Palo Alto, CA.

Yeh, G. T., and V. S. Tripathi, 1989, "A Critical Evaluation of Recent Developments in Hydrogeochemical Transport Models of Reactive Multichemical Components," *Water Resources Research*, vol. 25, pp. 93-108.

Yeh, T. C. J., L. W. Gelhar, and A. L. Gutjahr, 1985a, "Stochastic Analysis of Unsaturated Flow in Heterogeneous Soils, 1: Statistically Isotropic Media," *Water Resources Research*, v. 21, n. 4, pp. 447-456.

Yeh, T. C. J., L. W. Gelhar, and A. L. Gutjahr, 1985b, "Stochastic Analysis of Unsaturated Flow in Heterogeneous Soils, 2: Statistically Anisotropic Media with Variable α," *Water Resources Research*, v. 21, n. 4, pp. 457-464.

Yeh, T. C. J., L. W. Gelhar, and A. L. Gutjahr, 1985c, "Stochastic Analysis of Unsaturated Flow in Heterogeneous Soils, 3: Observations and Applications," *Water Resources Research*, v. 21, n. 4, pp. 465-471.

Youd, T. L., and D. K. Keefer, 1985, "Landslides Caused by Earthquakes: Discussion and Reply," *Geol. Soc. Amer. Bull.*, v. 96, n. 8, pp. 1091-1094.

Youd, T. L., and D. M. Perkins, 1978, "Mapping of Liquifaction-Induced Ground Failure Potential," Amer. Soc. Engineers, *Jour. Geotechnical Engineering Div.*, v. 104, no. GT4, pp. 433-446.

Youngs, R. R., and K. J. Coppersmith, 1985, "Implications of Fault Slip Rates and Earthquake Recurrence Models to Probabilistic Seismic Hazard Estimates," *Bull. Seismol. Soc. Amer.*, v. 75, n. 4, pp. 939-964.

Yucemen, M. S., 1973, "A Probabilistic Study of Safety and Design of Earth Slopes," Ph.D. Thesis, Univ. of Illinois, Urbana-Champaign, IL.

Zavoshy, S. J., P. L. Chambre, and T. H. Pigford, 1985, "Mass Transfer in a Geologic Environment," *in* C. M. Jantzen, J. A. Stone, and R. C. Ewing, *eds., Scientific Basis for Nuclear Waste Management VIII*, Proc. 44, Materials Res. Soc. Symp., North-Holland, New York, pp. 311-322.

Zellmer, J. T., G. R. Roquemore, and B. A. Blackerby, 1985, "Modern Tectonic Cracking Near the Garlock Fault, California," *Geol. Soc. Amer. Bull.*, v. 96, n. 8, pp. 1037-1042.

Zhen, Yu T., and P. Zu-Zeng, 1983, "Classification System of Fuzzy Sets for Rock Engineering," *Proc. Internat. Symp. Engineering, Geol. and Underground Construction*, Lisbon, pp. II1-II10.

Zienkiewicz, O. C., 1977, *The Finite Element Method*, 3rd ed., McGraw-Hill, London, England, 785 pp.

Zimmerman, Roger M., and Mark L. Blanford, 1985, "Evaluation of the Accuracy of Continum-Based Computational Models in Relation to Field Measurements in Welded Tuff," *Proc. Internat. Symposium on Fundamentals of Rock Joints*, Bjorkliden, Sweden, September 15-20, 1985, pp. 233-245.

Ziony, J. I., C. M. Wentworth, J. M. Buchanan-Banks, and H. C. Wagner, 1974, *Preliminary Map Showing Recency of Faulting in Coastal Southern California*, Miscellaneous Field Investigations Map MF-585, U.S. Geological Survey, (3 sheets, 1:250,000 scale, with explanatory text, 15 pp.).

Zoback, M. D., 1983, "Intraplate Earthquakes, Crustal Deformation, and In-Situ Stress," *in* W. W. Hays and P. L. Gori, *eds., A Workshop on The 1886 Charleston, South Carolina, Earthquake and Its Implications for Today*, Open-file Report 83-0843, pp. 169-179.

Zoback, M. D., and M. L. Zoback, 1981, "State of Stress and Intraplate Earthquakes in the United States," *Science*, v. 213, n. 4503, pp. 96-104.

Zoback, M. L., and G. A. Thompson, 1978, "Basin and Range Rifting in Northern Nevada: Clues from a Mid-Miocene Rift and its Subsequent Offsets," *Geology*, v. 6, n. 2, pp. 111-116.

Zoback, M. L., and M. D. Zoback, 1980, "State of Stress in the Conterminous United States," *Jour. Geophys. Research B*, v. 85, n. 11, pp. 6113-6156.

Author Index

Subject Index